International Studies in the History of Mathematics and its Teaching

The International Studies in the History of Mathematics and its Teaching Series creates a platform for international collaboration in the exploration of the social history of mathematics education and its connections with the development of mathematics. The series offers broad perspectives on mathematics research and education, including contributions relating to the history of mathematics and mathematics education at all levels of study, school education, college education, mathematics teacher education, the development of research mathematics, the role of mathematicians in mathematics education, mathematics teachers' associations and periodicals.

The series seeks to inform mathematics educators, mathematicians, and historians about the political, social, and cultural constraints and achievements that influenced the development of mathematics and mathematics education. In so doing, it aims to overcome disconnected national cultural and social histories and establish common cross-cultural themes within the development of mathematics and mathematics instruction. However, at the core of these various perspectives, the question of how to best improve mathematics teaching and learning always remains the focal issue informing the series.

Alexander Karp
Editor

Advances In The History Of Mathematics Education

Springer

Editor
Alexander Karp
Teachers College
Columbia University
New York, NY, USA

ISSN 2524-8022 ISSN 2524-8030 (electronic)
International Studies in the History of Mathematics and its Teaching
ISBN 978-3-030-95234-1 ISBN 978-3-030-95235-8 (eBook)
https://doi.org/10.1007/978-3-030-95235-8

Cover caption: Cover image is based on the cuneiform tablet, known as Plimpton 322 (collections of Columbia University, New York)

This Springer imprint is published by the registered company Springer Nature Switzerland AG
The registered company address is: Gewerbestrasse 11, 6330 Cham, Switzerland

Five years after (In Place of an Introduction)

The book being offered to the reader's attention grew out of materials presented at "The history of the teaching and the learning of mathematics" Topic Study Group (TSG) at the Fourteenth International Congress on Mathematics Education. This book, of course, does not represent the proceedings of that conference: far from all presentations are represented here,[1] and those that are represented are notably different from those heard at the conference, if only because they are many times greater in size. On the other hand, the present volume and the presentations of the TSG are not independent of one another, but pertain to the same ongoing studies. This book is structured on the same principles as the collection by Furinghetti and Karp (2018), which was published after the Thirteenth International Congress on Mathematics Education.

Because of the pandemic, 5 years passed between the conferences, instead of the customary four, and this relatively round number gives us occasion to reflect on what has happened to the scholarly field of the history of mathematics education over the intervening time. One may recall that in the course of these 5 years, two other conferences on the history of mathematics have taken place, and subsequently collections of the papers presented at these conferences have been published (Bjarnadottir et al. 2019; Barbin et al. 2020). One may also list other books on the subject that have been published, both as part of the series edited by Alexander Karp and Gert Schubring, to which the present book also belongs, and those published elsewhere (for example, as part of the series edited by McKenzie A. (Ken) Clements and Nerida Ellerton). One may note that in 2019 the Freudenthal medal, one of the highest honors in the field of mathematics education, was awarded to Gert Schubring, first and foremost for his work in the field of the history of mathematics education, which may be considered a sign of the recognition of the significance of this field as a whole.

[1]All presenters were invited to prepare articles based on their presentations, which then went through a peer review process, on whose outcomes the present collection was formed.

There have been no dramatic changes—scientific revolutions—over these years, of course. Using Thomas S. Kuhn's terminology (with all due reservations about the degree to which his theory applies to such a small scholarly field as the history of mathematics education, which is so clearly dependent on other fields), we may say that no new research paradigm has emerged (indeed, we should have been surprised by a more frequent emergence of such paradigms). Research has been accumulating within the frameworks of existing general approaches.

This, however, is also significant. The history of mathematics education, which was the subject of the first scholarly works on mathematics education, at least in the United States (recall the dissertations of Jackson 1906 and Stamper 1906), for a certain period found itself sidelined if not silenced altogether. *Normal* scholarly work on mathematics education was often expected to employ statistical methods, while history was regarded as something unrigorous, subjective, and above all, a thing of the past. At a certain point, the situation began to change, and it has continued to change, in the sense that history is coming back into the sphere of legitimate objects of research. This, in fact, constitutes the most important characteristic of the present time. It is becoming evident ever more broadly that without attempting to understand processes that have taken place, that is, history, it is impossible to understand the present.

Attention to history also brings with it new battles for history (to use an expression from Febvre 1953). The history of mathematics education may become distorted for various reasons. Sometimes, these are relatively innocent, as when researchers sincerely believe that what they have done is especially important and begin demanding that history be rewritten in a way that, say, allocates a special position to the materials they have collected.

Sometimes people wish to rewrite history so as to make it accord more with one or another general theory. Researchers of Russian history, for example, will recall how at one time, in order to strengthen and promote love for Russia, it was considered important to explain that it was not Felix Klein or other foreigners, but specifically Russians who had first come to the idea of reforms in mathematics (Karp 2019)—explanations that Russia, of course, did not need in the least. Such arguments can be heard today as well, in relation to different countries and different groups of people: the indisputable statement that all people are created equal must be confirmed for some reason by speculations that all people and nations simultaneously and independently created mathematics education of equal quality, while other people and nations caused nothing but harm to them.

In a certain sense, an opposing, "anti-ideological" tendency is also emerging, in which fear of charges of bias leads to the renunciation of any attempt to generalize what is observed in the history of mathematics education or to connect it with social history, which, indeed, itself comes under suspicion.

Finally, there are also writings that are dictated above all by ignorance, whose authors, unhesitatingly making the boldest generalizations, paraphrase what they have heard from someone else, while giving no thought to comparisons with other sources or even to the existence of such sources, to criticizing their own source, and so on. As a striking example of such an approach, I would name a course in the history of Russian mathematics education published 20 years ago (Kolyagin 2001), whose author believed that he "vividly demonstrates what level of preparation existed in secondary educational institutions"[2] in Russia before 1917 because he reproduced a copy of his aunt's graduation certificate (pp. 134, 136–137). Such writings appear not only in Russia, of course.

All of this, however, is characteristic not only of the history of mathematics education, but of many other historical disciplines as well, and however lamentable, is probably unavoidable. The response to shoddy work that has appeared in the last 5 years must be work that is conscientious and substantive, for which there are more than enough objects of research.

The history of mathematics education, in our view, differs from many other historical disciplines precisely by the abundance of unstudied and uncollected materials. Sometimes this has occurred because it has been difficult to document what took place, for example, in classrooms; sometimes, because it has not been customary to be interested in certain phenomena; and sometimes, because entire dimensions of mathematics education were not included in what mathematics education was officially considered to be (as, indeed, today all lessons with tutors or all kinds of preparatory classes are usually unaccounted for, although it is precisely in these contexts that learning very often takes place). In any event, it is important to expand the range of what is studied, to which end new generations of researchers must be drawn in—which, one must hope, will bring us a new understanding of the past (and therefore also of the present) of mathematics education.

The book being offered to the reader's attention is therefore interesting already simply by virtue of its chosen subject matter. Let us list the directions along which research is being conducted.

The most popular topic in the book is reforms in mathematics education. It is addressed in papers by Dirk De Bock, Elisabete Zardo Búrigo, Antonio M. Oller Marcén, which are devoted to various aspects of the reforms that began in the second half of the 1950s and went under different names in different countries (New Math, Mathématiques modernes, etc.). These reforms took place relatively recently, and are often seen as having ultimately failed, yet they undoubtedly exerted an enormous influence on the development of education around the world, and much about them remains unstudied and not understood.

The study of curricula is another traditional topic of historical research. The paper by Karolina Karpińska, however, is untraditional, because it is devoted to the study in Polish schools of a topic that lies at the intersection of several disciplines and is rarely studied elsewhere—the construction of sundials.

The paper by Vasily Busev and Alexander Karp analyzes the work of a major mathematician (P.L. Chebyshev) in mathematics education—a topic to which many studies have been devoted, but which is far from exhausted, not only because there are still no easily accessible studies of many mathematicians, but also because by

[2] All translations from Russian are by the author.

studying the role of research mathematicians, we may also better understand the system with which they interacted.

Sian Zelbo's paper belongs to the history of periodicals, a subject matter that has been gaining popularity in recent years. Indeed, by studying the journals of a historical period, one may reach a better understanding of what took place during that period, as well as to form a better conception of the people who wrote for and read these journals, that is, the actors in the unfolding process.

Mathematics education for women might have been expected to be one of the most studied topics for decades already. For various reasons (which, incidentally, deserve thinking about), this is not the case. Yana Shvartsberg's paper is devoted to women's education in the United States during the period of progressive education, when the foundations of mass-scale education in high schools were laid and many other principles and guidelines were established that remain important to this day.

The paper by Viktor Freiman and Alexei Volkov may be considered to belong to the history of ideas and intellectual influences in mathematics education, which has clearly not been sufficiently worked out. The paper is devoted to David Eugene Smith and his reception of German literature devoted to teaching arithmetic and such teaching in practice.

Lastly, the paper by Alexander Karp is devoted to entrance exams to higher educational institutions in prerevolutionary Russia and the preparation for such exams. Consequently, in it an attempt is made to investigate informal mathematics education—education outside official schools.

One might say that this book has one overarching subject: the formation and development of various aspects of mathematics education—ideas of teaching or the language of textbooks and what is expected of textbooks; the industry of exam preparation or attitudes about how mathematics education for women should be structured and what should be expected of it; the American mathematics community and the groups and people who implemented reforms; new types of textbooks for elementary schools, new approaches toward studying traditional topics and teaching nation-specific topics. This is natural: the processes observable in history can very often be characterized as processes (even if sometimes very slow ones) of formation and concurrent processes of obsolescence. We would like to hope that the papers collected in this volume will aid the study and better understanding of these processes.

Teachers College,
Columbia University,
New York, NY, USA

Alexander Karp

References

Barbin, Évelyne, Kristin Bjarnadottir, Fulvia Furinghetti, Alexander Karp, Guillaume Moussard, Johan Prytz, and Gert Schubring, eds. 2020. *"Dig where you stand" 6*. Munster: WTM.

Bjarnadottir, Kristin, Fulvia Furinghetti, Jenneke Krüger, Johan Prytz, Gert Schubring, and Harm Jan Smid, eds. 2019. *"Dig where you stand" 5*. Utrecht: Utrecht University – Freudenthal Institute.

Febvre, Lucien. 1953. *Combats pour l'histoire*. Paris: A. Colin.

Furinghetti, Fulvia, and Alexander Karp. 2018. *Researching the History of Mathematics Education. An international overview*. Cham, Switzerland: Springer.

Jackson, Lambert Lincoln. 1906. *Educational significance of sixteenth century arithmetic from the point of view of the present time. Doctoral dissertation*. Columbia University.

Karp, Alexander. 2019. The Russian National subcommission of ICMI and the mathematics education reform movement. *In National Subcommissions of ICMI and their Role in the Reform of Mathematics Education*, ed. Alexander Karp, 149–191. Cham, Switzerland: Springer.

Kolyagin, Yu.M. 2001. *Russkaya shkola i matematicheskoe obrazovanie. Nasha gordost'i nasha bol'*. [Russian school and mathematics education. Our pride and our pain]. Moscow: Prosveschenie.

Stamper, Alva Walker. 1906. *A history of the teaching of elementary geometry with reference to present day problems. Doctoral dissertation*. Columbia University.

Contents

About the Authors

Elisabete Zardo Búrigo is a full professor of mathematics education at the Mathematics and Statistics Institute of the Federal University of Rio Grande do Sul, Brazil. She graduated with a mathematics degree from the same university and received her PhD in education from the University of São Paulo in 2004. Her research interests include the history of mathematical education in primary and secondary schools, curricular policies and practices, and teacher training. She coordinated a research project on teacher training for teachers of mathematics in elementary school in Rio Grande do Sul, Brazil, covering the period from 1889 to 1970.

Vasily Busev chief editor of the electronic library *Mathematics Education* (https://mathedu.ru), graduated from Moscow State Pedagogical University, completed his PhD studies at the S.I. Vavilov Institute for the History of Natural Science and Technology of the Russian Academy of Sciences, taught mathematics in school, and worked as an editor for publishers of educational texts and subsequently at the K.D. Ushinsky Scientific Pedagogical Library. His scholarly interests include the history of Russian education, especially mathematics education, as well as the development of information systems for science and education. In addition, he has been actively involved in the description and publication of archival documents as well as the creation of a bibliography of scientific and methodological literature.

Dirk De Bock is a professor of mathematics in the Faculty of Economics and Business of the University of Leuven (Belgium). His major research interests are history of mathematics education, psychological aspects of teaching and learning mathematics, role of mathematics in economics and finance, and financial literacy. His recent research in the field of history of mathematics education focused on the role of Belgian mathematicians and mathematics teachers in the international modern mathematics reform movement of the 1960s, and it led to a monograph: De Bock, Dirk & Vanpaemel, Geert (2019). *Rods, sets and arrows: The rise and fall of modern mathematics in Belgium*. Cham, Switzerland: Springer. Ongoing research projects include the study of New Math or modern mathematics from an international perspective.

Viktor Freiman, PhD is a full professor in mathematics education at the Université de Moncton, Canada, and director of the CompeTI.CA (ICT competences in the Atlantic Canada) Partnership Network. His research interests and publications are in the field of teaching and learning algebra, problem-solving, use of technology, STEAM education, history of mathematics education, and mathematical creativity and giftedness. He is a president-elect of the International Group for Mathematical Creativity and Giftedness as well as co-organizer of Mathematics and its Connections to the Arts and Sciences symposia. He is co-editor of the book series *Mathematics Education in the Digital Era.*

Alexander Karp is a professor of mathematics education at Teachers College, Columbia University. He received his PhD in mathematics education from Herzen Pedagogical University in St. Petersburg, Russia, and also holds a degree from the same university in history and education. Currently, his scholarly interests span several areas, including the history of mathematics education, gifted education, mathematics teacher education, and mathematical problem-solving. He served as the managing editor of the *International Journal for the History of Mathematics Education* and is the author or editor of over 100 publications, including over 30 books.

Karolina Karpińska is an assistant professor in the Institute for the History of Science of the Polish Academy of Sciences. After completing her master's and doctoral studies in mathematics at the Nicolaus Copernicus University in Toruń, she received her PhD in the history of mathematics from the Institute for the History of Science of the Polish Academy of Sciences. Her scholarly interests are related to the history of mathematics education. Currently, she is working on a project—Mathematics Education in Prussian Poland in 1800–1914—which concerns the time when all Polish territories were under the occupation of Prussia, Russia, and Austria. She analyzes the ways in which the occupiers influenced the teaching of mathematics. She is a book review editor of the journal *Analecta: Studies and Materials for the History of Science* and is the author of over a dozen publications. Her last paper "'Denominate Numbers' in Mathematics School Textbooks by Stefan Banach" is in print in *Historia Mathematica.*

Antonio M. Oller-Marcén was born in Zaragoza (Spain) in 1981. He graduated in mathematics in 2004 and obtained a Master of Science degree in algebra in 2006 at the University of Zaragoza. He obtained a PhD degree in mathematics education at the University of Valladolid in 2012. He is currently an associate professor of mathematics at the Centro Universitario de la Defensa de Zaragoza. His research interests include both the history of mathematics education and its use as a tool, particularly in a teacher-training context. He is the principal investigator of the group "Investigación en Educación Matemática" and a member of the HPM Advisory Board.

Yana Shvartsberg is a clinical assistant professor at Pace University in New York City. She received her PhD in mathematics education from Teachers College, Columbia University, and also holds a Master of Science degree in applied mathematics from Hofstra University. Her research interests include the history of mathematics education, history of female mathematics education, impact of vocational context on mathematics education, and social factors that affect students' pursuit of mathematics knowledge in high school and college.

Alexei Volkov graduated from the Department of Mathematics and Theoretical Mechanics of Moscow State University and obtained his doctoral degree from the Institute of History of Natural Sciences and Technology of the Russian Academy of Sciences. He worked in research centers and universities in the Soviet Union/Russia, France, Hong Kong, and Canada; currently, he is a professor at the National Tsing-Hua University (Taiwan). His publications are devoted to the history of mathematics and mathematics education in East and Southeast Asia, in particular China and Vietnam.

Sian E. Zelbo received her JD from the University of Texas and her PhD from Teachers College, Columbia University. Currently, she teaches mathematics at The Brearley School in New York City and mathematics education at Stern College for Women, Yeshiva University. Her primary research interest is the history of American mathematics education. Her articles have examined both the formal and informal processes through which mathematical knowledge was disseminated in the nineteenth-century America.

[illegible] is a clinical assistant professor at Pace University in New York City. She received her PhD in mathematics education from [illegible], Columbia University and also holds a Master of Science degree in applied mathematics from [illegible] [illegible]

[illegible]

[illegible]

Contributors

Elisabete Zardo Búrigo Universidade Federal do Rio Grande do Sul, Porto Alegre, Brazil

Dirk De Bock KU Leuven, Leuven, Belgium

Vasily Busev Mathematical Institute, Russian Academy of Sciences, Moscow, Russia

Viktor Freiman Université de Moncton, Moncton, NB, Canada

Alexander Karp Teachers College, Columbia University, New York, NY, USA

Karolina Karpińska L. & A. Birkenmajer Institute for the History of Science, Polish Academy of Sciences, Warsaw, Poland

Antonio M. Oller-Marcén Centro Universitario de la Defensa de Zaragoza, Zaragoza, Spain

Yana Shvartsberg Pace University, New York, NY, USA

Alexei Volkov National Tsing-Hua University, Hsinchu City, Taiwan

Sian E. Zelbo The Brearley School, New York, NY, USA

Stern College for Women, Yeshiva University, New York, NY, USA

Contributors

[illegible]

Pafnuty Chebyshev and the Mathematics Education of His Time

Vasily Busev and Alexander Karp

Abstract This paper focuses on the work of the outstanding Russian mathematician Pafnuty Chebyshev in the Scientific Committee of the Ministry of Education of Russia, or more precisely, his reviews and comments on textbooks submitted to the Committee. These reviews were published in full only in 2021. An analysis of them makes it possible not only to learn more about Chebyshev, or more broadly, about what a major mathematician brings to mathematics education, but also to form a more complete picture of Russian mathematics education during the second half of the nineteenth century. One may draw the conclusion that Russian mathematics education was then still in a formative stage (including those aspects of it which pertain to expository language and style of presentation, its standards of accuracy and precision, and so on). P.L. Chebyshev's influence on this formative process was very substantial.

Keywords Chebyshev · Russian mathematics education · Scientific Committee of the Ministry of Education · Textbooks · Mathematics teachers · Reviews

1 Introduction

This paper is based on work on the publication of previously unpublished remarks by the outstanding Russian mathematician Pafnuty Chebyshev (1821–1894) on questions of mathematics education. Chebyshev was for a long time a member of the so-called Scientific Committee of the Ministry of Education of Russia, which oversaw among other things the approval of textbooks and the authorization of

V. Busev
Mathematical Institute, Russian Academy of Sciences, Moscow, Russia

A. Karp (✉)
Teachers College, Columbia University, New York, NY, USA
e-mail: apk16@columbia.edu

A. Karp (ed.), *Advances In The History Of Mathematics Education*,
International Studies in the History of Mathematics and its Teaching,
https://doi.org/10.1007/978-3-030-95235-8_1

handbooks for use in educational institutions. Over 250 reviews and comments by Chebyshev on the most varied topics have been preserved in the journals of the Scientific Committee's meetings. Vasily Prudnikov, a historian of mathematics and of mathematics education who devoted a number of papers to Chebyshev (for example, Prudnikov 1948, 1964, 1976), published and analyzed a certain portion of them—those publications constitute a part of the final, fifth volume of Chebyshev's collected works (Chebyshev 1951). Far from all of the surviving materials were published, however, and even in those that were published, cuts were made—certain passages, which were not of sufficient interest in the opinion of the publishers, were omitted.

At present, in connection with the bicentennial of Chebyshev's birth, a far more complete edition of his pedagogical remarks is being published.[1] It is on the analysis of these materials that the present paper is based.

As far as is known to the authors of this paper, Prudnikov's works have never been published in any foreign languages and hence are little accessible to researchers from countries other than Russia. However, the analysis of Chebyshev's work is of interest in the context of the study of what was done by other mathematicians in mathematics education. Without attempting even briefly to list all works devoted to this topic, let us mention by way of example the papers by Livia Giacardi (2006) on Luigi Cremona or Gert Schubring (2019) on Felix Klein.

Let us note at once that there exist quite varied views on the role and place of research mathematicians in the development of education. Confining ourselves to the recent history of Russian education, let us point to the paper by Sossinsky (2010), which assesses the contribution of mathematicians unequivocally positively, and the book by Kostenko (2013), which speaks quite a bit about the "absolute inability of scientists to put themselves in the child's place"[2] (p. 55). The topic of "the mathematician and mathematics education" clearly continues to attract attention, and the study of Chebyshev's work is unquestionably relevant to this topic.

However, the unpublished and even the published materials are also useful from other perspectives. Chebyshev worked under the conditions of a certain reality, and based on his reviews, as well as based on what was sent to him for review, conclusions may be drawn not only concerning Chebyshev himself, but about this reality as well. Chebyshev's reviews are a source on the history of Russian mathematics education with its general principles, the teachers then working, and the understanding of school mathematics. Prudnikov (1964) devoted an entire chapter to the discussion of the advantages and shortcomings of Chebyshev's methodological views from the modern perspective (that is, that of 1964). Less attention was devoted by him in this and other books about Chebyshev to the views held and problems faced by the teachers contemporary to Chebyshev. In the present paper, we will thus attempt to supplement and extend his research.

---

[1] While this article was being prepared for publication, a book by Busev and Rebrieva (2021) was published.

[2] All translations from Russian are by the authors.

Let us repeat that the main focus of analysis in the present paper are materials from the holdings of the Russian State Historical Archive, in referring to which we will indicate either the pertinent archival information or the relevant pages in Chebyshev (1951), if they have been published there.[3]

But first our discussion will turn to the Scientific Committee, to Chebyshev, and to the system of education that existed at the time of his tenure on the Committee.

2 The Scientific Committee of the Russian Ministry of Education

In order to understand what Chebyshev's pedagogical work consisted in, we must first tell about the Ministry's Scientific Committee. The actual date of its establishment can be reckoned in different ways. Georgievsky (1902) begins his account of the relevant activities practically with the Commission for the Establishment of Public Educational Institutions, which was established in 1782, and whose functions included much of what later came to be performed by the Scientific Committee. Similar objectives were also addressed by other commissions and committees established and dissolved by the Ministry (the Textbooks Committee of the Committee on the Organization of Educational Institutions, the Scientific Committee of the Main Academic Directorship, and others). In 1856, an organization bearing the official name the Scientific Committee of the Ministry of Education was established (or re-established, as is sometimes said, commemorating earlier organizations with the same functions) on the initiative of then-minister A. S. Norov (the reorganizations of the committee continued, the most significant of them taking place in 1863, for which reason the founding of the Committee is sometimes dated to that year).

The guidelines for the Committee's work formulated the following objectives:

(a) Assessing courses at educational institutions;
(b) Creating or reviewing programs for textbooks;
(c) Reviewing and assessing textbooks;
(d) Reviewing other books and manuscripts, and.
(e) Reviewing plans for presentations on academic, educational, and pedagogical topics and other similar projects (Georgievsky 1902, p. 19).

In order to carry out all of these functions, eminent experts in the main disciplines were included in the membership of the Committee. During the very first meeting of the newly established Committee, its members were announced by Minister Norov. They included Chebyshev, an adjunct at the Academy of Sciences, who was charged with overseeing the teaching of mathematics (naturally, along with fulfilling all his other existing duties). It should be pointed out that involving major

[3] All of these documents have now also been published in Busev and Rebrieva (2021).

mathematicians in the work of the Scientific Committee or organizations carrying out similar functions was something traditional—we might mention Dmitry Chizhov, who served in such a capacity during the first half of the nineteenth century, and Nikolay Sonin, who died shortly before the revolution of 1917 (Karp 2017, 2019). As a Committee member, P. L. Chebyshev performed various tasks, the primary ones being reviewing numerous textbooks and discussing various plans for reorganizing education, including the opening of new educational institutions and the development of educational curricula. Below, we will focus specifically on Chebyshev's textbook and manual reviews; Prudnikov's study (1964) contains information about other vectors of Chebyshev's activity as well.

3 Pafnuty Chebyshev and His Life

Already during his life, Chebyshev was recognized as an outstanding mathematician and the founder of a school of mathematics. The literature about him is substantial and includes both writings by research mathematicians who were in one or another respect his disciples (for example, Posse 1904 or Bernstein 1947), and works of a general biographical nature (Prudnikov's studies must be mentioned first and foremost, but they are not alone). Prudnikov (1964) describes in detail the genealogy of the Chebyshev family, nobility from Tula who served the tsar and their country mainly in the military arena. The slightly lame boy was unsuited for military service and was not even sent to an ordinary gymnasium. He studied at home and was taught mathematics by Platon Pogorelsky (Bogomolov 2006, Prudnikov 1956), who is remembered for making a new translation of the famous French textbook usually referred to in Russia as "Bellyaven" (Bellavène 1832–1834). The previous edition (Bellavène 1824–1825) clearly failed to satisfy Chebyshev's teacher, as is evident from the fact that many additions and changes in the book were made.

At the age of sixteen, Chebyshev enrolled in Moscow University, where his professors included Dmitry Perevoschikov, whose publications included textbooks for gymnasia (Perevoschikov 1837) and the so-called Small Encyclopedia (Ruchnaya…1826–1827). In 1841, Chebyshev graduated from the university, and in 1846 he received a master's degree there. Subsequently, Chebyshev moved to St. Petersburg, where he became an Adjunct professor (then this title had a bit different meaning than now) . In 1849, he was awarded a doctoral degree, becoming a professor extraordinarius shortly thereafter, and a professor ordinarius in 1860. In all, Chebyshev worked at the St. Petersburg university for 35 years. He was esteemed as an outstanding lecturer and the teacher of significant mathematicians, and concurrently wrote a number of important works on the theory of functions, the theory of probability, number theory, the theory of mechanisms, and other fields. His papers were actively published in the most prestigious and well-known journals of his time. Such concepts as the Chebyshev inequality, the Bertrand–Chebyshev theorem, Chebyshev polynomials, and Chebyshev bias have entered the international language of mathematics. Chebyshev was a member of the Russian and many other countries' academies of sciences. We should also note his academic connections with many well-known mathematicians of his time, which began already in 1852 when he took a long trip to Europe. Letters of the most important scientists of the time have survived (for example, Hermite), in which Chebyshev is recognized as one the greatest mathematicians in the world (Prudnikov 1964, p. 240).

Apart from strictly academic duties, Chebyshev also fulfilled many others, for example, he was a member of the artillery section of the so-called Military-Academic Committee and advised artillery experts. Likewise important was his work on the Scientific Committee, which is our subject in this paper. But first we must tell about the general organization of the Russian system of education during this period.

4 On the System of Russian Education in the Nineteenth Century

By 1856, when Chebyshev was appointed as a member of the Scientific Committee, and until the revolution of 1917, the system of Russian education was complex and multifaceted (and there was also the possibility of receiving a home education, as Chebyshev had done, in a certain sense outside this system). In effect, it had been largely created by the reforms carried out under Catherine II by Fyodor Yankovich de Mirijevo, whose plans were adopted in 1786 (Karp 2014). But subsequently the system continued to change and become restructured. Educational institutions were naturally divided into public and private ones (which were variously granted or not granted various rights), but even the public institutions could be subject to different agencies and ministries. Without going into all the details, we should say that there were educational institutions that were overseen by the Ministry of War—cadet corps and so on; by the Ecclesiastical Department—parochial schools, seminaries; by the Ministry of Finance or the Ministry of Industry and Trade (although with the participation of the Ministry of Education)—commercial schools; and lastly, institutions overseen by the Ministry of Education itself. This last category, in turn, included the following: gymnasia, the most privileged secondary educational institutions, which granted graduates the right to enroll in universities; real schools—secondary educational institutions based on the German Realschule model, which usually granted their graduates slightly lesser rights than gymnasia; as well as educational institutions that did not provide a full secondary education, but which could be used to prepare for admission to a gymnasium—uyezd (district) and municipal schools, which offered one-year, two-year, three-year, four-year, and even in special cases five-six-year programs. The concept of "public elementary schools"was also in use: their purpose, in the words of the regulations governing them (Polozhenie 1874), was "to inculcate religious and moral understanding in the people and to disseminate elementary useful knowledge," and they could include institutions affiliated with various agencies and even private and community ones. (Note that we have by no means listed all existing types of elementary and secondary educational institutions here. The collection Karp (2018a) contains more detailed materials about some, but far from all, types of prerevolutionary educational institutions.)

Each of these educational institutions relied on its own mathematics curriculum and its own textbooks. It is important to note that in spite of all departmental and other differences, approval of educational materials by the Ministry of Education was usually desirable. Consequently, such materials were sent for review to the Scientific Committee (which did not rule out their inspection by special committees of other agencies—for example, there was a special commission for reviewing educational literature at the Ministry of War, and Chebyshev was a member of this commission as well, which lies outside the scope of this paper).

In general, the Russian state strove for uniformity. The Russian Emperor Nicholas I plainly wrote as much in May 1826 in his rescript announcing the establishment of one of the predecessors of the Scientific Committee, the Committee for the Examination of Textbooks: "Reviewing with especial attention the organization of the educational institutions in which Russian youth is educated to serve the state, I regret to observe that they lack that necessary and indispensable uniformity which must be the foundation of both child rearing and instruction" (Perepiska 1826–1828, p. 1). To achieve the desired uniformity nonetheless proved impossible (at least when considered from a later, Soviet position).

Chebyshev, reporting to the Scientific Committee in 1863–1864 about the textbooks employed in gymnasia, names books that were used in all (at that time) seven school districts, only in six, five, four, and even only in three, two, or one (Chebyshev 1951, p. 330). At the same time, it is clear that by no means in all gymnasia of the same district the same books were used. Chebyshev notes that from the information submitted to him for review, it turns out that "most of these institutions [gymnasia] predominantly use books that were not only approved or recommended by the Ministry, but also published by the Department" (that is, a subdivision of the Ministry) (p. 329). However, from this description it is also clear that not all the textbooks being used fell into this category. A catalogue of recommended and approved publications was published (one such catalogue will be discussed below), which included several textbooks in each subject. We dispose of no information about the extent to which other textbooks were used in the teaching process.

The report concludes with the following important words by Chebyshev:

> As for improvements to textbooks in this subject, they are, of course, desirable, since the textbooks available today can hardly be considered perfect; but this, due to the nature of mathematical subjects, cannot be achieved by announcing competitions, since the writing of a course in these subjects, when carried out in haste, cannot be especially successful; such courses must be written over years and represent the outcomes of prolonged pedagogical activity.
>
> In order to arouse our pedagogues to such a labor, however, it is necessary merely constantly to follow new publications and permit everything to be used in gymnasia that proves worthy of such use, without granting any privileges in this respect to any books over others. (pp. 330–331).

This in fact was the program of Chebyshev's work, which will be our focus in what follows.

5 On Chebyshev's Work in the Scientific Committee

Chebyshev served on the Scientific Committee from 1856 (effectively from 1857, since he was away on a work trip before this) until 1873, when he retired from it. Prudnikov (1976) lists Chebyshev's main duties on the Committee:

> (1) Participation in developing regulations for elementary, secondary, and higher educational institutions. (2) Reviewing mathematics textbooks. (3) Continual monitoring to identify which mathematics textbooks schools needed for more successful teaching and finding the means for satisfying this need. (4) Writing and reviewing curricula for elementary and secondary schools. (5) Reviewing books with a physics-mathematics content designated to be presented to the tsar and members of his family as the most useful, or whose use among the public needed to be forestalled due to their defective character. (6) Recommending books with mathematics content that were intended for dissemination throughout the country's educational institutions. (pp. 113–114)

Naturally, all of these duties were in a definite way interconnected. For example, the report prepared by Chebyshev in 1867 "On Textbooks and Handbooks in Mathematics Used Today in Gymnasia and Elementary Public Schools" begins with the suggestion that M. Tsvetkov's algebra textbook (which was named among those being used in the aforementioned report of 1863–1864) be excluded from the catalogue of scientific textbooks and handbooks, and replaced with the textbook of A. Davidov. Chebyshev offers elaborate reasons for his view.

> I consider it necessary to exclude Mr. Tsvetkov's course from the catalogue because this course contains many topics that are poorly presented. Thus, the author begins section II on equations in the following manner: "Any problem whose subject is only subtraction can be expressed in the form of an equality between what is sought for and the totality of the unknown numbers" (p. 48). It is difficult even to surmise what exactly the author wished to say with these words.
>
> Section V begins with the words: "Equations are called indeterminate when the unknown has several solutions" (p. 80). This is wrong, not to mention the incorrectness of the formulation. Determinate equations can have several solutions.
>
> About logarithms, Mr. Tsetkov says, among other things, the following: "The logarithms of basic numbers are in general incommensurable logarithms, and therefore any such logarithm is expressed in a general form as any unmeasurable magnitude is expressed, for example, $\sqrt{a+x}$, $\sqrt[3]{a+x}$, $\sqrt[4]{a+x}$ is expressed by an infinite series of terms distributed along the degrees of the number x" (стр. 187). This is poorly expressed and wrong.
>
> On p. 79 we read: "Infinitely large solutions always indicate that the question is meaningless, except when the question pertains to higher geometry." Such an opinion is unfounded. (Chebyshev 1951, pp. 336–337).

Chebyshev's report thus includes a review of a textbook (we will refer to this review below). Nonetheless, we should say at once that as far as possible we will confine ourselves only to reviews of textbooks that are actually presented as such. Chebyshev put together quite a number of such reviews. Prudnikov indicates that in 1857, Chebyshev presented five reviews; in 1858, seven reviews; in 1859, ten reviews; in 1860, five reviews, and so on (Chebyshev 1951, p. 313). It must be said, however, that works reviewed by Chebyshev included ones that we would not consider textbooks or handbooks in mathematics today: "Handbook on Drawing According to the Rules of Perspective," "The Home Land Surveyor, or Handbook on How to

Measure and Compute a Sazhen [Fathom] of Different Ground Lines in a Simple Manner by Using Squares and How to Translate Them into Decimalized Proportions," or even "Handbook on Correctly Determining Advantages and Disadvantages for Economic Enterprises" (these titles appear in Chebyshev (1951, p. 313)). With a minor exception, we will not analyze reviews of such texts.

6 On Chebyshev's Reviews of Mathematics Textbooks

6.1 Some General Considerations

One of Chebyshev's best known students, Konstantin Posse (who subsequently served as the head of the Russian national subcommission of ICMI—Karp, 2017) noted that:

> the main contingent of those unhappy with Chebyshev consisted of authors of bad textbooks in elementary mathematics, toward which his attitude was one of relentless severity, knowing as he did all the harm these textbooks caused students and the benefit their authors derived from them when they received approbation (Posse 1904; also cited in Prudnikov, 1964, p. 227).

It should be said that the authors of the reviewed texts occupied the most varied positions: confining ourselves even to reviews written before 1862 (contained in the Russian State Historical Archive in fond 734, list 2), among the authors reviewed we see professors, such as the aforementioned Perevoschikov; Fyodor Busse, effectively the founder of Russian mathematics methodology, who for many years served as the director of a gymnasium; gymnasium teachers from Arkhangelsk, Belostok, Kazan, Kargopol, Minsk, St. Petersburg (the famous textbook author Konstantin Krayevich), Smolensk, Stavropol; the inspector of the Kazan gymnasium; a teacher at the Poltava Cadet Corps, Simashko (another well-known author of textbooks); apparently, a teacher at the Naval Cadet Corps; teachers from uyezd (district) and elementary schools from various regions of Russia; a teacher from a Vilna rabbinical school; a teacher from an orphanage for poor children; as well as a conductor of Marine Engineers Corps; a district land surveyor; and persons whose respective files indicate only their ranks—collegiate assessor, collegiate councillor, collegiate secretary, district secretary, or even simply nobleman, burgher, or mister. In addition, sometimes compositions published abroad and written by foreigners were reviewed. The authors of school textbooks were thus usually teachers themselves, although they could also be skilled professionals (such as the land surveyor).

Prudnikov (1964) formulates certain basic principles by which, in his view, Chebyshev was guided in writing reviews. They include conformity not only with the requisite scientific level, but also with the requirements of the school system, implying the absence of any fundamental deviations from the existing system of instruction in each of the mathematical subjects, linguistic precision and clarity, the prioritization of what is most important in the exposition, and so on. For example, one of the documents prepared by the Scientific Committee (that is, effectively by Chebyshev) states: "All explanations and proofs must be sufficiently rigorous, without superfluous details, and with an extremely circumspect use of words such as 'obvious,' 'self-evident,' and so on" (Prudnikov 1964, p. 230).

Agreeing with these general conclusions, below we will analyze certain examples, attempting to show with what exactitude Chebyshev demanded that curricula be adhered to, precisely which mistakes and laxities he pointed out, what objections he presented concerning the language of the exposition, and lastly, we will touch on some of his methodological views.

6.2 On Adhering to Curricula

It is fair to say that Chebyshev on the whole envisioned the system of textbooks as rather rigid, adhering to curricula in details, and devoid of any extravagances, including excessive numbers of textbooks. In one of his reviews, Chebyshev enunciates his understanding of the matter clearly:

> Arithmetic is taught in all elementary and secondary schools, but in keeping with these schools' course components, each of them—the parish school, the uyezd school, and the gymnasium—requires its own special textbook. (Chebyshev 1860).

Consequently, textbooks in arithmetic in general, for example, which combined the easy with the difficult, were rejected by him. "Does not correspond to the scope of arithmetic instruction in any of the various categories of the ministry's educational institutions" (Chebyshev 1861b)—such a remark was a weighty argument for denying a book approbation.

The course had to be presented in the generally established fashion; in the reviews, contrary approaches usually serve as evidence of deficiency: "Does not correspond to the generally established system for presenting geometry" (Chebyshev 1858b); "because [the text] deviates from the generally established system, the most simple and natural, and contains erroneous conclusions at its foundation, it cannot usefully serve as a textbook" (Chebyshev 1861a); "this course's deviations from the generally established way of presenting trigonometry are clearly to the detriment of the most significant qualities required of any textbook" (Chebyshev 1862a). The list of such quotes could be continued.

In the best cases, a book that contained more material than was prescribed by the curriculum could be recommended for purchase by school libraries. This was the case, for example, with Ritt's arithmetic, translated from the French by a certain Andreev (Chebyshev 1857b). It was noted that

> since this book presents certain topics in an elementary fashion that are not part of the course components in arithmetic (such as: progressions, measuring surfaces and volumes, and solving problems that require writing equations) but nonetheless constitute a good supplement to the course in arithmetic, making Mr. Andreev's books accessible to arithmetic teachers in elementary schools promises to be undoubtedly beneficial.

However, this was far from always the case.

Changing the sequence of topics in a course was undesirable. Thus, for example, Chebyshev writes about the curriculum put together by the teacher Zakharov as follows (Chebyshev 1951, p. 340):

> It differs from the currently accepted order in which arithmetic is presented in courses taught at uyezd schools mainly because it introduces fractions before nominative numbers. Such an order cannot be preferred to the long-accepted one in the course taught at uyzed school, in which nominative numbers, as a simpler subject, understandable by itself and particularly necessary in practice, are introduced before the chapter on fractions.

Prudnikov (1964, c. 229) expressed indignation at the fact that Chebyshev turned down such an important book in the history of Russian methodology as Pyotr Guryev's *Practical Arithmetic*. Indeed, Chebyshev wrote the following:

> Mr. Guryev has not been constrained by the requirements of the system of arithmetic instruction, a system universally accepted in our schools. Thus, between chapters on subtraction and the division of numbers not greater than 100 (p. 13), a chapter appears about the most useful measures of length, weight, etc. The same chapter, with certain additions, appears after a chapter on changes in the quotient that arise from various changes in the divisor and the dividend (p. 132). The same chapter with new additions appears yet a third time on pp. 216–222 (Chebyshev 1951, pp. 369–370).

Thus, catching Guryev in the act of a concentric presentation of arithmetic and noting certain other flaws as well, Chebyshev concluded that the book cannot be accepted among the number of textbooks used in schools.

To be sure, Chebyshev clearly believed that the number of such textbooks should not be large. The following conclusion in connection with Mudrov's textbook in arithmetic is indicative (Chebyshev 1857a):

> In the opinion of Mr. Chebyshev, the "Handbook" written by Mr. Mudrov does not deserve particular attention: because, although it does not contain significant defects, neither does it possess such virtues as might give it preference over textbooks in use in the Ministry's schools.

It should be recognized, however, that Cheybshev's severity, not always justified from today's point of view, was displayed at a time when those in positions of authority were in fact expected to be rigorously demanding, and a professor who was a member of the Scientific Committee was undoubtedly and correctly perceived as someone in a position of authority. Schools in Chebyshev's time and later were not directly subject to universities, but professors (by no means only members of the Committee) were encouraged to become involved in what in Soviet times was known as "guidance and oversight." Note that even after 35 years after Chebyshev left the Committee, an observer wrote that the teachers of gymnasia believe that "if a professor is sent to any teacher, then a punishment awaits [the latter] in the near future" (Istrin 1908, p. 63). In the period preceding Chebyshev's, Dmitry Chizhov, a university professor who also became involved in working with schools, likewise did not manifest any special kindheartedness, and not only in questions of mathematics, but also in strictly administrative matters (Karp 2019). Later, the author of a famous problem book in algebra, Nikolay Shaposhnikov (1906), published a special pamphlet violently attacking the work then being done by the Scientific Committee (more precisely, first and foremost attacking its member Boris Koyalovich); but even he saw fit to point out that at one time "official pedagogical criticism, represented by the Scientific Committee of the Ministry of Public Education, could be regarded as a more or less normal phenomenon" (p. 1). Very likely, this was precisely how Chebyshev's work was perceived by others.

6.3 Mistakes and Inaccuracies Identified by Chebyshev

The identification of mistakes and inaccuracies formed an important part of Chebyshev's work. Here it must be said that the mathematical caliber of the materials under review was by no means always exemplary. This was true not only, say, of the composition by Major General Poluektov, who had sent in his own method for dividing an angle into three equal parts (Chebyshev 1862b)—such contributions are received everywhere and at all times. Or even of the disquisitions of Collegiate Secretary Smyarovsky, about which Chebyshev made the following remark:

> As for the presentation, as one always observes in authors poorly acquainted with the subject of their compositions, it is distinguished by its obscurity and disconnectedness. But in Mr. Smyarovsky's case, these expository defects are additionally compounded by something specific that attests to the not altogether normal state of his mental faculties. (Chebyshev 1861c).

This, too, was clearly an isolated case.

Far more indicative as a characterization of the general level of mathematical literacy, in our view, even among people who use mathematics in their work, is the already mentioned composition "The Home Land Surveyor" (Chebyshev 1858a). Chebyshev even praises the author for his diligence and the laudable desire "to offer assistance to those landowners who are unable to measure their lands themselves due to their insufficient knowledge of the arithmetic and geometry necessary for this purpose." The trouble, however, lies in the fact that the author, "in calculating lands, pays attention exclusively to the lines that delimit them, and based on the length of these lines, he proposes to calculate their areas by using the tables included in his work." So that Chebyshev is forced to point out that shapes "can have different areas when the lines are the same."

Teachers were not always better. Chebyshev quotes Kirshteyn, a teacher at the Mtsensk uyezd school:

> From the first rule, he said, we conclude: if none of the factors is divisible by some number, then their product will also not be divisible by this number without a remainder" (p. 60). The inaccuracy of this conclusion is demonstrated by the simplest examples: thus, neither 6 nor 10 is divisible by 4, but the product of 6 and 10 is the number 60, which is divisible by 4. (Chebyshev 1859b)

This mistake was not made by Kirshteyn alone. In a review of the course prepared by Lieutenant Colonel Khristiani of the General Staff (Chebyshev 1863b), we read:

> Mr. Khristiani, after certain considerations, supposes that their first consequence is the following: "If a product is divisible by some number, then at least one of its factors must be divisible by this number." Using the simplest examples, we can see the groundlessness of this conclusion: for example, the product of 10 and 12, which is equal to 120, is divisible by 8, while neither 10 nor 12 are divisible by 8. Further, Mr. Khristiani writes: "If one number cannot be divided by another without a remainder, then its power will also not be divisible by this number." This is also wrong: 10, for example, is not divisible by 4, but all of its powers $10^2 = 100$, $10^3 = 1000$ and so on are evidently divisible by 4.

These, as Chebyshev immediately points out, are not the author's only mistakes in elementary number theory.

> With reference to factoring numbers, the author remarks that in searching for divisors, numbers should be tried out only until the divisor being tested is greater than half of the dividend (p. 37). For this, as is well known, a different, closer limit can be given.

In certain cases, the mistakes are such that one would like to assume that they are the results of common misprints; but on the other hand, these mistakes are repeated. Chebyshev (1866) writes: "In addition, it must be noted that Mr. Nikulin employs symbols incorrectly: thus, he writes: 5–12 = 7; 12–19 = 7 and so on (p. 117)."

And what certainly cannot be dismissed as a mere misprint is a theorem invented by Volens, the author of many textbooks. Chebyshev (1951, p. 377) writes as follows:

> In conclusion, I will note that on p. 88 we even find a proof that leads to a conclusion that is positively false. Here, the author proves, among other things, that in any polygon that is inscribed in a circle, the sides are equal if the angles are equal. That this is not always the case can be seen from the fact that about any rectangle, no matter what its sides are, a circle can be circumscribed.

The same Volens also make a more subtle mistake (pp. 376–377):

> Here is how the author, on pp. 28 and 29, deduces that from a given point, only one straight line can be drawn that is parallel to BC: "From point A, we can always drop a perpendicular AB to BC; then let us draw the perpendicular AD to the straight line AB at point A. The straight lines AD and BC will be parallel, since both are perpendicular to a third, AB. Consequently, since at point A, only one perpendicular to AB can be drawn, it stands to reason that from point A only one straight line parallel to BC can be drawn." But from the fact that only one parallel is obtained by means of the construction demonstrated here, one cannot yet conclude that another construction would not give another parallel.

It should be noted, however, that mistakes in the proofs of uniqueness are found in other textbooks as well, including even the famous *Geometry* by Kiselev.

The following, on the other hand, is a mistake of a philosophical nature, so to speak, although it reveals a serious lack of comprehension of the structure of geometry, from the book of another prolific author, V. I. Berens (Chebyshev 1951, p. 378):

> Proceeding to Euclid's famous postulate, the author writes as follows: "Now drawing some [other] straight line through the point A, it will be necessary to prove that it will meet with DC. No exact proof of this truth thus far exists; therefore, we will accept it without a proof." But lack of a strict proof alone cannot be a basis for accepting anything as indubitable.

Mistakes pertaining to general ideas can be encountered even in the writings of the famed author of arithmetic textbooks V. A. Yevtushevsky. Chebyshev (1951, p. 381) writes as follows:

> Here is what [the author] says about arithmetical operations in general (p. 128): "Addition is the fundamental arithmetical operation—all other arithmetical operations derive from it by way of the simplification of computations." This directly contradicts the idea of direct and inverse operations.

But let us not cite any more mistakes (which are indeed noted by Chebyshev in the presentation of all school subjects). Instead, let us now turn to Chebyshev's remarks concerning the expository language and style of the textbooks, about whose obscurity and disconnected he very frequently complains. And would that this pertained only to compositions of an explicitly philosophical character, such as E. Shneyder's "Genetically Organized Arithmetic"! Its author, according to Chebyshev, writes as follows:

> "In solving various problems, we notice that each of them can be solved by means of ratiocination, but that nonetheless from the sequence of various operations that is repeated in each problem, general rules for solving may be deduced; initially, problems may be put in the form of either two equal sums or two equal products" and the like (p. V).
>
> On the connections between the different parts of arithmetic, Mr. Shneyder expresses himself as follows: "in the arithmetical sciences, each new step must be nothing other than merely another form of a previously deduced truth" (p. VI; a judgment that is evidently false).
>
> This is how Mr. Shneyder formulates the idea of zero:
>
> "Zero expresses the fact that the idea exists, but without an object belonging to it. From this it follows that zero may also be taken as a number, but only in the form of a term in a numerical sequence" (p. 3). (Chebyshev 1951, p. 363).

But Chebyshev makes the following remarks about Malinin's and Burenin's arithmetic, a work that is usually considered a classic of Russian mathematics education:

> on p. 2 the following definition of a number is given: "A number is one unit or the totality of several uniform units." On p. 3, where a conclusion is drawn from the concept of number, § 4 begins as follows: "Because any number is the totality of several units, therefore..." and so on. (Chebyshev 1951, p. 360).

Chebyshev's review of Tsvetkov's widely used algebra textbook with its extremely awkward formulations has already been cited. And here is Chebyshev's remark about the course in geometry by Berens:

> The next paragraph § 3 begins as follows "Although we can always have surfaces, lines, and points on a body, we can also imagine them separately" (p. 2). 378. It is difficult to understand the real meaning of the first sentence: "to have points on a body," etc. (Chebyshev 1951, p. 378).

And here is Chebyshev's judgment on a text written by Zakharov, a teacher at a gymnasium in Stavropol:

> Thus, in § 96 we find the following passage, which is completely incomprehensible: "What is called the accuracy of a decimal? A vivid explanation of accuracy is as follows. You know very well from where A might receive money; therefore, when you find out that he has just now received money, you can guess how much he received. You say, for example, that he has 100 rubles, but he has, say, 105; therefore, you are mistaken by 5," etc. (Chebyshev 1951, p. 340).

Chebyshev speaks very often about the obscurity of the presentation. And this, of course, has to do largely with the lack of awareness of the specific nature of mathematical concepts on the part of the authors and their resulting desire to formulate and define these concepts "in a simple way" with a certain naivete. The already mentioned Volens wrote that "triangles that are identical in size and form are called equal; therefore, equal triangles, when superposed, will coincide." With regard to this, Chebyshev (1951, p. 376) remarked: "What exactly the author means by the definition 'identical in size and form' is not evident, and how the consequence that the author deduces from this follows is impossible to understand." Such pseudo-definitions are found in textbooks to this day (one cannot help recalling that many American textbooks state that geometrical figures are similar if "they have the same shape but are different in size"), demonstrating on the one hand the authors' lack of understanding of what constitutes a mathematical definition, and on the other hand, their pedagogical inability to circumvent a situation in which giving a precise definition is for any reason difficult.

We will confine ourselves to just one more example, which demonstrates such a naive approach. Here is what Chebyshev wrote about the course in arithmetic written by the uyezd land surveyor Boguslavsky:

> at the very beginning of his work, Mr. Boguslavsky, mixing up the concept of quantity in general with finite quantities—and the counting of objects with the measuring of quantities—writes as follows: "We call a finite quantity anything that is able in our imagination and in reality to increase and decrease, for example, a table, a blackboard, a pen, a notebook, and so on; from this we see that each finite quantity is subject to change or alteration; and next to the notion of a change in quantity stand numbers. Numbers are familiar to us from that time in our life when we began definitively to match or compare objects with one another and to distinguish between the quantity of one and the other, for example, a little babe says: 'I have one toy, you have two toys, he has three toys, etc." (Chebyshev 1861e).

Prolixity was seen as facilitating understanding, to which Chebyshev always objected, critically noting when a book was "replete with superfluous details" (Chebyshev 1859a). The language of school textbooks and the language of elementary mathematics in general had not yet taken shape and remained subject to different influences, from what may be described as the quasi-philosophical seminary tradition to not always literate borrowings from foreign literature. Below, we will cite examples of the language of textbooks from before Chebyshev's time, whose traditions in a certain sense Chebyshev was precisely trying to overcome. In general, the question of the formation of the language of school textbooks, in our view, requires further study. In any case, it is clear that Chebyshev contributed not a little to its development.

6.4 On the Rigor of Proofs in Textbooks

As we have already said, Chebyshev always insisted on rigor and completeness in explanations. For example, he commented the course in arithmetic by Mel'gunov as follows (Chebyshev 1861d):

> In Mel'gunov's course, the most fundamental topics are presented without the requisite explanations, such as: the rules of addition, subtraction, and the like. The author considers it sufficient to say how various arithmetical operations are carried out, without explaining the meaning of various techniques that are used in the process, without which no successful teaching of any mathematical subject is possible. Here, for example, is what the author limits himself to when presenting the multiplication of ordinary fractions (p. 39): 'In order to multiply one fraction by another, both the numerators and the denominators must be multiplied by one another. For example: 5/7 × 2/3=10/21.' Most of the topics in Mr. Mel'gunov's course are presented in this way.

And here is Chebyshev's remark concerning the manuscript of Moscow gymnasium teacher Polyakov:

> in its presentation, Mr. Polyakov's arithmetic hardly meets the requirements of the gymnasium course, since many proofs of especially important propositions are presented in it in an unsatisfactory manner. In order to show to what extent Mr. Polyakov's arithmetic suffers in this respect, I will quote here the proofs offered by the author in connection with various propositions.
>
> "The invariability of the product when changing the order of the factors; 5 × 4 means to repeat 5 four times by adding, that is, 5 + 5 + 5 + 5 = 20. 4 × 5 means to repeat 4 five times by adding, that is, 4 + 4 + 4 + 4 + 4 = 20. Thus, it does not matter whether it is 5 × 4 or 4 × 5; therefore, from changes in the order of the factors, the product does not change..."
>
> It is not hard to see that the conclusions drawn here by the author lack proper rigor, and many of the proofs offered by Mr. Polyakov suffer from the same defect. (Chebyshev 1951, pp. 374–375).

Remarks of this kind are quite numerous. For example, about a book on arithmetic by a Tver gymnasium teacher, Pol'kovsky, Chebyshev (1863a) writes as follows: "The inadequacy of Mr. Pol'kovsky's course in this respect may already be seen from the fact that he leaves a large part of his most significant propositions without proofs, or limits himself only to mentioning what might serve as a basis for such proofs." About the course in geometry, which was intended not even for gymnasia, but for uyezds and cities public schools, Chebyshev (1951, p, 323) wrote that "nothing must be presented without a proof."

What is noteworthy in connection with the aforementioned Muscovite Polyakov, however, is that here we have the opportunity to observe a methodological discussion. Pyotr Polyakov sent an explanatory note to the Scientific Committee, which contained the following paragraph (cited in Chebyshev 1951, p, 388):

> In my previous explanatory note, submitted along with the textbook, I myself admitted this defect [insufficient rigor] and cited the following reason for it: the first part of the practical arithmetic and the beginning of the second (the section on divisors) contain not entirely precise definitions and proofs because strict scientific definitions and proofs are inaccessible to children of the age at which arithmetic is usually covered. One must perforce sacrifice rigorous precision in return for clarity and the accessibility of the subject. Yet indeed, such a necessary sacrifice cannot in any way harm matters, because arithmetic, when it is repeated by students in the seventh grade of a gymnasium, is presented in a rigorously scientific form, using textbooks that are completely unlike those used in the lower grades.

Chebyshev remained completely unconvinced by this argument. He responded as follows:

> Such a view of the teaching of arithmetic is positively incorrect. Proofs lacking requisite rigor can bring nothing but harm. Without even mentioning the purposeless loss of time spent on studying such proofs, unrigorous proofs exert a harmful influence on students' mental abilities, accustoming them to seeing sufficient reasons where there are none. If something cannot be proved rigorously, students must be told about this directly, and not be led into deception by being offered an unrigorous proof. An example of this was given to us by Euclid himself in his geometry. Mr. Polyakov is likewise mistaken when he claims that when arithmetic is repeated in the seventh grade in the gymnasium, it is presented using textbooks entirely different from those used in the lower grades. Without even mentioning the fact that it is impossible to cover algebra successfully with students who have not fully mastered arithmetic, the teaching of arithmetic in gymnasia would require too much time if it were covered twice using two different textbooks, with different proofs of the same truths.

The Scientific Committee supported Chebyshev, of course, and rejected Polyakov's book, which, however, did not prevent it from going through numerous editions (for example, Polyakov 1897).

6.5 Digression: An Example of a "Pre-Chebyshev" Textbook

However, it is impossible not to ask: so how, then, should the fact that the multiplication of positive integers is commutative have been explained? We will not attempt to compare all textbooks that existed by that time, of course, but confine ourselves to just one: the textbook by Bellavène, the very one that Chebyshev himself likely used as a student (the quotes below are from the textbook on arithmetic by Bellavène 1832–1834).

Here, we first learn what multiplication is. It is defined as the addition of several identical numbers. The conclusion of the argument is summed up as follows:

> Thus: Multiplication is an operation in which one number is taken as many times as a unit is taken to compose another. But it is evident that this definition will have no meaning when the second number is a fraction; therefore, we will express it in general as follows: multiplication is an operation in which the number sought is composed from a given first number in such a way as the second is composed from a unit (pp. 12–13).

Here it must be noted that the concept of the fraction was presented earlier: a number is called "*fractional* if [it is composed] of whole numbers and parts of a unit; or simply a *fraction* if it is composed of parts alone" (p. 4). However, the explanation of multiplying by a fraction will hardly appear comprehensible to a modern reader.

Further, there comes a discussion of commutativity:

> Although it is evident that multiplying 4 by 3 or 3 by 4, we will obtain identical products, the same must be understood of any two factors; many writers, however, have deemed it necessary to prove this, as for example, Legendre in *Number Theory*. In order to obtain an understanding of the proof proposed by him, suppose that we must multiply 8 by 3. First, $8 \cdot 3 = (3 + 5) \cdot 3 = 3 \cdot 3 + 5 \cdot 3$; next, $5 \cdot 3 = (3 + 2) \cdot 3 = 3 \cdot 3 + 2 \cdot 3$; but $2 \cdot 3 = 2 \cdot (2 + 1) = 2 \cdot 2 + 2 \cdot 1$; and finally $2 \cdot 1 = (1 + 1) \cdot 1 = 1 \cdot 1 + 1 \cdot 1$. If we combine these various parts of the product, we will obtain $8 \cdot 3 = 3 \cdot 3 + 3 \cdot 3 + 2 \cdot 2 + 1 \cdot 1 + 1 \cdot 1 = 9 + 9 + 4 + 1 + 1 = 24$. Now suppose that we must multiply 3 by 8: first, $3 \cdot 8 = 3 \cdot (3 + 5) =$

> $3 \cdot 3 + 3 \cdot 5$; next, $3 \cdot 5 = 3 \cdot (3 + 2) = 3 \cdot 3 + 3 \cdot 2$; but $3 \cdot 2 = (2 + 1) \cdot 2 = 2 \cdot 2 + 1 \cdot 2$; and finally, $1 \cdot 2 = 1 \cdot (1 + 1) = 1 \cdot 1 + 1 \cdot 1$. Thus, $3 \cdot 8 = 3 \cdot 3 + 3 \cdot 3 + 2 \cdot 2 + 1 \cdot 1 + 1 \cdot 1 = 9 + 9 + 4 + 1 + 1 = 24$.
>
> Consequently, the proposed proof consists in factoring an overall product into partial products, which should be composed of equal factors. And since the partial products here are in both cases the same, the overall product does not change from a change in the order of the factors. (pp. 13–14).

The modern reader will likely doubt the persuasiveness of the proposed argument, if only because the distributivity of multiplication over addition has not been discussed anywhere earlier.

The courses by Bellavène—not his Arithmetic, however, but his Algebra and Geometry—were included in the catalogue of the Ministry, endorsed by Chebyshev (Chebyshev 1951, pp.338–339), which was referred to above; nonetheless, we can by no means claim that Chebyshev required something specifically like what was in the quoted textbook. Another "pre-Chebyshev" textbook, by Bunyakovsky (1849), which likewise was included in the Ministry's catalogue, in its explanation of multiplication, very early on introduced the multiplication table and explained (or rather, illustrated) commutativity by using this table (p. 21). In the textbook by Guryev (1832), which was a component of a larger manual reviewed by Chebyshev, the commutativity of multiplication is presented as a rule, illustrated by an example, but not explained at all (p. 50). As has already been said, Chebyshev did not approve Guryev's textbook, but wrote nothing about this particular point.

Most likely, what Chebyshev did not like in Polyakov's textbook was precisely the fact that a student might accept a single example as a proof, while the author took no measures to preclude this. It should be borne in mind, however, that the very understanding of what constitutes a proof in a school textbook did not emerge by itself and all at once, nor did the understanding of what needs to be proved: as we saw above, an author could reason that something might seem obvious, yet some people consider it necessary to prove it, so we might as well prove it also. In this regard, too, we observe a certain developmental process, in which Chebyshev took part.

6.6 About Some of Chebyshev's Methodological Positions

Everything discussed above effectively pertains to methodological views. We may also specifically emphasize the fact that Chebyshev clearly attributed particular significance to the computational side. He himself plainly wrote as much:

> I think that it is more useful not to introduce new sciences into real gymnasia, but to devote particular attention to exercises in computation: pupils in real gymnasia who wish to enroll in schools of applied sciences or to devote themselves to practical activity directly will derive greater benefit from knowledge of techniques that reduce computations and eliminate mistakes than from a superficial acquaintance with analytical geometry and descriptive

> geometry, since at the aforementioned schools these sciences are studied from the very beginning, while in practical life these sciences can be applied in any way whatsoever only given a complete knowledge of them. (Chebyshev 1868b).

Along the same lines, we can mention Chebyshev's resistance to five-digit tables (for example, Chebyshev 1951, p. 365). He insisted on using seven-digit tables everywhere.

In this section, we would like mainly to tell about Chebyshev's attitude toward problems in textbooks on arithmetic. Here, he closely monitors the gradual increase in difficulty—based on the principle that difficult problems cannot appear before easy ones. Let us cite a long, but illustrative passage from a review of a collection of problems in arithmetic by Yevtushevsky (Chebyshev 1951, p. 374):

> As for the specific positions of problems within each section, they are done strictly in accordance with the magnitude of the numbers; but the problems' levels of difficulty do not always follow one another in the proper order. Thus, for example, comparing problems 3, 9, 11, and 24 in the first section of the first part, we find that problem 11—"A family consists of a father, a mother, a son, and a daughter. How many people are in this family, and how many men and women?"—turns out to be easier than problem 9, which also involves the number 4, and even easier than problem 3, which states: "A boy going to school received three kopeks from his father to buy breakfast; for one kopek, he bought an apple, and for the remaining two, pretzels. How much did he pay for each pretzel?" Problem 24 is as follows: "Two boys bought three apples; one bought one apple and the other bought two apples, for the same price each; the second paid two kopeks more for his apples than the first. How much did all three apples cost?" while the next problem (the 25th) involves simply counting days of the week, and so forth...

Such remarks are quite numerous. Chebyshev clearly believed that sequencing problems in keeping with an increasing level of difficulty was important in teaching arithmetic. Note that such remarks are practically nonexistent with reference to other subjects.

Methodology as organization and sequencing in terms of the level of difficulty was only beginning to develop, especially when dealing with problems (Karp 2015). In the teaching of arithmetic, however, problems and their positioning received more attention. The full (or rather, almost full) title of the book by Guryev (1832), cited above, is: "Arithmetical Sheets, Gradually Sequenced from the Easiest to the Hardest." Therefore, the idea of sequencing problems in terms of their difficulty belongs, of course, not to Chebyshev. But it was an idea that he adhered to.

In this regard, however, we should also note that Chebyshev did not like violations of order in general. For example, he proposed that the following problem be excluded from a problem book in arithmetic: "A boy said: If you give me another 40 nuts, I will have as many nuts as my brother; and if you give me 90 nuts, I will have twice as many nuts as he. How many nuts did he and his brother have?" (Chebyshev 1951, p. 357). The reason for this was that this problem, as Chebyshev put it, "actually belongs to algebra." The fact that in this particular case no algebra is needed to solve the problem, which, of course, he could not fail to see, clearly had no significance for him: the problem was not where it should be, since it involved two unknowns.

7 Conclusion

Pafnuty Chebyshev was a great mathematician, and our aim here is not at all either to praise or to criticize his pedagogical work. It is more important to document specifically what he did and to try to understand under what conditions and why he did this. At the beginning of this paper, we spoke about the study of the role of mathematicians in mathematics education and about certain possible typical aspects of their work. If one wants to, one can simply rebuke Chebyshev for insufficient attention to age specifics and excessive attention specifically to strictly mathematical details, which might be said to be typical of mathematicians—but Chebyshev was summoned to do specifically the work that he did, including searching out mistakes. And during times far closer to ours, too, it has usually been expected that the mathematician would search out mistakes, rather than engage in discussions about the general perception of mathematics that can be derived from a textbook. Chebyshev in fact exceeded the limits of this assignment; it is a separate issue that his conception of what the content of a course in mathematics should be was rather conservative, and his reasoning was based on the supposition that the image of mathematics that schools should convey to students was already determined.

It is a separate issue, too, that many mistakes identified by Chebyshev astonish us by the fact that it required Chebyshev to identify them. And this was the case, even though the authors of the submitted textbooks were by no means always neophytes. In certain cases, one might attempt to suppose that the reviewer exaggerated, presented as a mistake what was obviously a misprint, and so forth, but this is clearly not the only and not the main explanation.

Russian mathematics education was still in its formative stages. We have already mentioned that the language of textbooks was only taking shape. But much else was taking shape as well, including requirements for students and teachers, and techniques for interactions between them. The late nineteenth and the early twentieth centuries became a time of a remarkable blossoming of Russian methodology, but it was prepared by a prolonged preceding period (or rather, periods), when among other things the very profession of the mathematics teacher took shape (Karp 2018b). Recognition of the fact that the teaching of mathematics was a separate and special field had not yet been fully achieved by Chebyshev's time (to be sure, it has probably not been fully achieved even now)—the number of colonels, lieutenant colonels, land surveyors, landowners, and other *educated people* prepared to write textbooks, as we have seen, was great.

We can ask: what influenced the processes taking place? It is natural to think that the growing number of students was important in this respect. It was important for the development of the teaching profession, if only because it was accompanied by a growing number of teachers, who consequently began to think more about their work, but of course this was not the only reason. It became clear that it was necessary to teach differently. For example, as has already been noted, the role of problems changed.

Naturally, while algebra, and even more so, geometry, were seen as compendia of wisdom (even if such an assertion requires various caveats) that needed to be assimilated, but that no student was required to develop on his own to any degree or even specifically to apply to anything by solving problems, arithmetic was definitely oriented toward problem solving. Arithmetic, indeed, was taught to far more people than algebra or geometry. And the working teacher's "ingenuous" approach—that students had to solve certain problems, and that therefore in his course he had to show students how they were solved and train them to solve others like them—was giving way to an approach that was more systematic, with a gradual transition from one problem to another. Note that something like a methodological revolution (Karp 2015), when problems in algebra and geometry also began to be organized in a special manner, also occurred when greater numbers of people began to be taught.

The style of explanation, and in general the understanding of what an explanation was, was changing. It was changing in mathematics itself as well. But in mathematics education, in this case, too, of course, the change and expansion of the audience played its role. Chebyshev (1868a) wrote about what was accessible and what was inaccessible to "commoners," but since it was recognized that something was inaccessible, a different form of presentation and explanation had to be looked for, since more and more commoners had to be taught.

There emerged a kind of market in mathematics textbooks, which was indeed controlled to a certain extent by the Scientific Committee. Its influence (and thus Chebyshev's) was considerable, but was only a part of the influence exerted on the teaching of mathematics by the general system of requirements and conditions in the country. How a mathematics teacher lived, studied, worked, and was monitored by his superiors also inevitably exerted an influence on how he taught, just as the system of exams for students also inevitably exerted an influence on the process of mathematics education (see the paper in this volume).

There were, of course, other sources of influence as well, for example, education in other countries, whose outcomes and principles in one way or another reached Russia (we have seen that Chebyshev reviewed foreign publications and their translations, and we also see them among texts used in teaching—take, for example, the same Bellavène). It is important to emphasize that an important process was taking place, in which Chebyshev's word by no means necessarily turned out to be final, but in which it undoubtedly had an important significance. Reading Chebyshev's reviews, we see the underside of Russian mathematics education, and we can better understand what changed, how, and why, and this constitutes the value of the surviving materials, which cannot be reduced to their value for studying Chebyshev's biography.

We should also note, in conclusion, that for the contemporary reader, it is interesting to observe the genesis or early stages of various discussions which subsequently continued literally for centuries, such as, for example, the discussion about whether it was necessary to cover sections in school that were traditionally studied later on, like analytic geometry. Chebyshev's reviews make it possible once again to grasp the complexity and disputability of what occurred in the history of mathematics education.

References

Bellavène [Bellyaven], Jacques-Nicolas. 1824–1825. *Kurs chistoi matematiki, sostavlennyi dlia voennykh uchilishch po porucheniiu Bellyavenya [Course in pure mathematics written for military schools on the orders of Bellavène]*. St. Petersburg: Voennaya tipografia Glavnogo shtaba.

———. 1832–1834. *Kurs chistoi matematiki, sostavlennyi po porucheniiu Bellyavenya* [Course in pure mathematics written on the orders of Bellavène]. Moscow: Tipografia Lazarevykh Instituta vostochnykh yazykov.

Bernstein, Sergey N. 1947. Chebyshev, ego vliyanie na razvitie matematiki [Chebyshev, His Influence on the Development of Mathematics]. *Uchenye Zapiski Moskovskogo Gosudarstvennogo Universiteta 91*: 35–46.

Bogomolov, Nikolay V. 2006. *Ocherki o rossiiskikh pedagogakh-matematikakh [Sketches of Russian Teacher-Mathematicians]*. Moscow: Vysshaya shkola.

Bunyakovsky, Victor Ya. 1849. *Matematika. Arifmetika [Mathematics. Arithmetic]*. St. Petersburg: Tipografiya voenno-uchebnykh zavedenii.

Busev, Vasily M., and Tatiana B. Rebrieva. 2021. *P.L.Chebyshev v Uchenom Komitete Ministerstva Narodnogo Prosveschecheniya* (1856–1873). [P. L. Chebyshev in the Scientific Committee of the Ministry of Education (1856–1873)]. Moscow: MTsNMO.

Chebyshev, Pafnuty L. 1857a. O rukopisi Mudrova "Rukovodstvo k arifmetike" [About Mudrov's manuscript "Handbook on Arithmetic"]. Russian State Historical Archive, f. 734, op. 2, d. 2, p. 25.

———. 1857b. O knige Ritt G. Novaya arifmetika dlya pervonachal'nogo obucheniya [About G. Ritt's book "New Arithmetic for Elementary Education"]. Russian State Historical Archive, f. 734, op. 2, d. 2, pp. 276–277.

———. 1858a. O sochinenii Petushkova "Domashnii zemlemer" [About Petushkov's composition "The Home Land Surveyor"]. Russian State Historical Archive, f. 734, op. 2, d. 3, pp. 290–294.

———. 1858b. O rukopisi perevoda uchitelya Kargopol'skogo uezdnogo uchilischa Sokolova "Sobraniya geometricheskikh zadach" G. Ritta [About the manuscript of the translation of G. Ritt's "Collection of Geometry Problems" by Sokolov, a teacher at the Kargopolsky uezd school]. Russian State Historical Archive, f. 734, op. 2, d. 3, pp. 321–322.

———. 1859a. O rukopisi Zhibrata "Arifmetika, sostavlennaya po rukovodstavam Busse, Bunyakovskogo, Bourdona i dr" [About Zhibrat's manuscript, "Arithmetic, Based on Handbooks by Busse, Bunyakovsky, Bourdon, and Others"]. Russian State Historical Archive, f. 734, op. 2, d. 4, pp. 10–12.

———. 1859b. O rukopisi Kirshteyna "Rukovodstvo k arifmetike"[About Kirshteyn's manuscript, "Handbook on Arithmetic"]. Russian State Historical Archive, f. 734, op. 2, d. 4, pp. 338–342.

———. 1860. O rukopisi Rummelya "Arifmetika" [About Rummel's manuscript, "Arithmetic"]. Russian State Historical Archive, f. 734, op. 2, d. 5, pp. 387–388.

———. 1861a. O sochinenii Geeka "Rukovodstvo nachal'noy geometrii dlya gimnazii" [About Geek's composition, "Handbook on Elementary Geometry for Gymnasia"]. Russian State Historical Archive, f. 734, op. 2, d. 6, pp. 13–14.

———. 1861b. O knige Nikulin A. "Arifmetika" [About A. Nikulin's book, "Arithmetic"]. Russian State Historical Archive, f. 734, op. 2, d. 6, pp. 130–131.

———. 1861c. O rukopisi Smyarovskogo "Polnaya natural'naya teoriya logarifmov [About Smyarovsky's manuscript, "Complete Natural Theory of Logarithms"]. Russian State Historical Archive, f. 734, op. 2, d. 6, pp. 311–313.

———. 1861d. O rukopisi konduktora korpusa korabel'nykh inzhenerov Mel'gunova "Kurs arifmetiki" [About the manuscript of "Course in Arithmetic" by Mel'gunov, conductor of the corps of naval engineers"]. Russian State Historical Archive, f. 734, op. 2, d. 6, pp. 314–316.

———. 1861e. O knige: Boguslavsky A. "Matematika: Arifmografiya i arifmetika" [About the book: A. Boguslavsky, "Mathematics: Arithmography and Arithmetic"]. Russian State Historical Archive, f. 734, op. 2, d. 7, pp. 72–74.

———. 1862a. O rukopisi Duperrona "Trigonometriya ploskaya i sfericheskaya i osnovnye nachala analiticheskoy geometrii" [About Duperron's manuscript, "Plane and Spherical Trigonometry and Elementary Analytic Geometry"]. Russian State Historical Archive, f. 734, op. 2, d. 7, pp. 77–81.

———. 1862b. O sposobe deleniya ugla na tri ravnye chasti, pridumannom general-mayorom S.S.Poluektovym [About the method for dividing an angle into three equal parts invented by Major General S.S. Poleuktov]. Russian State Historical Archive, f. 734, op. 2, d. 7, pp. 155–158.

———. 1863a. Ob "Arifmetike", sostavlennoy uchitelem Tverskoy gimnazii Pol'kovskim [About "Arithmetic," written by Pol'kovsky, a teacher at a Tver gymnsasium]. Russian State Historical Archive, f. 734, op. 3, d. 1, pp. 27–32.

———. 1863b. O sostavlennom g. podpolkovnikom General'nogo Shtaba Khristiani "Polnom kurse elementarnoy matematiki" [About the "Complete Course in Elementary Mathematics," written by Mr. Khristiani, Lieutenant Colonel of the General Staff]. Russian State Historical Archive, f. 734, op. 3, d. 1, pp. 156–160.

———. 1866. O rukovodstvakh po arifmetike, sostavlennykh Nikulinym i Volensom [About Handbooks on Arithmetic written by Nikulin and Volens]. Russian State Historical Archive, f. 734, op. 3, d. 5, pp. 976–978.

——— 1868a. Mnenie Chebysheva: 1) Schetnyi samouchitel,' 3 chasti; 2) Arifmetika Fogelya. Kratkoe izlozhenie pervykh 4-kh pravil; 3) Tseringera. Shveytsarskii uchebnik arifmetiki (uprazhnenie v umstvennom schislenii) [Chebyshev's opinion: 1) Self-Teaching Manual on Calculation, 3 parts; 20 Fogel's Arithmetic. Brief formulation of the first four rules; 3) Tseringera. Swiss Textbook in Arithmetic (exercise in mental calculation)]. Russian State Historical Archive, f. 734, op. 3, d. 9, pp. 1180–1184.

———. 1868b. Mnenie gg. Chlenov Galakhova, Chebysheva, Georgievskogo, Khodneva i Shteinmana o programmakh predmetov kursa v Nikolaevskoy gimnazii, s nekotorymi poyasneniyami, vsledstvie soobrazhenii Uchenogo Komiteta o takovykh zhe programmakh, pervonachal'no predstavlennykh [Opinion of the member, Messrs. Galakhov, Chebyshev, Georgievsky, Khodnev, and Shteinman about the subject curricula of the course at the Nikolaevskaya gymnasium, with certain elucidations, following the considerations of the Scientific Committee about the same curricula as originally submitted]. Russian State Historical Archive, f. 734, op. 3, d. 9, pp. 1223–1241.

———. 1951. *Polnoe sobranie sochinenii. Tom 5 [Complete Works. Volume 5]*. Moscow-Leningrad: Izdatel'stvo Akademii nauk SSSR.

Georgievsky, Alexander I. 1902. *K istorii Uchenogo Komiteta Ministerstva Narodnogo Prosvescheniya [Toward a History of the Scientific Committee of the Ministry of Public Education]*. St. Petersburg: Senatskaya tipographiya.

Giacardi, Livia. (2006). From Euclid as Textbook to the Giovanni Gentile Reform (1867-1923). Problems, methods and debates in mathematics teaching in Italy. *Paedagogica Historica. International Journal of the History of Education* 42(4/5): 587–613.

Guryev, Pyotr S. 1832. *Arifmeticheskie listki [Arithmetical Sheets]*. St. Petersburg: Imperatorskaya Akademiya nauk.

Istrin, V. 1908. Po povodu otchetov o pis'mennykh rabotakh na ispytaniya zrelosti [On Reports about Written Solutions for Graduation Exams]. *Zhurnal Ministerstva narodnogo prosvescheniya* February, 57–95.

Karp, Alexander. 2014. Mathematics Education in Russia. In *Handbook on the history of mathematics education*, ed. Alexander Karp and Gert Schubring, 303–322. New York: Springer.

———. 2015. Problems in old Russian textbooks: How they were selected. In *"Dig where you stand" 3*, ed. K. Bjarnadottir et al., 203–218. Uppsala: Uppsala University.

———. 2017. On the Russian national subcommission of the ICMI. In *"Dig where you stand" 4*, ed. K. Bjarnadottir et al., 149–166. Rome: Edizioni Nuova Cultura.

———, ed. 2018a. *Otchety Rossiiskoy Natsional'noy Podkomissii Mezhdunarodnoy Komissii po Matematicheskomu Obrazovaniu [Reports of Russian National Subcommission of ICMI]*. St. Petersburg: Lema.

———. 2018b. Russian mathematics teachers, 1830–1880: Toward a group portrait. In *Researching the History of Mathematics Education*, ed. Fulvia Furinghetti and Alexander Karp, 107–130. Springer.
———. 2019. Dmitry Chizhov and the examination of mathematics textbooks in Russia during the 1820s-1830s. In *"Dig where you stand"5*, ed. K. Bjarnadottir et al., 205–220. Utrecht: Utrecht University – Freudenthal Institute.
Kostenko, Igor' P. 2013. *Problema kachestva matematicheskogo obrazovaniya v svete istoricheskoy retrospektivy [The Problem of the quality of mathematics education in the light of a historical restrospective]*. Moscow: RGUPS.
Perepiska ob uchrezhdenii Komiteta po ustroistvu uchebnykh zavedenii. Zhurnaly zasedanii Komiteta. [Correspondence on the Establishment of the Committee for the Organization of Educational Institutions. Journals of Meetings of the Committee]. 1826–1828. Russian State Historical Archive, f. 738, op. 1, d. 1.
Perevoschikov, Dmitry M. 1837. *Gimnazicheskii kurs chistoy matematiki, soderzhaschii arifmetiku, osnovaniya algebry geometrii, pryamolineinuiu trigonometriu i konicheskie secheniya* [Gymnasium Course in Pure Mathematics, Containing Arithmetic, Elementary Algebra, Geometry, Linear Trigonometry, and Conic Sections]. Moscow: Tipografiya Avgusta Semena.
Polozhenie o nachal'nykh narodnykh uchilischakh [Regulations for Elementary Public Schools]. 1874. Accessed on March 12, 2021 http://www.detskiysad.ru/raznlit/istped007.html
Polyakov, Pyotr N. 1897. *Rukovodstvo k arifmetike dlya srednikh uchebnykh zavedenii [Handbook on arithmetic for secondary educational institutions]*. Moscow: V.Dumnov.
Posse, Konstantin A. 1904. Chebyshev. In *Kritiko-bibliograficheskii slovar'*, ed. Semen A. Vengerov, v. 6, 8. St. Petersburg: Semenovskaya tipo-litografiya.
Prudnikov, Vasily E. 1948. Pedagogicheskoe nasledie P.L. Chebysheva [The Pedagogical Legacy of P. L. Chebyshev]. *Matematika v shkole 6*: 28–33.
———. 1956. *Russkie pedagogi-matematiki XVIII-XIX vekov [Russian Mathematics Educators of the XVIII-XIX Centuries]*. Moscow: Uchpedgiz.
Prudnikov, Vasily E. 1964. *P.L. Chebyshev. Uchenyi i pedagog [P. L. Chebyshev. Scientist and Teacher]*. Moscow: Prosveschenie.
———. 1976. *Pafnutiy L'vovich Chebyshev*. Leningrad: Nauka.
Ruchnaya matematicheskaya entsiklopedia [Small Mathematical Encyclopedia] 1826–1827. Moscow: Universitetskaya tipografiya.
Sossinsky, Alexey B. 2010. Mathematicians and mathematics education: A tradition of involvement. In *Russian mathematics education: History and world significance*, ed. Alexander Karp and Bruce Vogeli, 187–222. Hackensack, NJ: World Scientific Publishing.
Schubring, Gert. 2019. Klein's Conception of 'Elementary Mathematics from a Higher Standpoint'. In *The Legacy of Felix Klein*, ed. H.-G. Weigand et al., 169–180. New York: Springer.
Shaposhnikov, Nikolay A. 1906. *Nasha uchebno-uchenaya kritika. Razbor stat'i professora Koyalovicha* [Our Educational-Scientific Criticism. An Analysis of Prof. Koyalovich's Article]. Moscow: Universitetskaya tipographiya.

Frédérique Papy-Lenger, the Mother of Modern Mathematics in Belgium

Dirk De Bock

Abstract Belgian mathematician Frédérique Papy-Lenger's work in math education spans roughly four periods. Having been introduced to the CIEAEM community, Lenger made the study of student knowledge a main theme in her work during the 1950s. Toward the end of that decade, the modernization of the teaching of mathematics in secondary schools became central to her thinking. In 1967, the "experiment Frédérique," aimed at preparing for reform at the primary level, was developed and brought her into contact with the Comprehensive School Mathematics Program in the United States. From the 1980s onward, in the last phase of her career, Lenger focused on working with disabled children.

Keywords Belgium · Caleb Gattegno · CIEAEM · CSMP · Frédérique Lenger · Georges Papy · GIRP · Minicomputer · Modern mathematics · Primary education

1 Introduction

With few exceptions, the emergence of women on the international mathematics education scene is a post-World War II phenomenon. In Europe, the *Commission Internationale pour l'Étude et l'Amélioration de l'Enseignement des Mathématiques* (CIEAEM)/*International Commission for the Study and Improvement of Mathematics Teaching* has played an important role in this development. The CIEAEM was an informal group of people, created by Caleb Gattegno in 1950 and officially founded at the group's fourth meeting in La Rochette par Melun (1952). Meetings of the CIEAEM were held in a friendly atmosphere, in principle on an

D. De Bock (✉)
KU Leuven, Leuven, Belgium
e-mail: dirk.debock@kuleuven.be

A. Karp (ed.), *Advances In The History Of Mathematics Education*,
International Studies in the History of Mathematics and its Teaching,
https://doi.org/10.1007/978-3-030-95235-8_2

annual basis, with a focus on the study of pedagogical aspects of mathematics education with the common objective of improving the quality of this education. For that purpose, experts from various related scientific disciplines (mathematics, epistemology, logic, psychology, educational sciences, ...) were invited along with experienced mathematics teachers. From the very start, this latter group included a number of women who played an important role within the CIEAEM and in the international debates on mathematics education from the 1950s onward. The list of founding members of the CIEAEM mentions three women: the Italian Emma Castelnuovo, the French Lucienne Félix, and the Belgian Frédérique Lenger (Bernet and Jaquet 1998). From the late 1950s, a fourth woman, the Polish Anna Zofia Krygowska, entered the international arena (she joined the CIEAEM in 1958 and was its president in the period 1970–1975). Krygowska developed into a leading researcher in the then emerging scientific discipline of mathematics education.

As Furinghetti (2008) argued, the history of women in mathematics education is a rather neglected subject of research. While the lives and work of Castelnuovo, Félix, and Krygowska have been documented in the international literature on the history of mathematics education (see, e.g., Barbin 2013; Brasseur 2011; Furinghetti and Menghini 2014; Menghini 2015; Nowecki 1992), Frédérique Lenger's contribution to research and her implementation of innovations is still underexposed in this literature. There are several possible causes for this: (1) Lenger communicated her ideas mainly through her developmental work and was not a "theorist," (2) she worked partly in the shadow of her husband Georges Papy, the radical Belgian reformer of the 1960s, and (3) she was a representative of the modern mathematics movement which fell into disfavor from the early 1970s onward. In this chapter we aim to fill the gap in the existing literature by analyzing from a critical distance Lenger's role throughout her career as a mathematics teacher and as a developer and researcher in mathematics education from the 1950s to the 1980s. More specifically, we aim at answering the following research question: "What is the specific contribution of Lenger in shaping mathematics education in the second half of the previous century, particularly with regard to the modern mathematics movement of the 1960s and early 1970s?" To answer this question, we have carefully examined Lenger's publications and the reactions to them by her contemporaries as well as certain other relevant documents, such as teaching materials that we found in her personal archive, which is preserved at the University of Mons-Hainaut (Belgium).

This chapter is part of a line of research into the history of the modern mathematics movement in Belgium (De Bock and Vanpaemel 2019). Through additional studies we hope to shed light on the role and international influence of a number of key actors and events in this recent history. The contribution of Lenger had not yet been the subject of detailed historical research, nor had a systematic biography been published before. However, a website was recently created by former collaborators and friends of Lenger, notably Robert Kennes (www.rkennes.be/frederique.htm). This website, published in French, provides some biographical milestones, a bibliography and a selection of Lenger's papers, and some testimonials. An unsigned obituary in honor of Lenger appeared in 2005 in *Mathématique et Pédagogie* 154: 5–6.

2 Lenger and the Study of Student Knowledge

Frédérique Lenger was born on December 8, 1921 to an intellectual family in Arlon (Belgium); her father was a lawyer. She passed away on September 1, 2005 in Molenbeek-Saint-Jean, a municipality of the Brussels-Capital Region.

Frédérique Lenger studied at the *Lycée Royal* of her hometown, and in 1939 she went on to study mathematics at the *Université libre de Bruxelles*. In 1941, however, the university was officially closed to prevent its takeover by the German occupying forces. Lenger's studies continued underground and examinations were administered by the *Jury Central*. She graduated in 1943 and obtained a teacher's certificate in 1946. From 1947 to 1950, Lenger served as an assistant at the same university's geometry department, which was led by the renowned professor Paul Libois. The latter proposed that she study Élie Cartan's principle of triality and Girard Desargues' *Brouillon projet d'une atteinte aux événements des rencontres du cone avec un plan* (1639), which led to her first (and probably only) research paper in geometry (Lenger 1950). Simultaneously, Lenger was a mathematics teacher at the *École Decroly*, a comprehensive school following the principles of Ovide Decroly's "reform pedagogy." The *École* was directed by Lucie Libois-Fontaine, who was also a graduate in mathematics, and Libois' wife. In 1950, Lenger started to teach at the *Lycée Royal* in Arlon where she was appointed *préfète des études* in 1957 and at the same time became director of the *École Normale de l'État pour filles* (teacher training college for girls) in Arlon.

Lenger's early work in mathematics education is inextricably linked to the actions of the CIEAEM of which she was a founding member and "one of the faithful since the third meeting"[1] (Félix 1986, p. 71) until her resignation in 1970 (Fig. 1). It is likely that Libois, who participated in the second meeting in Keerbergen (Belgium) in 1951, brought her into contact with Gattegno and his group. It is her connection with them that shaped Lenger's entire career. In particular, Gattegno's "pedagogy of situations" (see, e.g., Gattegno 1963) would have a lasting influence on her work ("All my teaching has been inspired by Gattegno's pedagogy of situations," Frédérique. 1976, p. 95). In this approach to mathematics education, the role of the teacher consists of selecting and proposing situations which are easy to approach and sufficiently interesting and problematic to elicit and sustain investigations by learners. From this perspective the teacher does not transfer knowledge directly, but becomes a coordinator and a catalyst of the learning process. Gattegno regularly applied this pedagogical method himself during the early CIEAEM meetings (Félix 1986). Particularly in the 1970s, Lenger gave this method her own interpretation and made it the core of her didactic approach (see below).

We have little or no direct sources for identifying Lenger's contribution to the early debates within the CIEAEM. She did not contribute to either of the two collective volumes published by the group during the 1950s (Gattegno et al. 1958; Piaget et al. 1955). A historical overview by Lucienne Félix (Félix 1986), which

[1] All translations were made by the author.

Fig. 1 Left: Louis Jeronnez, Caleb Gattegno, and Frédérique Lenger at the CIEAEM meeting in La Rochette par Melun, 1952. Right: Frédérique Lenger, Lucien Delmotte, and Louis Jeronnez at the CIEAEM meeting in Bellano, 1955 (photos from Lucien Delmotte, collection Guy Noël)

covers the period 1950–1984, mentions Lenger only sporadically and provides no details of her interventions. Because the early CIEAEM meetings were not conferences in the contemporary sense, there are no proceedings available either; these only appeared regularly from 1974 onward (i.e. after Lenger's involvement in the group ended). A source that does shed some light on Lenger's fields of interest during the 1950s is *Mathematica & Paedagogia*, the journal of the Belgian Society of Mathematics Teachers, founded in 1953 with Lenger as a member of the first board. During the 1950s, *Mathematica & Paedagogia* became a main forum of ideas and proposals related to the teaching of mathematics, in particular for CIEAEM members (De Bock and Vanpaemel 2015). The journal had a recurrent structure with clearly distinguishable sections. Two of these sections are of particular importance in the context of this chapter: "Knowledge of the students" and "Contacts." The idea behind "Knowledge of the students," which was not a common theme in teachers' journals of the time, was that successful teaching of mathematics not only depended on the teachers' expertise in mathematics, but also on their knowledge about students' psychological development and reasoning abilities. Lenger published five articles in this section (Lenger 1953a–1954, 1953b–1954, 1954a–1955, 1954b–1955, 1955–1956), focusing on middle school students' misconceptions in geometry, their systematic calculating errors in algebra, and the mathematical potential of "less gifted" students. In the "Contacts" section, Lenger published (non-official) reports of the 6th, 7th, and 8th CIEAEM meeting, held respectively in Calw, Germany

(1953); Oosterbeek, The Netherlands (1954); and Bellano, Italy (1955) (Lenger 1953c–1954, 1954c–1955, 1954d–1955).

Lenger's papers of the 1950s strongly reflect the influence of Gattegno and the CIEAEM. As mentioned in an action plan of this group dated 1952 (Bernet and Jaquet 1998), one of the objectives was that teachers would design some of their lessons as "psychological experiments" to reveal the students' mental structures. Lenger (1953a–1954) describes such an experiment involving middle school students in which a rectangle was deformed so that it becomes a parallelogram. Students were asked which properties of the original rectangle were preserved under that deformation and which were not. Most students were convinced that although the figure's height decreased, its area remained unchanged, and their arguments disclosed their thinking (e.g. "the figure's sides do not change," or "what the area loses in height is compensated by what it wins in length," p. 18). In Lenger's (and Gattegno's[2]) view, errors should be framed as something natural and positive, as learning opportunities, both for the student and for the teacher:

> We believe that the errors regularly made by successive generations of students, far from being a constant source of irritation or despair for the teacher, can be, on the contrary, an interesting topic for reflection. It seems to us that such errors can open up one of the most reliable avenues of access to juvenile mathematical thinking. (Lenger 1953b–1954, p. 15)

We finally notice that as early as the mid-1950s, Lenger's ideas were influenced by "modern mathematics" and that she was thinking about its implementation in education:

> Modern mathematics [...] is the result of an awareness of structures and the relationships between these structures. The thinking of the modern mathematician is relational. And it seems to me that relational mathematical thinking can be recreated by a child or adolescent if the teacher is aware of it and if he presents the appropriate situations. Mr. Gattegno's experiments in many schools in England and on the Continent bear witness to this. (Lenger 1954c–1955, p. 58)

3 Toward Modern Mathematics at the Secondary Level

By the end of the 1950s, the time for action had come. In August 1958, in the margin of the 12th CIEAEM meeting in Saint Andrews (Scotland, UK), Frédérique Lenger and Willy Servais developed an experimental program for the teaching of mathematics based on set theory and topology (Le programme B des écoles normales gardiennes 1958–1959). A first experiment, based on the Lenger-Servais program, was run during the school year 1958–1959 in two schools for future kindergarten teachers in the French-speaking part of Belgium. It was led by Frédérique Lenger herself and Madeleine Lepropre. Participants were 15–16-year-old students who certainly did not belong to the top tracks in mathematics education; some of them

[2] See, e.g., Gattegno (1954, p. 11): "It is man's privilege to make mistakes; only through experience, experience that is often painful, does man learn and acquire some degree of wisdom."

even showed fundamental hostility toward mathematics and the people who teach it (Lenger and Lepropre 1959). The experiment was considered successful; students reacted lively and actively to the new material, and their hostility toward mathematics completely disappeared (for more details about this experiment, we refer to De Bock and Vanpaemel 2018). After this first year of experimentation, Lenger and Servais realized that they needed the assistance of a research mathematician to aid with follow-up experiments. Georges Papy, a professor of algebra at the *Université libre de Bruxelles*, was asked to collaborate, and he accepted. Lenger moved from Arlon to Brussels where she was appointed professor of mathematics at the *École Normale Berkendael*, a teacher training college of the state. In *Berkendael* Papy, in collaboration with Lenger, set up his own experiment involving future kindergarten teachers. In this period, Lenger became Papy's partner, both professionally and personally (they married on October 1, 1960). During the next decades, Georges and Frédérique Papy (or: Frédérique Papy-Lenger[3]) formed a complementary team permeated by the same ideas and with a firm determination to implement modern mathematics at all school levels (Noël 1993) (Fig. 2).

To coordinate the experimental trajectory and other initiatives related to intended reform, Papy created on May 24, 1961 the *Centre Belge de Pédagogie de la Mathématique* (CBPM), and Lenger was one of the founding members. The *Centre* brought all main actors in the reform movement together, both from universities and

Fig. 2 Georges and Frédérique Papy, 1970s (collection Robert Kennes)

[3] After she was married, Frédérique Lenger no longer used her maiden name but her husband's last name (and thus signed her publications as "Frédérique Papy"). From the end of the 1960s, she usually signed her publications only with her first name "Frédérique." Just like "Papy," Frédérique became a "brand" in reform-minded circles.

from secondary schools. In September 1961, Papy and his CBPM launched a new experiment aimed at teaching modern mathematics from the first years of general secondary education (12–13-year-olds). Twelve schools representing some thirty classes were involved. Lenger was responsible for one of these classes. The experiment expanded year after year, both within the original schools as well as in others, which joined later. However, at the upper secondary level (15–18-year-olds), the experimental efforts lost direction and vigor. Only the class taught by Lenger went through the entire six-year experimental trajectory. Although the experiment proved a success (see, e.g., Stievenart 1968)—Lenger was a very gifted teacher—few general conclusions could be drawn from it. Yet, the Papys did not monitor replication on a larger scale. At the end of the 1960s, they considered the reform in secondary education to be essentially accomplished and began to focus entirely on a reform in primary education (see below).

Parallel to and based on the experiments, Papy published his ground-breaking textbook series *Mathématique Moderne* "in collaboration with Frédérique." Intended for the teaching of modern mathematics to students aged 12–18 (G. Papy 1963–1967), the series presented a totally new conception of secondary school mathematics by basing it on the unifying themes of sets, relations, and algebraic structures. A central idea was to construct the Euclidean vector plane in the lower grades (12–15-year-olds) and then take this structure as a starting point (a "unique axiom") for the further build-up of the mathematical edifice from the beginning of the upper grades (15–18-year-olds). Although Papy's discourse was mathematically rigorous, it was interwoven with various didactic innovations, such as multi-colored graphs, non-verbal proofs by filmstrips, and playful drawings (De Bock et al. 2019). During the 1960s, Papy's approach, as reflected in his textbooks, inspired many reformers worldwide. It is difficult to determine the exact contribution of Frédérique Lenger to this project. In the preface to the third volume of the series, we read that "the reader owes Frédérique the picture-book character of this rigorous and intuitive presentation of geometry. Thanks to her, children have always been present in the minds of the authors" (G. Papy 1967b, p. xvi). Lenger may also have contributed on another level. Friends of the Papys testified that without Lenger's inspiration, strictness, and organizational talent, Papy might never have been able to realize his *magnum opus* (Matthys 2005).

Lenger was also a driving force behind the *Journées d'Arlon*, a series of annual in-service teacher training courses held between 1959 and 1968 and aimed at preparing Belgian mathematics teachers for the arrival of modern mathematics in their classrooms. She launched the series; the first course was given in July 1959 in her hometown of Arlon and was an opportunity to present and discuss the experiment carried out in the preceding school year (G. Papy 1959). Year after year, the success of the initiative was growing, with about 600 participants at its peak in 1965–1966 (Holvoet 1968). Most participants were Belgian mathematics teachers, but there were also some university professors, inspectors, political officials, and foreign guests. At the 8th and 9th meetings, held in 1966 and 1967, a modern approach to analysis, intended for 16–18-year-old students in general secondary education, was proposed (G. Papy 1966, 1967a). This approach was entirely based on Lenger's experimental classes and consisted of defining and studying, from an extensive topological framework, the concepts of continuity and limit (as a local continuous extension), and some basic elements of differential and integral calculus (with a focus on the concepts of derivability and integrability). In September 1968, shortly after the 10th and last edition of the *Journées*, modern mathematics was introduced in the first year of the general divisions of Belgian secondary schools, mandatorily and generally (De Bock and Vanpaemel 2019).

4 Lenger's Modernization Efforts at the Primary Level

4.1 The Experiment Frédérique

In September 1967 the CBPM started the *experiment Frédérique* to prepare for reform at the primary level (F. Papy 1970). The experiment was led by Frédérique Lenger herself, assisted by Danielle Incolle, and initiated in two classes of 6–7-year-olds in the primary section of the *École Normale Berkendael*. It was continued year after year, until the end of primary education. Lenger was not bound by any official program and could therefore give free rein to her creative talents. And she did so abundantly. About *L'Enfant et les Graphes* (Frédérique and Papy 1968), the first publication resulting from the experiment and describing the first ten lessons in which Lenger introduced multi-colored graphs ("*papygrammes*"), André Deledicq (2012) [4]wrote:

> As for myself, it is by opening *L'Enfant et les Graphes* that I learned what could be a "mathematics in color," that is to say a mathematics for which color is not just an aesthetic ornament but a significant component of the pleasure of searching and the reasoning it illustrates. (p. 266)

The main goal Lenger set herself was to build, with the help of the children, a "house of mathematics," a space of unity and harmony in which creativity could flourish, something completely different from "traditional" mathematics instruction. In Frédérique (1976) she phrased this metaphor as follows:

> I associate traditional mathematics with an American supermarket [...]. A building without poetry, without color, without windows, offering consumers a range of useful products to meet the most immediate needs of their daily lives, organized in a series of departments with few links between them, providing a range of tools that are certainly needed, but which only allow local troubleshooting or constructions without scope. [. . .] The house of mathematics of our teaching is not a futuristic construction but a comfortable, colorful home, offering places of meeting and exchange, corners for meditation and reflection, laboratories where some simple, well selected machine tools are available. For the teacher, this framework allows a better grasp of the path and perspectives of his teaching. For the student, the

[4]André Deledicq (born 1943) is a French mathematician and former lecturer at the University of Paris VII.

unitary structure offers security and comfort, elements that favor the development of intelligence and knowledge in mathematics. (pp. 70–71)

More down to earth, the *experiment Frédérique* aimed at (1) introducing children into the relational world of modern mathematics, and (2) progressively initiating them in calculation techniques in "sets of ever richer types of numbers" (F. Papy 1970, p. 95). Tools or teaching aids to realize these goals were Venn diagrams and *papygrammes*, to be drawn with felt-tip pens on large sheets of plain paper, Cuisenaire's numbers in color, the Logiblocs of Dienes, and "minicomputer."

Minicomputer was not a computer or calculator, but a new teaching aid created by G. Papy (1968b). It was a two-dimensional abacus with plates that were subdivided into four square sections, each colored according to the coding system of the Cuisenaire rods (Fig. 3). In these plates, numbers were represented in a binary way by counters that could be moved up and down, corresponding to the operations of doubling and halving. This representation was inspired by a hexadecimal computing system developed by the Belgian mathematician and astronomer Georges Lemaître (see, e.g., Lemaître 1954). Papy's minicomputer, however, was primarily designed to help young children to acquire an understanding of the decimal numeration system and a thorough knowledge of the basic operations of counting, addition, and subtraction in a concrete and playful way. Later in the

Fig. 3 A child at work with Papy's minicomputer, late 1960s (International Visual Aid Center, Brussels)

Fig. 4 Anna Zofia Krygowska, Hans-Georg Steiner, Frédérique Papy-Lenger, and Zoltan Dienes at the first ICME congress in Lyon, 1969 (Courtesy of Gert Schubring and Livia Giacardi)

experiment, minicomputer was re-used, this time for gaining insight in the operations of division and multiplication, and for working with decimal numbers and fractions. In 1969, Lenger was invited to present the minicomputer method at the first International Congress on Mathematical Education (ICME1) in Lyon (France) (F. Papy 1969) (Fig. 4).

In the early 1970s, Lenger made the annotated accounts of her experimental classes available in a luxuriously published four-volume book series entitled *Les Enfants et la Mathématique*[5] (Frédérique 1970–1976). The different volumes document the experiment in the first four years of primary education. Fifth and sixth volumes were announced (see, e.g., Frédérique. 1976) but the series ultimately remained unfinished. The first volume also included Lenger's doctoral dissertation (F. Papy 1968a), which she defended on December 8, 1968 and so became the first person in Belgium to obtain a doctorate in mathematics education.

Both in terms of content and style—with many colorful figures that elicit a mathematical idea or line of thought—the series resembles that of Papy's *Mathématique Moderne*. Through a careful examination of the different volumes as well as the published reviews (Cuypers 1970; Gille 1971; Goffree 1970–1971; Vredenduin 1970–1971, 1972–1973, 1975–1976, 1978–1979), we noted some key

[5] Volumes of this series were translated into English (*Mathematics and the Child*), German (*Kinder und Mathematik*), and Italian (*I bambini e la matematica*).

characteristics of Lenger's approach. First, it was mainly the mathematical topics themselves that were used to motivate the students (besides, of course, the nice materials, the good class atmosphere, and the enthusiasm of the teacher). Second, no attention was paid to procedural knowledge about numbers and repeated practice of technical computational skills. Instead, students were confronted with various "calculational experiences" (with real numbers in different representations, with sets and relations, with vectors, ...). Third, the mathematical background of the experiment was quite ambitious. To mention just one example: 6–7-year-olds were expected to learn to deal with negative numbers and to solve equations in the group of integers, obviously by means of concrete examples and with the help of arrow graphs. As Vredenduin (1975–1976) noticed: "Frédérique's great didactic talents enable her to achieve results with young children that will be unattainable for many" (p. 165). Above all, Lenger lays foundations, develops intellectual facilities in children, and prepares concepts and methods in an informal way without ever making them mathematically explicit (Vredenduin 1972–1973). Apparently, for Lenger, preparing children for modern secondary school mathematics was a main goal of mathematics education at the primary level. However, her views could not fully convince the Belgian educational authorities. When, in the second half of the 1970s, modern mathematics was introduced in Belgian primary schools, the implementation did not mirror the CBPM proposals, but, rather, those of the so-called "Waterloo group" (Jeronnez and Lejeune 1972a, b). Mathematical reflection was central to this competitive group's model as well, but they also paid a lot of attention to numbers, operations, and arithmetic skills that were deliberately trained.

4.2 *Lenger and the* Comprehensive School Mathematics Program

With the presentation at ICME1, Frédérique Lenger's work caught the attention of modern mathematics circles outside Belgium, such as, Burt Kaufman's team in the United States of America. Kaufman headed the *Comprehensive School Mathematics Program* (CSMP), a second generation New Math project (Kilpatrick 2012) established in 1966 and originally affiliated with Southern Illinois University in Carbondale, Illinois. The goals of the CSMP were to modernize both the content and teaching methods of mathematics, although in reality more emphasis was put on mathematics content than on pedagogy (Hayden 1981; Walmsley 2003). The project stood for an integrative approach to mathematics education and stressed logic, mathematical structure, and precision in language (Kaufman 1968). In order to achieve its objectives, curriculum materials, teachers guides, and assessment tools were developed that "reflect recent insights in the didactics of mathematics" (Kaufman and Steiner 1969, p. 319). CSMP desired that children of various ages, from kindergarten up to the age of 18, would not only learn arithmetic skills, but also, by using a "pedagogy of situations," develop a deeper understanding of numerical concepts and become involved in broader mathematical thinking (Kaufman and Haag 1977). Four non-verbal languages were employed for that purpose: Papy's minicomputer, arrows (to represent relations), strings (for classification tasks), and the upcoming electronic calculators (mainly as a tool for discovering patterns) (Comprehensive School Mathematics Program 1978). Despite the widespread use of its materials, the CSMP has never become a dominant reform project in the United States of America.

Soon after their first encounter in France in 1969, Kaufman invited Frédérique Lenger for several short term stays in Carbondale, to advise the project team and to give a number of demonstration lessons[6] based on *Les Enfants et la Mathématique* (Un groupe de professeurs Américains 1970). In 1973 Kaufman appointed Lenger as director of research of the CSMP, a role she fulfilled until 1978 (Fig. 5). In the United States of America, with a much richer environment and technical support than in the CBPM, Lenger's creativity exploded (IMACS 2006). In addition to the usual work—guiding the direction of the development work and teaching in experimental classes—she wrote profusely, including *Stories by Frédérique*, a series of 24 tales intended as a source of inspiration for lessons to children aged 5 to 14

Fig. 5 Frédérique Lenger at CSMP in discussion with Kevin Saunders (left), Burt Kaufman (middle), and an unidentified person (right), mid-1970s (collection Robert Kennes)

[6]The *Comprehensive School Mathematics Program Preservation Project* has digitized several films of Frédérique Lenger's classes and made them freely accessible at http://stern.buffalostate.edu/Movies/index.html.

(Frédérique 1974–1978). Each story, told with a minimum of words, was illustrated with cartoon-like drawings and incorporated discrete mathematical topics. Lenger invited children to enter a world where numbers live, act, have emotions, a world where children could become the friend of a number (IMACS 2006)[7].

5 Lenger and the Action of the GIRP

In 1970, after several years of internal tensions, Georges Papy, who was president of CIEAEM since 1963, felt that the Commission no longer had a role to play and proposed its dissolution (Bernet and Jaquet 1998). It is likely that Papy was no longer followed in the direction in which he had wanted to steer the Commission: Research on *his* modern mathematics pedagogy. But the proposed dissolution was rejected, Papy resigned as president and, along with his wife and a few other like-minded people, left the CIEAEM. Krygowska became his successor and from then on, under the impetus of Hans Freudenthal and Guy Brousseau, among others, CIEAEM's priority focus was on renewing the didactics of mathematics based on authentic problem situations and applications.

In 1971, Georges and Frédérique Papy, with some loyal followers, founded the Groupe International de Recherche en Pédagogie de la Mathématique (GIRP) (1971)[8]. As a rationale for founding the GIRP, the initiators stated that the intention was to bring together people who were fully focused on research in the field of (modern) mathematics education. The objectives of the GIRP were specified in its Articles: "The association aims to promote human and professional exchanges between mathematicians, psychologists, educators, and users of mathematics from all countries, and to organize international meetings, allowing researchers in mathematics education to exchange the results of their work" (p. 7). The focus of the GIRP was primarily on applied research into innovative educational experiments and practices:

> Its preference will go to the relations between current or recently completed experiments. It will focus on the preparation of experiments that can be carried out in different countries by the mathematical and methodological elaboration of new topics that can be integrated into education. (p. 5)

At the first meeting of the GIRP in Luxembourg (July 19–28, 1971), an executive committee was elected with Frédérique Lenger (Belgium) as president, Robert

[7] Lenger's materials are still available at http://stern.buffalostate.edu/CSMPProgram/Storybooks/bygrade.html.

[8] Along with Georges and Frédérique Papy, the following founding members are listed: Joana De Oliveira Bender (Brazil), Jacqueline Glardon (Switzerland), Peter Hilton (U.S.A.), Julien Constantin and Claude Lemaire (Canada), Ignacije Smolec (Yugoslavia), Jean Drabbe, Roger Holvoet, Paul Van Praag, Jean-Claude Matthys (Belgium), and Robert Dieschbourg (Luxembourg) (Groupe International de Recherche en Pédagogie de la Mathématique 1971, p. 4).

Fig. 6 Teresa Dutra (Portugal), Georges Papy, and Frédérique Papy-Lenger at the GIRP meeting in Dubrovnik, Easter 1977 (collection Robert Kennes)

Dieschbourg (Luxembourg) as secretary, and Chris De Munter (Belgium) as treasurer. Lenger will serve as president of the GIRP until 1981.

The Group organized one international meeting every year, each time in a different place and far away from the big cities (Fig. 6). The meetings were residential and lasted about ten days. The number of participants was deliberately limited (to a maximum of 50). The proceedings of the first four GIRP meetings (1971–1974) were published in the 10th, 13th, 16th, and 19th issues of *Nico*, the journal of the CBPM, which also served as a publication outlet for the GIRP and its members during the 1970s. These proceedings provide us with some information about the initial points of interest of the GIRP: (1) The reform of mathematics education at the primary level, as shaped by Lenger's experiments, (2) the use of the results of scientific research in the pedagogy of mathematics by in-service teachers in non-experimental classes, (3) the mathematical-logical foundation of modern mathematics education, and (4) modern mathematics as a therapeutic and educational tool for supporting mentally or physically disabled children, a theme that would become predominant in the last decade of Lenger's professional life (see the next section). Lenger's interventions at these meetings dealt with the final stages of her Belgian experiment at the primary level and the first materials she had developed for the CSMP. The GIRP survived until the late 1990s[9], but its original mathematics education ideology and

[9]The last meeting of the GIRP was held in Bardonecchia, Italy, in 1999 (Kristien De Bruyn, personal communication, May 14, 2021).

related rituals did not. The Italian mathematics educator Bruno D'Amore, the Group's president during 1992–1995, noticed:

> The critique that put an end to the tools created by Papy [*papygrammes*, minicomputer, ...] came from the same group he created (the GIRP, *Groupe International de Recherche en Pédagogie de la Mathématique*). Nowadays no one talks about those methods anymore. However, at the time they deceived the world with the trap called "new mathematics" or "modern mathematics." (D'Amore 2016, p. 25)

6 Lenger's Work with Disabled Children

Since the first meeting, the GIRP community had shown an interest in teaching modern mathematics to children with an intellectual disability or personality disorder (Dieschbourg 1971; Lowenthal 1971). The Group members shared a strong belief that the methodology Lenger had developed for primary education was also beneficial for special education:

> The impressive results obtained in teaching retarded children reveal, in a general and often surprising way, that Frédérique's method can be adapted to special education as long as it follows a slower tempo and decomposes some of the steps that normal children accept from the outset. [...] Children who were known to be totally impervious to certain mathematical concepts changed their attitudes completely thanks to the use of modern pedagogical means used in a more nuanced and refined way than in general education. The results obtained constitute a benefit for disabled children and sometimes allow a better understanding of the phenomenon of acquisition in normal children. (Actes de GIRP2. Prologue 1973, p. 5)

While she worked for the CSMP, Lenger had already started to work with "slow learners," often children with intellectual disabilities or severe behavioral problems. The CSMP intended to provide supplementary materials for this specific group. This lead to works such as *Math Play Therapy* (Frédérique. 1977–1978) and *Barry* (Frédérique 1984). Using games, puzzles and detective stories translated into the language of Venn diagrams, colored graphs, and minicomputer, Lenger developed activities to support disabled children in understanding basic mathematical ideas. According to Lenger, these tools were in line with these learners' natural mode of communication which is more likely to be oral and physical than written ("The burden of verbal ornamentation falls on the children, who do not stint themselves, thereby showing the intensity with which they are caught up in it," F. Papy 1983, p. 36). Consequently, paper-and-pencil exercises were avoided and, instead, the use of pocket calculators was promoted. Between 1981 and 1992, Lenger continued this work in the special education section of the *École Normale de l'État* in Nivelles (Frédérique and Papy 1987). She was able to show that her method, which requires little verbal support, was helpful for students with severe intellectual disabilities and has a positive influence on the behavior of these students. This work had a great influence on the later research on nonverbal communication by Francis Lowenthal, Laurent Lefebvre, and Christiane Vandeputte (University of Mons-Hainaut) (Lowenthal and Lefebvre 2014; Lowenthal and Vandeputte 2006). Lenger's final work included studies with autistic children (F. Papy et al. 1995); in particular she conducted research with Thierry Bordignon (University of Mons-Hainaut) on young adults with Asperger's syndrome (Bordignon 2004a, b).

7 Concluding Remarks

We described the life and career of Frédérique Papy-Lenger, one of the prominent pioneer women on the international scene of mathematics education. After her first steps within the CIEAEM, alongside her husband Georges Papy, she became the mother of modern mathematics in Belgium and received international recognition in modern mathematics circles during the 1960s and 1970s. In the last decades of her career, Lenger developed various materials based on the language of modern mathematics with which to study and improve the mathematical knowledge and skills of disabled children.

Lenger's most important contribution to the modern mathematics movement of the 1960s and 1970s was at the level of primary education. Although her approach to primary education shows some parallels with that of other leading reformers of that time in Europe, such as Nicole Picard in France, Geoffrey Matthews in England, and Bent Christiansen in Denmark (see, e.g., Moon 1986), she was not directly influenced by these models. Lenger was more radical in her objectives and the way she shaped them. In her experimental teaching Lenger tried to lower the age for the introduction of fundamental concepts such as sets, relations, functions, groups, and real numbers (Servais 1975). To reach that goal, Lenger relied on the tools of modern mathematics: Venn diagrams, arrow graphs, and minicomputer; she did not encourage children to build alternative representations of problem situations. Lenger, however, did not only pursue purely cognitive objectives. With her "pedagogy of situations" she also wanted to stimulate children's enjoyment in using mathematics, and much more:

> Teaching mathematics also means contributing to the development of the human personality of the teacher and the student, making them both aware of their creative potential; giving them responsibility for their achievements; learning how to listen to each other; opening up the path to freedom: freedom of expression, of decision, of observation, of analysis, of criticism of the society in which they live; enabling them to reach the highest level of communication. (Frédérique. 1976, p. 68)

Very early on, Lenger was aware of the importance of the affective component to the learning of mathematics, including the inhibiting influence of mathematical anxiety. She concluded her "intellectual testament" (Frédérique. 1976) with a discussion on how teachers can help students overcome fear of mathematics. Lenger advises them to (1) speak little and above all listen carefully, (2) recognize errors as something natural and positive, (3) not judge and refrain from comparing students, (4) nip the spirit of rivalry in the bud, but create a climate of emulation, self-surpassing, and mutual aid, (5) continue to learn the most varied mathematics possible, and (6) not underestimate the importance of a certain form of humor and cheerfulness in the child-adult relationship.

Acknowledgement The author thanks Francis Lowenthal, Guy Noël, and Christiane Vandeputte, as well as two anonymous reviewers, for their feedback on a draft of this paper.

References

Actes de GIRP2. Prologue. 1973. *Nico* 13: 2–6.

Barbin, Évelyne. 2013. Le genre « ouvrage d'initiation »: l'Exposé moderne des mathématiques élémentaires de Lucienne Félix (1959–1961). In *Les ouvrages de mathématiques dans l'histoire. Entre recherche, enseignement et culture*, ed. Évelyne Barbin and Marc Moyon, 117–129. Limoges: Presses universitaires de Limoges.

Bernet, Théo, and François Jaquet. 1998. *La CIEAEM au travers de ses 50 premières rencontres*. Neuchâtel: CIEAEM.

Bordignon, Thierry. 2004a. Est-il possible de développer des habiletés numériques chez des personnes présentant des troubles autistiques: Étude d'un cas. *APEPA 89*: 7–11.

———. 2004b. Utilisation « d'outils structurés » pour favoriser le développement cognitif d'un adulte autiste. *Bulletin Scientifique de l'Arapi* 12: 38–40.

Brasseur, Roland. 2011. Lucienne Félix, 1901–1994. *Bulletin de l'Union des Professeurs de Spéciales* 84 (236): 20–27.

Comprehensive School Mathematics Program. 1978. *CSMP in action*. St. Louis, MO: CEMREL Inc.

Cuypers, Karel. 1970. Review of *Les enfants et la mathématique 1* by Frédérique. *Persoon en Gemeenschap* 23 (1): 51–52.

D'Amore, Bruno. 2016. Preface (translated from Italian by Maura Iori). In *Mathematics and mathematics education. In occasion of the 70 years of Bruno D'Amore*, ed. Maura Iori, 17–30. Bologna: Pitagora Editrice.

De Bock, Michel Roelens Dirk, and Geert Vanpaemel. 2019. *Mathématique moderne*: A pioneering Belgian textbook series shaping the New Math reform of the 1960s. In *"Dig where you stand" 5. Proceedings of the Fifth International Conference on the History of Mathematics Education*, ed. Kristín Bjarnadóttir, Fulvia Furinghetti, Jenneke Krüger, Johan Prytz, Gert Schubring, and Harm Jan Smid, 129–145. Utrecht: Freudenthal Institute.

De Bock, Dirk, and Geert Vanpaemel. 2015. The Belgian journal Mathematica & Paedagogia (1953–1974): A forum for the national and international scene in mathematics education. In *Proceedings of the Seventh European Summer University on the History and Epistemology in Mathematics Education*, ed. Évelyne Barbin, Uffe Thomas Jankvist, and Tinne Hoff Kjeldsen, 723–734. Copenhagen: Aarhus University, Danish School of Education.

———. 2018. Early experiments with modern mathematics in Belgium. In *Advanced mathematics taught from childhood? In Researching the history of mathematics education: An international overview*, ed. Fulvia Furinghetti and Alexander Karp, 61–77. Cham: Springer.

———. 2019. *Rods, sets and arrows. The rise and fall of modern mathematics in Belgium*. Cham: Springer.

Deledicq, André. 2012. Hommage à Frédérique et Georges Papy. *Bulletin de l'Association des Professeurs de Mathématiques de l'Enseignement Public* 499: 265–266.

Dieschbourg, Robert. 1971. Un enseignement de la mathématique moderne pour des enfants mentalement handicapés. *Nico* 10: 34–63.

Félix, Lucienne. 1986. *Aperçu historique (1950–1984) sur la Commission Internationale pour l'Étude et l'Amélioration de l'Enseignement des Mathématiques (CIEAEM). 2ième édition revue et augmentée*. Bordeaux: l'IREM de Bordeaux.

Frédérique. 1970–1976. *Les enfants et la mathématique 1, 2, 3, 4*. Brussels, Belgium-Montréal, Canada-Paris, France: Marcel Didier.

———. 1974–1978. *Stories by Frédérique (24 Volumes)*. St. Louis, MO: CEMREL Inc.

———. 1984. *BARRY, Une thérapie par le calcul, avec une préface de PAPY*. Bruxelles, Belgium: Commission française de la Culture de l'agglomération de Bruxelles.

Frédérique, and Georges Papy. 1987. *Un enfant spécial et la mathématique*. Brussels, Belgium: CBPM.

Frédérique, and Georges Papy. 1968. *L'enfant et les graphes*. Brussels, Belgium-Montréal, Canada-Paris, France: Marcel Didier.

Frédérique. 1976. Liberté créative dans l'enseignement de la mathématique. *Nico* 20: 67–127.

———. 1977–1978. *Math play therapy*. Vol. 1 and 2. St. Louis, MO: CEMREL Inc.

Furinghetti, Fulvia. 2008. The emergence of women on the international stage of mathematics education. *ZDM: The International Journal of. Mathematics Education* 40 (4): 529–543.

Furinghetti, Fulvia, and Marta Menghini. 2014. The role of concrete materials in Emma Castelnuovo's view of mathematics teaching. *Educational Studies in Mathematics* 87 (1): 1–6.

Gattegno, Caleb. 1954. Mathematics and the child, III: The use of mistakes in the teaching of mathematics. *The Mathematical Gazette* 38 (323): 11–14.

———. 1963. *Algèbre et géométrie pour les écoles primaires*. Neuchâtel: Delachaux et Niestlé.

Gattegno, Caleb, Willy Servais, Emma Castelnuovo, Jean-Louis Nicolet, Trevor J. Fletcher, Lucien Motard, Luigi Campedelli, Alphonse Biguenet, Jack W. Peskett, and Pedro Puig Adam. 1958. *Le matériel pour l'enseignement des mathématiques*. Neuchâtel: Delachaux et Niestlé.

Gille, Arthur. 1971. Review of *Les enfants et la mathématique 1* by Frédérique. *Bulletin de la Société Mathématique de Belgique* 23 (1): 73–74.

Goffree, Fred. 1970–1971. Review of *Les enfants et la mathématique 1* by Frédérique. *Euclides* 46 (5): 194–195.

Groupe International de Recherche en Pédagogie de la Mathématique (GIRP). 1971. *Nico* 10: 3–11.

Hayden, Robert W. 1981. *A history of the "new math" movement in the United States* (Dissertation Iowa State University). Iowa State University Digital Repository. Retrospective Theses and Dissertations. 7427. Retrieved January 6, 2021, from http://lib.dr.iastate.edu/rtd/7427.

Holvoet, Roger. 1968. De werkgroepen van het Centrum. In *Arlon 10*, ed. Georges Papy , 90–111. Brussels: CBPM.in collaboration with Paul R. Burgraeve, Roger Holvoet, Frédérique Papy, and Andrée Terfve

IMACS. 2006. *Burt Kaufman: An appreciation*. Retrieved November 14, 2020, from https://www.imacs.org/home/about-us/burt-kaufman/.

Jeronnez, Louis, and Isabelle Lejeune. 1972a. L'expérience de Waterloo d'un enseignement moderne de la mathématique à l'école primaire. *Mathematica & Paedagogia* 53–54: 69–80.

———. 1972b. L'expérience de Waterloo d'un enseignement moderne de la mathématique à l'école primaire (II). *Mathematica & Paedagogia* 55: 211–221.

Kaufman, Burt A. 1968. *The CSMP approach to a content oriented highly individualized mathematics education*. Carbondale, IL: CEMREL-CSMP.

Kaufman, Burt A., and Vincent H. Haag. 1977. New math or old math?—the wrong question. *The Arithmetic Teacher* 24 (4): 287–292.

Kaufman, Burt A., and Hans-Georg Steiner. 1969. The CSMP approach to a content-oriented, highly individualized mathematics education. *Educational Studies in Mathematics* 1 (3): 312–326.

Kilpatrick, Jeremy. 2012. U.S. mathematicians and the new math movement. In *"Dig where you stand" 2. Proceedings of the Second International Conference on the History of Mathematics Education*, ed. Kristín Bjarnadóttir, Fulvia Furinghetti, José Manuel Matos, and Gert Schubring, 251–261. Lisbon: New University of Lisbon.

Le programme B des écoles normales gardiennes. (1958–1959). *Mathematica & Paedagogia* 16: 70–75.

Lemaître, Georges. 1954. *Calculons sans fatigue*. Louvain, Belgium: E. Nauwelaerts.

Lenger, Frédérique. 1950. La notion d'involution dans l'œuvre de Desargues. In *IIIe Congrès National des Sciences, Bruxelles 30 mai–3 juin 1950*, 109–112. Liège: Desoer.

———. 1953a–1954. Enseigner en apprenant. *Mathematica & Paedagogia* 1:16–19.

———. 1953b–1954. Les erreurs mathématiques. *Mathematica & Paedagogia* 3:15–17.

———. 1953c–1954. Rencontre internationale de professeurs de mathématiques 20 au 26 juillet 1953, Calw (Baden-Wurttemberg, All.). *Mathematica & Paedagogia* 1: 55–56.
———. 1954a–1955. Les erreurs mathématiques des élèves. *Mathematica & Paedagogia* 4:13–15.
———. 1954b–1955. Réflexions sur l'enseignement des mathématiques aux enfants dits 'peu doués'. *Mathematica & Paedagogia* 5:20–23.
———. 1954c–1955. Rencontre internationale de professeurs de mathématiques. *Mathematica & Paedagogia* 4:57–59.
———. 1954d–1955. VIIIe rencontre internationale de professeurs de mathématiques. *Mathematica & Paedagogia* 6: 86–88.
———. 1955–1956. Témoignage. *Mathematica & Paedagogia* 8: 22–24.
Lenger, Frédérique, and Madeleine Lepropre. 1959. Initiation aux mathématiques fondamentales dans une classe de première normale gardienne. In *Journées d'études 8: L'initiation aux mathématiques—Problèmes psycho-pédagogiques*, 17–24. Brussels: Ministère de l'enseignement public, Secrétariat général de la réforme de l'enseignement moyen et normal.
Lowenthal, Francis. 1971. Enseignement de la mathématique à deux groupes d'enfants caractériels. *Nico* 10: 69–86.
Lowenthal, Francis, and Laurent Lefebvre. 2014. Nonverbal communication devices, language, cerebral flexibility, and recursive exercises. In *Language and recursion*, ed. Francis Lowenthal and Laurent Lefebvre, 39–55. New York, NY: Springer.
Lowenthal, Francis, and Christiane Vandeputte. 2006. *Une remédiation cognitive pour l'acquisition des prérequis de la lecture et du nombre. In Actes du Congrès X-fragile–Europe*, 22. Belgium: Liège.
Matthys, Jean-Claude. 2005. [Funeral speech in honor of Frédérique Lenger, dated September 7, 2005]. Retrieved November 14, 2020, from http://www.rkennes.be/Hommage/JCLM1.htm.
Menghini, Marta. 2015. Emma Castelnuovo's commitment to creating a new generation of mathematics teachers. In *"Dig where you stand" 3. Proceedings of the Third International Conference on the History of Mathematics Education*, ed. Kristín Bjarnadóttir, Fulvia Furinghetti, Johan Prytz, and Gert Schubring, 263–278. Uppsala: Uppsala University, Department of Education.
Moon, Bob. 1986. *The "new maths" curriculum controversy: An international story*. London: The Falmer Press.
Noël, Guy. 1993. La réforme des maths moderne en Belgique. *Mathématique et Pédagogie* 91: 55–73.
Nowecki, Bogdan J. 1992. Anna Zofia Krygowska. *Educational Studies in Mathematics* 23 (2): 123–137.
Papy, Frédérique. 1968a. *Sur le premier enseignement de la mathématique et une méthodologie de la formation continue des enseignants*. Unpublished doctoral dissertation, Faculté des Sciences, Université Libre de Bruxelles, Brussels, Belgium.
———. 1969. Minicomputer. *Educational Studies in Mathematics* 2 (2-3): 333–345.
———. 1970. L'expérience Belge à l'école primaire. In *New trends in mathematics teaching – Tendances nouvelles de l'enseignement des mathématiques*, ed. André Revuz, vol. II, 95–111. Paris: Unesco.
———. 1983. When children fail to learn arithmetic in primary school, what are the roots of the failure? What approaches can be taken towards a cure? In *Proceedings of the Fourth International Congress on Mathematical Education*, ed. Marilyn J. Zweng, 36–37. Boston, MA: Birkhäuser.
Papy, Frédérique, Georges Papy, and Adriana Schuler. 1995. *La pensée hors langage: À la rencontre d'adolescents autistes*. Paris: Bayard.
Papy, Georges. 1959. *Aperçu général de la théorie des ensembles. Quelques notions essentielles de topologie (Report of Arlon 1, July 4–6, 1959)*. Brussels: Presses Universitaires de Bruxelles.
———. 1963–1967. *Mathématique moderne 1, 2, 3, 5, 6*. Brussels, Belgium-Montréal, Canada-Paris, France: Marcel Didier.
———. 1966. *Arlon 8. Premières leçons d'analyse mathématique par Frédérique*. Brussels, Belgium: CBPM.

———. 1967a. *Arlon 9. Nouvelles leçons d'analyse mathématique par Frédérique*. Brussels: CBPM.

———. 1967b. *Mathématique moderne 3. Voici Euclide*. Brussels, Belgium-Montréal, Canada-Paris, France: Didier.

———. (in collaboration with Frédérique)1968b. *Minicomputer*. Brussels: IVAC.

Piaget, Jean, Evert Willem Beth, Jean Dieudonné, André Lichnerowicz, Gustave Choquet, and Caleb Gattegno. 1955. *L'enseignement des mathématiques*. Neuchâtel: Delachaux et Niestlé.

Servais, Willy. 1975. Continental traditions and reforms. *International Journal of Mathematical Education in Science and Technology* 6 (1): 37–58.

Stievenart, Pierre. 1968. La guerre des math aura-t-elle lieu? *Spécial* 153: 16–19.

Un groupe de professeurs Américains. 1970. Frédérique dans une classe Américaine. *Nico* 7: 63–70.

Vredenduin, Piet G.J. 1970–1971. Review of *Les enfants et la mathématique 1* by Frédérique. *Euclides* 46 (5): 194–195.

———. 1972–1973. Review of *Les enfants et la mathématique 2* by Frédérique. *Euclides* 48 (2): 72–73.

———. 1975–1976. Review of *Les enfants et la mathématique 3* by Frédérique. *Euclides* 51 (4): 163–165.

———. 1978–1979. Review of *Les enfants et la mathématique 4* by Frédérique. *Euclides* 54 (6): 248–249.

Walmsley, Angela L.E. 2003. *A history of the 'new mathematics' movement and its relationship with current mathematical reform*. Lanham, Maryland: University Press of America.

The Discarding and Endurance of the Rule of Three: Changes and Stability in Arithmetic Teaching in France and Brazil

Elisabete Zardo Búrigo

Abstract In the 1960s, the modern mathematics movement brought about changes in school curricula in different parts of the world. It was not focused on secondary education only, the movement also reached mathematics education in the early grades of various countries, including France and Brazil. In both countries, until the 1950s, the rule of three occupied a prominent place in the early grades. It was the culmination of arithmetic study and had multiple applications in everyday life. In both countries, the emergence of the modern mathematics movement combined with broadening access to secondary education leading to changes in mathematics education goals in the early grades. The combined effect of modernizing ideas and changes in school pathways affected the teaching of the rule of three. In France, the study of functions was introduced in elementary school to replace the mechanical application of the rule of three; in Brazil, this proportionality study was postponed until the later grades. A decade later in France, and two decades later in Brazil, new reforms rehabilitated the rule of three. Presenting a narrative of discarding and restoring the rule of three in both countries, this article attempts to contribute to studies on change and permanence in school arithmetic.

Keywords Arithmetics teaching · Elementary school · Problems resolution · School programs · Rule of three

E. Z. Búrigo (✉)
Universidade Federal do Rio Grande do Sul, Porto Alegre, Brazil
e-mail: elisabete.burigo@ufrgs.br

A. Karp (ed.), *Advances In The History Of Mathematics Education*, International Studies in the History of Mathematics and its Teaching,
https://doi.org/10.1007/978-3-030-95235-8_3

1 Introduction

What remains and what changes in school teaching of arithmetic over time? Bjarnadóttir (2014), in a retrospective on education in Europe and the United States, shows that in the 19th and 20th centuries the teaching of arithmetic in primary school oscillates between practical instruction, preparation for work and everyday life, and development of logical thinking. Educational trends as well as changes in purpose, in the training of teachers, in new technologies, and in reorganization of the education system all imply changes in methodologies and teaching content.

This article focuses on the discarding of the rule of three from elementary school programs. The rule can be stated nowadays as a tool to find the unknown fourth term of a proportion when the other terms are given, namely, multiplying the second by the third term, and dividing the result by the first one. It was spread by commercial arithmetic books in Europe since the 15th century as a technique to solve various problems and was later progressively incorporated into textbooks and programs in several countries (Swetz 1992). According to Kristin Bjarnadóttir "The rule survived well into the twentieth century, but was practically eradicated by the school mathematics reforms of the 1960s" (Bjarnadóttir 2014, p. 442).

The study deals with two particular cases of curricular change. In France and Brazil, until the 1950s, solving arithmetic problems through the rule of three was the culmination of arithmetic study in elementary school. Discarding the teaching of the rule of three had different outcomes in the two countries. In France, the theme of proportionality was linked to the study of linear functions; in Brazil, it was associated with the study of rational numbers and postponed to a later level of schooling. In 1985, in France, the rule of three was reinstated into the programs; and in 1998, it was brought back as a middle school study topic in Brazil. So in both cases, the discarding was not precisely an eradication.

The case of France is particularly interesting because, for decades, the rule of three issue was explicitly and repeatedly mentioned in pedagogical texts and official documents (d'Enfert 2015). The contrast between the two countries is appealing because French teaching had been, from the 19th century until the late 1950s, a reference for the programs, pedagogical manuals, and textbooks produced in Brazil (Búrigo 2017). In Brazil's case, in light of the variations between local legislation, the study focuses on the reforms in São Paulo, the wealthiest and most populous state, and Rio Grande do Sul, where the theme was subject to detailed regulation.

The study takes as its main sources the elementary school mathematics programs in force in both countries from the 1950s to the present as well as textbooks from the same period. The narrative of the debates that preceded and justified the curricular changes is anchored in previous research and takes into account the processes of broadening access to secondary education which, at that time, had an impact on the purposes assigned to elementary school. We seek to understand how the intertwining between those changes and pedagogical movements affected the programs in the two cases, thus contributing to the understanding of change and permanence in school arithmetic.

2 The Rule of Three as a Preparation for Practical Life and the Exams

In the 1940s and 1950s in France and Brazil, primary school was the only school accessible to the majority of the population. In both countries, primary schools' teaching of arithmetic was centered on calculations with whole numbers, decimal numbers or fractions, and arithmetic problem-solving. According to official guidelines, arithmetic should be eminently practical.

In France, elementary education was reorganized after the liberation from German occupation in 1945. It lasted for five years, with fourth and fifth grade making up the *cours moyen* (9–11 year olds). At this stage, the rule of three, as well as the study of decimal numbers, percentages, and fractions, was associated with solving "problems of everyday life" (MEN 1945).

The 1945 instructions advocated for approaching the rule of three using the method of "reduction to unity." This approach became widespread in France from the 1850s and was widely accepted because it avoided an "overly mechanical use of proportions" (d'Enfert 2003, p. 30) as well as the theory of ratios and proportions (Galisson 2008). Given two different quantities, the "unit value" was obtained as an intermediate ratio to be calculated: "price per unit of length, distance, surface, volume, capacity, or time," "distance or volume per unit of time," "weight per unit of length or volume" (MEN 2015a, p. 401). The desired value would then be obtained by the formula "total value = unit value × number of units." The instructions presented the rule of three as a procedure that summarized the method of reduction to unity based on the argument that multiplication could precede division.

Figure 1 presents the calculation of a monthly heating expense, given the expense for a six-day period in French francs (F). It is taken from a textbook which was written by Ardiot, Wanaud, and Budin (1963, p. 271) for the *cours moyen.* The wording reads: "In a heating apparatus, a tank of butane which costs 12,80 F lasted 6 days. What is the cost of heating for a month of 30 days?". The solution follows as: "For one day heating costs (12,80 F)/6. We can perform this division. We obtain an approximate quotient of 2,13 F. But one can also keep this quotient in its fractional form. Heating for 30 days costs: (12,80 F × 30)/6 = 64 F. A lesson is drawn from the example and underlined: "Starting with the multiplication, **the rule of three allowed us to find an exact result**" (bold as in the original text). One ending remark points out: "Let's keep in mind that the number placed at the head of the rule of three and the result are of the same nature: (12,80 F × 30)/6 = 64 F".

In Brazil in the 1950s, primary school teaching programs varied by state. In the southernmost state of Rio Grande do Sul, the arithmetic program was established by Decree 8.020 in 1939, which prescribed the study of proportions by using the rule of three and its applications for the fifth grade (11–12 year olds). Much like in France, the rule of three was to be presented following the method of reduction to unity. The Primary Course Experimental Mathematics Program, enacted in 1959, maintained the rule of three (Rio Grande do Sul 1959). In São Paulo state's primary school programs, the rule of three was not included. However, the method of

La règle de trois CM

Dans un appareil de chauffage, la bouteille de butane qui coûte 12,80 F a duré 6 jours.
Quel est le prix de revient du chauffage pendant un mois de 30 jours?

Pour un jour, le chauffage coûte : $\frac{12{,}80\ F}{6}$

On peut effectuer cette division. On obtient un quotient approché : 2,13 F. Mais on peut aussi garder ce quotient sous cette forme fractionnaire.

Le chauffage en 30 jours coûte : $\frac{12{,}80\ F \times 30}{6} = 64\ F.$

Commencée par la multiplication,
la règle de trois nous a permis de trouver un résultat exact.

Retenons — Le nombre placé en tête de la règle de trois et le résultat sont de même nature : $\frac{12{,}80\ F \times 30}{6} = 64\ F$

Fig. 1 Introducing the rule of three based on the unit reduction method

reduction to unity was recommended for the fourth and fifth years in solving problems involving fractions of whole quantities, interest, and conversions between currencies or measurement systems (São Paulo 1950a, b).

In both countries, published rules advocated use of the rule of three in arithmetic problems. French guidelines recommended that these problems should be "feasible, as the ones which the pupil has seen or will see around him," and "reasonably present themselves in practice" (MEN 2015a, p. 400). The directives of the state of Rio Grande do Sul, in Brazil, stated that problems should be "capable of frequently occurring in practical life and in the same way as they ordinarily present themselves, that is, formulated with the difficulties they offer in reality and using real data" (Rio Grande do Sul 1939, p. 291).

Solving arithmetic problems by applying the rule of three quickly and accurately was an objective of teaching arithmetic as preparation for work and everyday life. Thus, in French textbooks of the 1950s, an initial presentation of the rule was followed by a list of word problems, aiming at training for efficiency in solving.

These arithmetic problems follow a familiar standardized format: a brief description of an everyday situation containing the numerical data that should be used in the solution, followed by a question, to be answered with a number. While similar to those of Renaissance commercial arithmetics, these problems were both abbreviated and standardized in school, as shown in Fig. 1. They became "teachable" (Chervel 1988, p. 90), that is, written so that children could learn to read, recognize the tasks involved, and solve them.

In the mid-1950s, several types of tests evaluated the students' ability to solve arithmetic problems in both countries: tests applied at the end of each year to decide on the advancement of each student, exams to obtain the certificate of completion of primary school in France, and selective entrance processes for the first cycle of secondary school in both countries. Problems involving proportionality made up the entrance exams to the *sixième* grade (11 year olds) in France and the *ginásio* (11–15 year olds) in Brazil.

Teaching and learning to use the rule of three, therefore, prepared students not only for life but also for exams which determined their school paths. In France, there were many manuals dedicated to preparing for arithmetic tests. A typical example of these manuals is the book *1300 problèmes*, by Pierre Jorez and Georges Koëll, published in several editions since 1952. The rule of three item takes up a few pages of the book, which are filled with dozens of problems. In Brazil, the fifth year of primary school centered on preparing for the exam. Much like in France, preparatory manuals offered dozens of problems focused on training and using the rule of three (Segundas noções 1955, p. 197).

In France, the teaching of arithmetic problems had been questioned at least since the 1920s. Inspectors of education denounced training for examinations, asserting that it led to exhaustive solving of similar problems without challenging students to interpret them and identify the operations involved (Sarrazy 2003). In the 1950s, the teaching of the rule of three was also subject to questioning.

Thus, the textbook author François Brachet advocated that the proportional magnitudes concept should be addressed beginning in elementary school, arguing that its use "will gain in efficiency if one dares to speak its name, without going beyond the bounds of possibility" (Brachet 1955, p. 87). In this approach, the rule of three would be one of the consequences of proportionality study and no longer a topic in the curriculum.

Henri Canac, assistant principal of the Saint-Cloud Normal School, questioned the practical nature of arithmetic problems: "schoolwork does not consist in and cannot claim to provide that the child routinely solve all the particular difficulties that life may offer to him." The utilitarian character of arithmetic teaching was already under debate: "would the best preparation for life not be, in the end, the most methodical intellectual exercise?" (Canac 1955, p. 117).

3 Mathematical Thinking from Pre-school to University

In 1956, France abolished the entrance exams to the *sixième* grade, deflating the dynamics of training for the tests. In 1959, the Berthoin Reform unified school paths and instituted a two-year "observation cycle," extending the stage of schooling which was common to all students. All the *sixième* and *cinquième* classes were made to follow the same math program (d'Enfert 2011). Several polls and studies accompanied the change.

At the beginning of the 1958–1959 school year, the educational research department of IPN (National Pedagogical Institute) conducted a survey on mathematical knowledge with 324 students from six schools in the Paris region. The results were considered inadequate, especially for problems involving the rule of three. Only

17% of the students correctly solved the following problem: "Mom bought a suitcase labeled 6500 francs. She paid only 6175 francs. Calculate the percentage of the reduction granted to her" (Dabout 1960, p. 9). According to the investigators, students' poor performance was due to miscalculations but, above all, to "reasoning" failures (Dabout 1960, p. 29).

In a nationwide survey conducted by INETOP (National Institute of Study of Labor and Vocational Guidance), researchers found that only 55% of the students in second-year *Cours Moyen* classes (fifth grade) correctly solved the following problem: "if a cyclist rides 5 kilometers in 20 minutes, how far will he ride in 50 minutes?" (Gloton 1960, p. 14).

The entrance of a new contingent of students in the now unified *sixième* highlighted the existing gap between prescribed and assimilated knowledge at the end of elementary school. According to Bishop and Dorison (2015), the evaluation of program inadequacy gradually replaced the idea of student inadequacy. In 1964, some arithmetic topics in the *cours moyen*, including the rule of three, were considered "premature" or "dispensable" and thus became optional (MEN 1964, p. 1795).

Actors engaged in the modernization of mathematics education also questioned the programs. The APMEP (Association of Public Education Mathematics Teachers) had been advocating for the renewal of mathematics education in secondary school since the 1950s. International events on the teaching of mathematics, such as those promoted by CIEAEM (International Commission for the Study and Improvement of Mathematics Teaching) since 1950, or by the OECD (Organization for Economic Co-operation and Development), such as the renowned Royaumont Seminar in 1959, favored the spread of modernizing proposals and the push for reform. In the early 1960s, the notion that mathematics education should form a coherent whole spread and was translated into the catchphrase "from pre-school to university" (d'Enfert 2011).

The rule of three was one of the targets of the movement's renewal. Mathematician Gustave Choquet published an article that proposed discarding the expression "rule of three," and instead focusing on proportional magnitudes through linear functions (Choquet 1961). The secondary school teacher and APMEP leader Gilbert Walusinski (1963) supported this proposal. The functional approach was also taken on by a Commission ministerial meeting on May 22, 1965, chaired by Inspector General Marius Beulaygue (Walusinski 1966).

A functional approach to proportionality was developed and applied in primary education classes in Paris as part of an experiment under the guidance of the IPN (National Pedagogical Institute), beginning in 1964. Nicole Picard, inspired by Jean Piaget and Zoltan Dienes's works, coordinated the experience, which involved children's engagement in various activities, interspersed with discussions. In the second-year *cours élementaire* classes (third grade), operator and function notions were introduced by the simulation of "machines" capable of effecting specific transformations on objects or numbers. The discussion about the machines involved, for example, identifying the machine that does the opposite of "multiply by 2" or the machine that "does nothing" (Picard 1967).

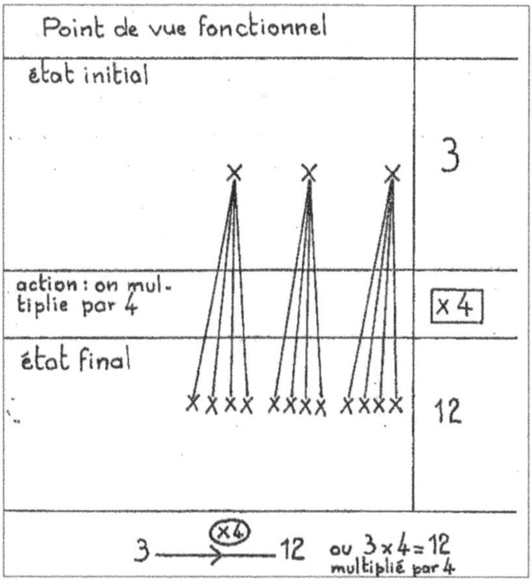

Fig. 2 Illustration of "functional point of view" on multiplication

Figure 2 shows a tree representation of the multiplication machines used by the third graders (Picard 1967, p. 27). From an "initial state of 3," and applying the "action: multiplying by 4," one gets the "final state of 12."

From 1965 on, the experiment was extended to several regions of France. Reports on teaching in *cours moyen* classes (fourth and fifth grades) show the exploration of various representations of functions, including tables and cartesian graphs, in the study of proportionality (Touyarot 1968). Figure 3 represents a graph plotted on a plywood board (Vandendriessche 1968, p. 53). The author explains that by placing matches at points corresponding to ordered pairs and stretching a plastic string passing by two of them, one can find the others and, therefore, solve problems on proportional magnitudes.

The CEEM (Commission for the Study of Mathematics Teaching) was created in 1966 by a ministerial decision to propose reforms of mathematics teaching in *collège*. From 1969 on, discussion encompassed the development of a new mathematics program for elementary education. According to Renaud d'Enfert (2011), the commission discussed two projects. The first one, which had been developed within the scope of APMEP and with the participation of primary teachers, primary inspectors, and teachers of normal schools, grounded the new program decreed in 1970 (MEN 1970). The other version, proposed by the IPN, was more audacious and was forseen to be applied from 1973 on, after teachers' adequate training.

In the 1970 program, the "rule of three" item gave way to the topic of "proportionality." Comments published in a ministerial circular proposed proportionality as a particular case of "numerical relations" to be studied using graphs and tables. Using a "multiply by *a*" or "divide by *b*" operator, two lists of numbers could be matched, as in Fig. 4. The *Circular* stated that "all the problems dealt with by the

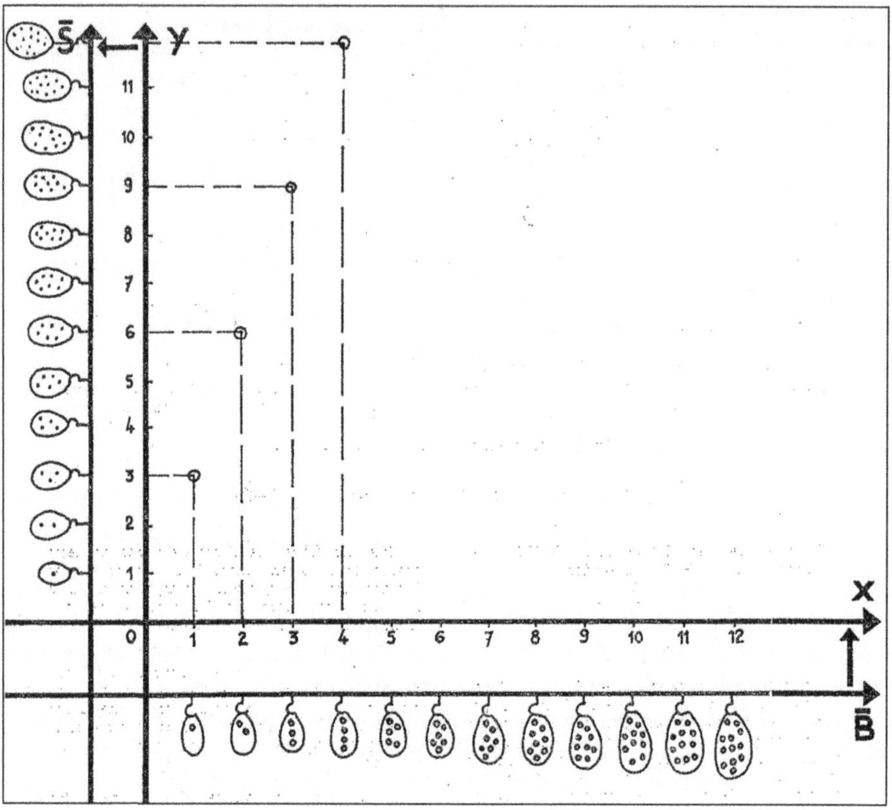

Fig. 3 Graphic representation of a relationship between numbers of chocolates and price

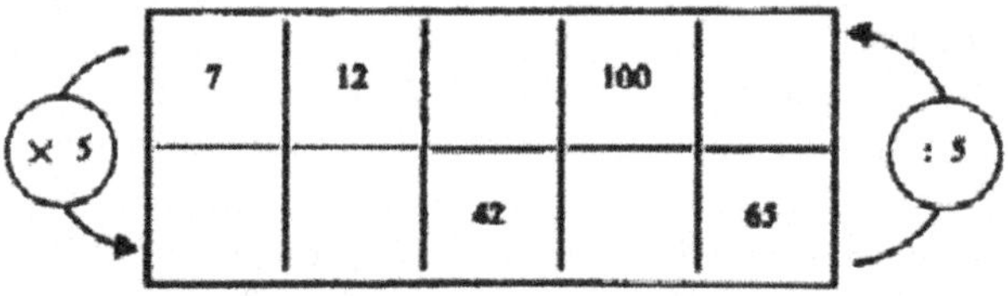

Fig. 4 *Circular*'s example of a table illustrating proportionality

'rule of three' arise from the previous mathematical model" and that "it is one and the same problem, which should be explained in new terms" (MEN 2015b, p. 519).

In the new approach, the students' task was no longer to follow a sequence of steps but to "learn to move from a situation to a diagram that describes it" (MEN 2015b, p. 531). Through the graph, a student could discover the operator that would relate two lists of values of various magnitudes rather than the value corresponding to a given unit.

Distance sur la carte, unité : cm	5	18	
Distance sur le terrain, unité : km	10	.	32

Fig. 5 Scheme for solving the problem involving scale

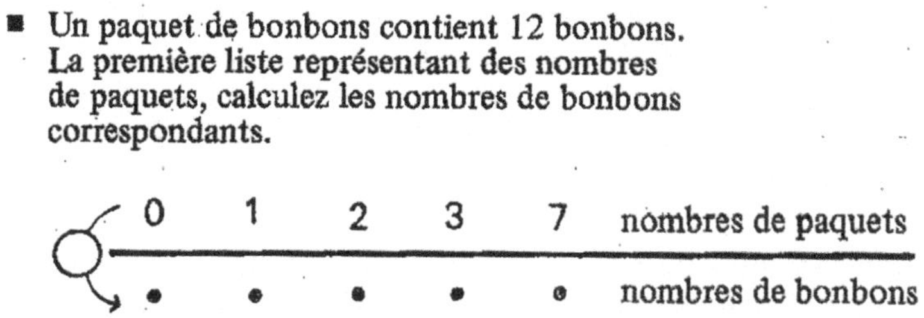

Fig. 6 Graph for representing packages and corresponding numbers of chocolates

As an example, we have the statement: "5 cm on a map represent 10 km on the ground. How far in the field is a distance of 18 cm on the map? What distance on the map represents a distance of 32 km on the field?" (MEN 2015b, p. 520).

From a scheme like the one in Fig. 5, where the upper line records distances on the map given in centimeters and the lower line records distances on the site in kilometers, the student should identify that operators moving from one line to another are "multiplying by 2" and "dividing by 2" and then obtain the missing data.

We see that the decree and the *Circular*, therefore, made the functional approach to proportionality official, even if represented only by two-line tables. Authors published new textbooks following the *Circular*, replacing the "rule of three" section with the topic "proportionality" or "proportional magnitudes." Some of them presented the rule of three after problem solving activities using tables and operators (Denise et al. 1970). In Fig. 6, we can see the similarity between the *Circular*'s guidelines and a textbook example about package and numbers of chocolates. The wording reads: "A package of chocolates contains 12 chocolates. Since the first list represents numbers of packages, calculate the corresponding numbers of chocolates (Goergler et al. 1971, p. 162).

It is important to note that these early 1970s books contain activities in varied formats in addition to these graphical representations. These activities include the traditional lists of arithmetic problems, either in their usual form or with longer statements, including more than one question or diagram. These lists also correspond to the guidelines of the *Circular*, which describes problem solving as a privileged activity and explains: "There is a problem if, knowing a certain amount of information about a situation, one proposes to deduce from this information that which was not initially explicit." (MEN 2015b, p. 530).

4 The Delay of the Study of Proportions in Brazil

In Brazil, ideas about modernizing mathematics circulated locally as early as the 1950s. At that time, at the General Flores da Cunha Education Institute in Porto Alegre, in the state of Rio Grande do Sul, primary school teacher trainers already took courses on set theory.

The creation of the GEEM (Mathematics Teaching Study Group) in São Paulo in 1961 was a founding landmark. In 1962, at the IV Brazilian Congress on Mathematics Teaching, the GEEM presented a proposal of a new curriculum for secondary school. From then on, in different parts of the country, teachers organized into groups that disseminated modern mathematics ideas. Different books, authors, and experiments from the United States and Europe were taken as reference and combined in the elaboration of new textbooks and teacher training courses (d'Ambrosio 1987).

In Brazil, in the early 1960s, only two out of every ten children who entered school completed primary school, and one of ten continued to study beyond that stage (Cunha 1980; Gouveia and Havighurst 1969). Evoking Jean Piaget's (1955) statements about correspondences between mental and mathematical structures, advocates of modernization promised more attractive and easier mathematics lessons (Búrigo 2009).

As in France, the modernizing movement's initial focus was secondary education, but the innovation proposals quickly expanded to primary education (d'Ambrosio 1987; d'Enfert 2011). In 1963, a group of GEEM members toured several cities in different regions, teaching courses for primary school teachers. In the same year, they launched the book *Introduction of Modern Mathematics in Primary School*, disseminating guidelines (Medina 2007). From 1966 on, the General Flores da Cunha Education Institute in the state of Rio Grande do Sul offered a one-and-a-half-year specialization course entitled Didactics of Modern Mathematics for the Primary School.

During the 1960s, the dissemination of modernizing ideas came through teachers' groups. Only in the state of São Paulo was there a slight curricular reform in 1969, with the main goal to produce a "simple," "basic and common" program (São Paulo 1969, p. 8). Having a mathematics degree and being a GEEM member, Manhucia Liberman participated in the program's writing, inserting modern mathematics features, such as the emphasis on sets and operations with sets (Medina 2007). In this "simple" program, the unit reduction method was no longer mentioned. There were only implicit references to proportionality, as in the item "percentage" and "application to simple problems," to be addressed in the fourth series (São Paulo 1969, p. 78). The theme of proportionality was postponed until middle school (11–15 year olds).

Within the reorganization of education in 1971, New Education Law established a single compulsory education stage, now with an extended duration of eight years. As earlier in France, the primary school lost its terminal character through the extinction of the entrance exam to secondary education.

The national guidelines prescribed a curriculum organized according to areas, with Mathematics being a sub-area of Sciences and aiming at achieving "the development of logical thinking" and "the experience of the scientific method and its applications" (CFE 1971, p. 191). In this conception of curriculum, "the integration of related contents" and "the situations of experience" should be valued as much as "systematic knowledge" (CFE 1971, p. 191).

Adapting to the new legislation, states developed new curricular proposals for the eight-year cycle called "first degree"; technical concepts of school curriculum were in vogue. Those proposals were then organized according to well-defined, detailed, and operational objectives so that their achievement could be measured.

In the state of Rio Grande do Sul, the Department of Education published new curricular guidelines in 1972 (Rio Grande do Sul 1972). In line with the new law's spirit, the document was not organized around content but presented a list of objectives to be achieved. The document was organized in two parts. In the text addressed to the first three school grades (7–10 year olds), we can see modern mathematics marks, such as the valuation of operations with sets, order, and equivalence relations, as well as other themes. Proportionality is implicitly evoked in the study of reductions, extensions, and equivalence between fractions. Still, there are no references to the rule of three nor to the idea of a multiplicative operator.

For the second stage of the first degree, fourth to eighth grade (10–15 year olds), the document was organized into three areas: Communication and Expression, Social Studies, and Science, the latter covering Mathematics. Zilá Maria Guedes Paim was responsible for the mathematics area at the Department of Education and coordinated the production of the guide's Science section. In the list of objectives, the theme of proportions appears explicitly associated with the construction of the rational numbers set and problem solving in the group (Q, ·), that is to say in the algebraic structure resulting from considering the rational numbers set and the multiplication in this set. Reasons and proportions, scales of reduction and enlargement, percentage, rates, interest, and exchange are some of the contents listed.

In the state of São Paulo, only one team was in charge of writing guidelines for teaching mathematics over the eight years of the first degree. The invited authors—Almerindo Marques Bastos, Anna Franchi, and Lydia Condé Lamparelli—had their degrees in mathematics. Medina (2007) explains that, not by chance, every one of them was engaged in the modern mathematics movement. It was the participants in the movement who were most keenly interested in discussing curriculum.

Emphasizing the first-degree cycle's continuity and unity, the Mathematics section of the São Paulo Curricular Guide (São Paulo 1973) covered each of the themes throughout the eight series. The theme of proportionality was associated with the study of rational numbers. The topic of percentages was moved to the fifth grade (11–12 year olds), while proportional quantities, including their use in the solution of simple problems, remained in the sixth grade (12–13 year olds). According to the curricular guide, at this level, a ratio should be presented as a quotient between two rational numbers and a proportion should be presented as an equality between rational numbers. That knowledge should be used in solving simple problems involving proportional magnitudes. "Relations and functions" should be addressed from the first grade (7–8 year olds), and in the sixth grade this study should include the relations "to be a multiple of" and "to be a divisor of," but linear functions would be studied only in the eighth grade (14–15 year olds).

The documents from the states of São Paulo and Rio Grande do Sul which were produced between 1969 and 1973, therefore, no longer dealt with the rule of three as a mechanism to be learned and applied by young children in preparation for solving adult problems. Moreover, in both documents, the initial study of proportionality did not include, as it did in France, the study of operators and was disconnected from functions. Both documents prioritized the presentation of the number sets and followed the logic of a progressive extension - natural numbers, integers, rational numbers, and then real numbers.

Some textbooks were produced and published in line with the new programs. The collection *Modern Course of Mathematics for the Elementary School*, written by GEEM members and other authors, had a wide circulation. In the book intended for the fifth grade, problems involving proportionality were solved by using fractions, percentages, and graphing schemes, and the rule of three was not mentioned (Sanchez and Liberman 1972).

The sixth-grade book presented the ratios and proportions topic through various situations, representations, and tasks. Figure 7 shows an extract of the teachers' guide (Averbuck et al. 1975, p. 197). The first image represents a set A of girls, containing subsets of three elements, and a set B of boys, containing subsets of two elements. The first task is to write the cardinal of each set. Item *b* asks, "By means

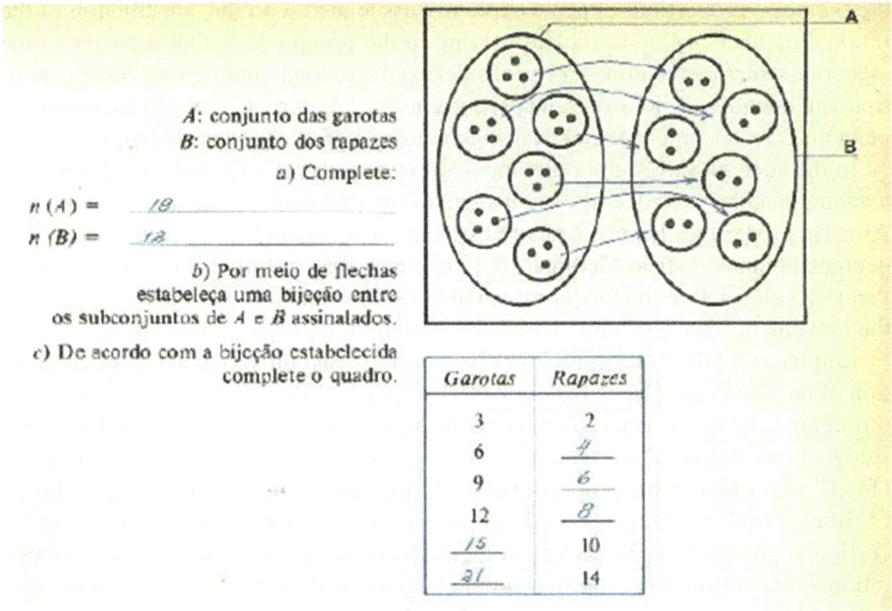

Fig. 7 Representations of a proportionality relationship using sets

of arrows establish a bijection between the subsets of A and B denoted." Similarly to the French books, item *c* asks the student to complete the table, with numbers of girls in the first column and corresponding numbers of boys in the second.

Although absent from official documents, the approach to proportionality from the point of view of functions was present in an experiment coordinated by GEEMPA (Study Group on Mathematics Teaching in Porto Alegre), a group of school and university teachers created in 1970 in the state of Rio Grande do Sul. In 1972, GEEMPA organized an experiment with pilot classes, rearranging curricular contents according to the "psychogenetic stage of cognition, enabling real learning by discovery" and "the need to respect the mathematical sequence of content" (GEEMPA 1975, p. 26). According to the experiment plan, operators were present from the second and third grades in the "machines" format proposed by Nicole Picard. The multiplicative operators should be explored beginning with the fourth grade (10–11 year olds). Still, the rule of three was not to appear until the seventh grade (13–14 year olds) and was associated with the study of rational numbers, proportions, and geometric transformations.

The modernization proposals tried out in French elementary schools were known in Brazil, as shown in the GEEMPA experiment. However, curricular reforms and textbooks published in the states of São Paulo and Rio Grande do Sul followed another logic. In the early grades (7–11 year olds), modernization focused on presenting elementary notions of sets and relations. The *Synopses for Modern Secondary School Mathematics* (OEEC 1961), drawn up by a group of experts after the Royaumont seminar, advocated that proportional magnitudes be approached through functions in middle school. The book was translated and published in Brazil on the initiative of GEEM, but its path was not followed either. The solving of problems involving proportionality was postponed to the final years (11–15 year olds) and connected to the study of rational numbers. This connection and the stepwise approach to the natural numbers, integers, rational numbers, and real numbers were aligned with the North American School Mathematics Study Group (SMSG) proposal. The SMSG middle school book series had been translated and published by the IBECC (Brazilian Institute of Education, Science, and Culture), with the participation of Lydia Lamparelli. It is interesting to note that the French middle school reform did not adopt this sequence, instead teaching real numbers prior to the rational numbers set (Búrigo 2019).

5 Rehabilitation

The modernizing impetus cooled down at different paces in France and Brazil from the mid-1970s on.

While the APMEP (Association of Public Education Mathematics Teachers) advocated a more profound reform at all schooling stages, the National Union of Primary School Teachers opposed any abrupt change (d'Enfert 2017). In September 1970, some months after the publishing of the new programs and guidelines, a new circular tried to calm fears of radical modifications, stating that this was "the program of 1945 lightened and presented in a more rational way…whose essential goal is to approach notions already known with a new spirit" (MEN 2015c, p. 538). Still, the program was subject to various kinds of criticism. The controversies and tensions related to the reform of mathematics teaching at the *collège* (middle school) caused a ceasing of the changes and the end of CEEM's work in 1973 (Alix et al. 2015). The project of comprehensive teacher training and the more profound reform that was to follow were cast aside.

New programs for French elementary education were published gradually from 1977 on. In the text written for the *cours moyen* and published in 1980, a significant section concerns "problem situations," stating that "the teacher will avoid stereotyping the formatting of the approach or results" (MEN 2015d, p. 584). The section dedicated to numerical functions covers proportionality, but the guidelines no longer recommended that the functions involved be studied and explained: "they will be discovered and used, without formalization, during the study of appropriate situations," which cover "problems previously solved by the rule of three" (MEN 2015d, p. 595).

In 1985, a new program for elementary education reconciled the study of functions and the rule of three. Students were expected to recognize and use the numerical functions given by "n → n + a and n → n × a," solve problems related to these functions as well as those "more particularly of proportionality (example of the rule of three)" (MEN 2015e, p. 607). According to d'Enfert (2017), it was part of a "back to basics" turn, and one of the outcomes considered essential to the preparation for the *collége*, now accessible and mandatory for all young people. The rule of three is again mentioned in the 1991 programs as one among many valid strategies to solve proportionality problems: "use of graphs, numerical tables, linearity properties, possibly the rule of three" (MEN 2015f, p. 620). From 1995 on, it was no longer recommended nor repudiated. On the other hand, proportionality remains a privileged topic and includes the solution of traditional verbal statement problems. In recent textbooks, we find statements that evoke the book lists of the 1950s: "To make 5 hazelnut cakes, you need 300g of flour. How much flour is needed to make 36 hazelnut cakes?" (Peltier et al. 2009, p. 169).

In Brazil, the 1970s were a time of rapid expansion of education. In a time of decentralized and non-mandatory curricular guidelines, when abbreviated teacher training predominated, textbooks began to play the role of teaching guides. In a context of criticism and progressive oblivion of modern mathematics, some innovations were left aside, others were incorporated, and some traditions persisted. Most sixth-grade textbooks kept a chapter dedicated to ratios and proportions, with the traditional lists of verbal problems and no concern for exploring different representations and activities.

The new Brazilian National Education Law, approved in 1996, granted autonomy to schools in the construction of curricula. However, the Ministry of Education decided to produce a document to guide teachers' actions and, above all, to subsidize government policies. The PCNs (National Curricular Parameters) for the initial and the final four years of elementary education were published, respectively, in

1997 and 1998. The document dedicated to the early years restores proportionality as a theme to be treated early on in the study of multiplication and division (MEC 1997). The volume devoted to the final years preserves that tradition, including in the program of the fifth and sixth year "Solving problem situations that involve the idea of proportionality, including calculations with percentages, by the use of non-conventional strategies" (MEC 1998, p. 71). The seventh- and eighth-grade program recalls the rule of three: "Solving problems involving directly proportional or inversely proportional magnitudes using various strategies, including the rule of three" (p. 87). The problem statements are similar to those of the 1950s: "If 8 meters of fabric cost 5.80 *reais*, how much will I pay for 16 meters of fabric?" (p. 109).

In the 1990s' curricula, therefore, the rule of three reappears as one of the valid resources for problem solving, alongside other strategies. What are the reasons for the comings and goings of the rule of three, in both countries, after being dismissed as something inconsistent with teaching oriented towards mathematical thinking development? Many reasons could be raised, such as the pressure from textbook authors or teachers' tendencies to teach the topic as they learned it, or even negotiations between the different pedagogical orientations of the specialists involved in writing the programs. After being taught in school for so long, it is even possible that the rule of three has become knowledge mobilized in different cultural practices, regardless of its intentional transmission by the school. It would, in that case, be a component of what André Chervel (1988, p. 100) calls "school acculturation."

We can also conjecture that the persistence of the rule of three in the official texts—even if it is not as prominent as in other times—is related to the enduring character of school arithmetic problems involving proportional quantities. The place of the problems varies: some textbooks, from different eras, predominantly treat problems as applications of a studied topic while others present problems as an introduction to a new subject. The expected solutions to problems also vary in predictability, uniformity, and level of detail. What seems to be persistent in France, Brazil, and elsewhere, is the strong association between problems and the content that enables students to solve them "quickly and accurately," as was intended in the 1950s.

Whether accepted or rejected, the rule of three always appears linked to the theme of proportionality and the problems associated with it. It is interesting to examine the persistence of this theme in the arithmetic of the early school years. Even when the functional approach—supposedly more generic or more generalizing—was predominant in France, linear functions continued to be privileged. Most recent textbooks present tasks in which the students are asked to decide whether the magnitudes involved are proportional or not. Still, the discussion revolves around proportionality and doesn't provoke a more exploratory study of the relationships between numerical variables, which may or may not be expressed algebraically.

More recently, we can see a valorization of the study of proportionality in international assessments such as PISA (Programme for International Student Assessment).

The 2022 PISA Mathematics test includes the topic "Percents, ratios and proportions: Numerical description of relative magnitude and the application of proportions and proportional reasoning to solve problems" (OECD 2018, p. 28). Although the experts responsible for evaluating mathematical literacy state that the tests are inspired by the idea of mathematical modeling (Stacey and Turner 2015), the items already released have a format similar to textbooks, with brief statements and previously selected data. See for example this item of the PISA 2012 tests: "Normally, a penguin couple produces two eggs every year. Usually the chick from the larger of the two eggs is the only one that survives. With rockhopper penguins, the first egg weighs approximately 78 g and the second egg weighs approximately 110 g. By approximately how many percent is the second egg heavier than the first egg?" (OECD 2013, p. 54). We see that considerations about the survival of chicks are dismissible, but the rule of three can help answer the question. Considering the equality between two reasons and writing the rate as $x/100$, one calculates 110 g − 78 g = 32 g and obtains $x/100 = 32\text{g}/78\text{ g}$. Applying the rule of three, that is, multiplying the second by the third, and dividing that product by the other given term, one gets $x = (32\text{ g} \times 100) \div 78\text{ g}$, a value close to 41. An increase of approximately 41% is the expected answer in this question that is claimed to assess the ability to "Calculate with percentage within a real context" (OECD 2013, p. 54).

6 Final Considerations

The case of the rule of three illustrates how the modernization movement in education, combined with the expanded access to education, affected curricula in France and Brazil. It also helps us to perceive permanences that coexist with changes.

In both countries, until the 1950s, an important elementary school task was to solve problems using the rule of three. With the extension of schooling in both countries, elementary school was no longer under pressure to prepare for adult life; therefore, the utilitarian character of arithmetic problems lost importance. It was also no longer important for elementary students to solve problems quickly to succeed in examinations once these had been eliminated.

In France, concerns about coordinated mathematics education at all stages of schooling justified the anticipation of the study of functions in 1970 and the "back to basics" movement in 1985. The rule of three was depreciated and later rehabilitated as one of the possible ways to solve problems involving proportionality.

In Brazil, in the early 1970s, the study of proportionality was associated with the study of the set of rational numbers and postponed to the more advanced grades. The rule of three was readmitted in 1998, provided it was preceded by solving proportionality problems through other methods.

The movement that became known as modern mathematics encompassed many strands, with some common slogans. The emphasis on generalizing concepts and on understanding these concepts, as opposed to memorizing techniques, motivated the rejection of the teaching of algorithms such as the rule of three, with different consequences in the two countries. Meanwhile, the precedence of proportionality over other numerical relationships has remained almost untouched in both countries. The returns of the rule of three are associated with these permanences.

Why does the study of proportionality, in both countries and so many others, precede the study of all other types of functions, and especially those that an algebraic expression cannot depict convey? Because of its applicability in various situations, in and out of school? Because it is one of those topics that the school has learned to teach, and whose learning can be assessed through test questions?

References

Alix, Philippe, Renaud d'Enfert, and Hélène Gispert. 2015. Des commissions pour réformer les disciplines (1945-1980) [Commissions to reform the disciplines (1945-1980)]. In *Réformer les disciplines. Les savoirs scolaires à l'épreuve de la modernité (1945-1985)*, ed. Renaud d'Enfert and Joël Lebeaume, 85–111. Rennes: Presses Universitaires.

Ardiot, Rémy, Wanaud, André, and Budin, Bernard. 1963. *Calcul. Livre unique pour les cours élémentaire et moyen* [Calculation. Single book for elementary and middle school]. Paris: Hachette.

Averbuck, Anna, Franca Cohen Gottlieb, Lucilia Bechara Sanchez, and Manhucia Perelberg Liberman. 1975. *Curso Moderno de Matemática Para o Ensino de 1° grau - Guia do professor, sexta série*. [Modern Course of Mathematics for the Elementary School—Teacher's guide, 6th grade]. São Paulo: Companhia Editora Nacional.

Bishop, Marie-France, and Catherine Dorison. 2015. Genèse et enjeux du problème de la liaison du cours moyen 2 à la sixième, 1959-1980 [Genesis and Issues of the Fifth to Sixth Grade Linkage Problem, 1959-1980]. *Le français aujourd'hui* 189: 13–24.

Bjarnadóttir, Kristín. 2014. History of teaching Arithmetic. In *Handbook on the history of mathematics education*, ed. Alexander Karp and Gert Schubring, 431–457. New York: Springer.

Brachet, François. 1955. Fractions et règle de trois [Fractions and the rule of three]. In *L'enfant et le nombre. Éléments pour une pédagogie du calcul élémentaire*, ed. François Brachet, Henri Canac, and Eugène Delaunay, 79–99. Paris: Didier.

Búrigo, Elisabete Z. 2009. Modern mathematics in Brazil: the promise of democratic and effective teaching. *International Journal for the History of Mathematics Education* 4 (1): 29–41.

———. 2017. Les problems de la vie à l'école: discours pédagogiques au Brésil dans les années 1950 (le cas du Rio Grande do Sul) [The problems of life in school: pedagogical discourses in Brazil in the 1950s (the case of Rio Grande do Sul)]. In *Les mathématiques à l'école élémentaire (1880-1970). Études France-Brésil*, ed. Renaud d'Enfert, Marc Moyon, and Wagner R. Valente, 187–207. Limoges: Presses universitaires de Limoges.

———. 2019. Real numbers in school: 1960s experiments in France and Brazil. In *Researching the History of Mathematics Education. An international overview*, ed. Alexander Karp and Fulvia Furinghetti, 23–43. Cham: Springer.

Canac, Henri. 1955. Les problèmes dits "pratiques" [The so-called "practical" problems]. In *L'enfant et le nombre. Éléments pour une pédagogie du calcul élémentaire*, ed. François Brachet, Henri Canac, and Eugène Delaunay, 113–117. Paris: Didier.

CFE. (Conselho Federal de Educação) [Brazilian Federal Education Council]1971. Parecer n° 853/71, do CFE. Núcleo comum para os currículos do ensino de 1° e 2° graus. A doutrina do currículo na Lei 5692. [Ruling 853/71, from the Federal Education Council. Common core for 1st and 2nd grade curricula. The curriculum doctrine in Law 5692]. *Documenta* 132: 166–195.

Chervel, André. 1988. L'histoire des disciplines scolaires: réflexions sur un domaine de recherche [The history of school subjects: reflections on a research field]. *Histoire de l'éducation* 38: 59–119.

Choquet, Gustave. 1961. L'enseignement de l'arithmétique à l'école primaire et à l'école secondaire [Arithmetic teaching in elementary and secondary schools]. *Bulletin de l'APMEP* 215: 365–372.

Cunha, Luiz A. 1980. *Educação e desenvolvimento social no Brasil*. [Education and social development in Brazil]. Rio de Janeiro: Francisco Alves.

D'Ambrosio, Beatriz S. 1987. *The dynamics and consequences of the Modern Mathematics reform movement for Brazilian Mathematics education*. Doctoral dissertation, Indiana University.

D'Enfert, Renaud. 2003. Introduction. In *L'enseignement mathématique à l'école primaire de la révolution à nos jours. Textes officiels. Tome 1: 1791-1914*. [Mathematics education in elementary school from the revolution to the present days. Official documents. Volume 1: 1791-1914], ed. Renaud d'Enfert, 13–44. Paris: Institut national de recherche pédagogique.

———. 2011. Une réforme ambigüe: l'introduction des 'mathématiques modernes' à l'école élémentaire (1960-1970) [An ambiguous reform: the introduction of 'modern mathematics' in elementary school (1960-1970)]. In *Le temps des réformes. Disciplines scolaires et politiques éducatives sous la Cinquième République: les années 1960*, ed. Renaud d'Enfert and Pierre Kahn, 53–73. Grenoble: Presses Universitaires de Grenoble.

———. 2015. *L'enseignement mathématique à l'école primaire de la Révolution à nos jours. Textes officiels. Tome 2: 1915-2000*. [Mathematics education in elementary school from the revolution to the present days. Official documents. Volume 2: 1915-2000]. Limoges: Pulim.

———. 2017. Quelles mathématiques pour l'école élémentaire? Une perspective historique (1945-début XXIe siècle) [Which mathematics for the elementary school? A historical perspective (1945-early 21st century)]. *Education & Formations* 94: 7–22.

Dabout, Marcel. 1960. Recherche sur l'acquisition et l'utilisation des connaissances et mecanismes de base en mathématiques à l'entrée en sixième [Research on the acquisition and use of basic knowledge and mechanisms in mathematics at the beginning of the sixth grade]. *Courrier de la recherche pédagogique* 11: 2–33.

Denise, Henri, André Joly, and Madeleine Thévenon. 1970. *Calcul cours moyen deuxième année. Nouvelle édition completée et mise à jour*. [Calculus for the fifth year. New complete and updated edition]. Paris: Delagrave.

Galisson, Marie-Pierre. 2008. Génèse d'une discipline scolaire: l'exemple de l'arithmétique primaire [Genesis of a school discipline: the example of primary arithmetic]. *Spirale* 42: 95–107.

GEEMPA. (Grupo de Estudos em Ensino de Matemática de Porto Alegre) [Study Group on Mathematics Teaching in Porto Alegre]1975. *Reformulação metodológica do ensino da matemática no 1° grau [Methodological reformulation of mathematics teaching in school]*. Porto Alegre: GEEMPA.

Gloton, Robert. 1960. L'articulation entre le cours moyen et la classe de sixième: sélection et orientation [The articulation between the fifth and the sixth grade: selection and guidance]. *Enfance* 13 (1): 13–22.

Goergler, Bernard, André Viala, and Robert Andrieu. 1971. *L'univers mathématique Cours Moyen*. [The mathematical world. Fourth and fifth grades]. Paris: Éditions de l'école.

Gouveia, Aparecida Joly, and Robert James Havighurst. 1969. *Ensino médio e desenvolvimento*. [High School and Development]. São Paulo: Melhoramentos.

MEC. (Ministério da Educação) [Brazilian Department of Education]1997. *Parâmetros curriculares nacionais: Matemática [National Curricular Standards: Mathematics]*. Brasília: MEC.

———. (Ministério da Educação) [Brazilian Department of Education]1998. *Parâmetros curriculares nacionais: Matemática [National Curricular Standards: Mathematics]*. Brasília: MEC.

Medina, Denise. 2007. *A produção oficial do movimento da matemática moderna para o ensino primário do estado de São Paulo (1960-1980)* [The official production of the modern mathematics movement for primary education in the state of São Paulo (1960-1980)]. Master thesis, Pontifícia Universidade Católica de São Paulo.

MEN. (Ministère de l'Éducation nationale) [French Department of Education]1945. Arrêté du 17 octobre 1945 [Order of October 17, 1945]. *Bulletin Officiel de l'Éducation nationale* 52: 3485–3491.

———. (Ministère de l'Éducation nationale) [French Deparment of Education]1964. Circulaire sur le calcul et la grammaire au CM2 [Circular on numeracy and grammar in fifth grade]. *Bulletin Officiel de l'Éducation nationale* 30: 1795–1798.

———. (Ministère de l'Éducation nationale) [French Department of Education]1970. Arrêté fixant le programme de mathématiques de l'enseignement élémentaire [Order establishing the mathematics program for elementary education]. *Bulletin Officiel de l'Éducation nationale* 5: 347–348.

———. (Ministère de l'Éducation nationale) [French Department of Education]2015a. Instructions relatives à l'application de l'arrêté du 17 octobre 1945 fixant les horaires et les programmes des écoles primaires. 7 décembre 1945 [Instructions concerning the application of the decree of October 17, 1945 fixing the schedules and programs of elementary school. December 7, 1945]. In *L'enseignement mathématique à l'école primaire de la Révolution à nos jours. Textes officiels. Tome 2: 1915-2000*, ed. Renaud d'Enfert, 388–404. Limoges: Pulim.

———. (Ministère de l'Éducation nationale) [French Department of Education]2015b. Circulaire sur l'enseignement des mathématiques à l'école élémentaire. 2 janvier 1970 [Circular on the teaching of mathematics in elementary school. January 2, 1970]. In *L'enseignement mathématique à l'école primaire de la Révolution à nos jours. Textes officiels. Tome 2: 1915-2000*, ed. Renaud d'Enfert, 506–537. Limoges: Pulim.

———. (Ministère de l'Éducation nationale) [French Department of Education]2015c. Circulaire sur l'enseignement des mathématiques à l'école élémentaire. 4 septembre 1970 [Circular on the teaching of mathematics in the elementary school. September 4, 1970]. In *L'enseignement mathématique à l'école primaire de la Révolution à nos jours. Textes officiels. Tome 2: 1915-2000*, ed. Renaud d'Enfert, 537–538. Limoges: Pulim.

———. (Ministère de l'Éducation nationale)2015d. Objectifs, programmes et instructions pour le cycle moyen de l'école élementaire (arrêté). 16 juillet 1980 [Aims, programs and instructions for the fourth and fifth grades of the elementary school (decree). July 16, 1980]. In *L'enseignement mathématique à l'école primaire de la Révolution à nos jours. Textes officiels. Tome 2: 1915-2000*, ed. Renaud d'Enfert, 572–603. Limoges: Pulim.

———. (Ministère de l'Éducation nationale) [French Department of Education]2015e. Programmes et instructions pour l'école élémentaire (arrêté). 15 mai 1985 [Elementary school programs and instructions (order). May 15, 1985]. In *L'enseignement mathématique à l'école primaire de la Révolution à nos jours. Textes officiels. Tome 2: 1915-2000*, ed. Renaud d'Enfert, 603–608. Limoges: Pulim.

———. (Ministère de l'Éducation nationale) [French Department of Education]2015f. Les cycles à l'école primaire. 1991. Programmes et instructions pour l'école élémentaire (arrêté). 15 mai 1985 [Cycles in the elementary school. 1991. Programs and instructions for elementary school (decree). May 15, 1985]. In *L'enseignement mathématique à l'école primaire de la Révolution à nos jours. Textes officiels. Tome 2: 1915-2000*, ed. Renaud d'Enfert, 615–622. Limoges: Pulim.

OECD. (Organisation for Economic Co-operation and Development). 2013. *PISA 2012. Mathematics items*. http://www.oecd.org/pisa/test/PISA%202012%20items%20for%20release_ENGLISH.pdf.

———. 2018. *PISA Mathematics Framework*. (Organisation for Economic Co-operation and Development). https://pisa2021-maths.oecd.org/.

OEEC. (Organisation for European Economic Co-operation)1961. *Synopses for modern secondary school mathematics*. Paris: OEEC.

Peltier, Marie Lise, Joël Briand, Bernadette Ngono, and Danielle Vergnes. 2009. *Euromaths. CM2*. Paris: Hatier.

Piaget, Jean. 1955. Les structures mathématiques et les structures opératoires de l'intelligence [The mathematical structures and the operative structures of intelligence]. In *L'enseignement des mathématiques*, ed. Jean Piaget, Edward W. Beth, Jean Dieudonné, André Lichnerowicz, Gustave Choquet, and Caleb Gattegno, 11–34. Neuchâtel: Delachaux et Niestlé.

Picard, Nicole. 1967. L'enseignement des mathématiques modernes (Enseignement élémentaire et préscolaire) [Modern Mathematics Teaching (Elementary and Preschool)]. *Le Courrier de la Recherche Pédagogique* 31: 5–41.
Rio Grande do Sul. 1939. Programa de Matemática [Mathematics Program]. *Revista do Ensino* 4: 284–292.
———. Secretaria de Educação e Cultura. Centro de Pesquisas e Orientação Educacionais [Center for Educational Research and Guidance of the state of Rio Grande do Sul]1959. *Programa Experimental de Matemática [Mathematics Experimental Program]*. Porto Alegre: Imprensa Oficial.
———. Secretaria de Educação e Cultura [Education Department of the state of Rio Grande do Sul]. 1972. *Brasil 71/74, em tempo de construir. Ensino de 1° grau no Rio Grande do Sul* [Brazil 71/74, times of construction. Primary Education in Rio Grande do Sul]. Porto Alegre: Secretaria da Educação e Cultura.
Sanchez, Lucilia, and Manhucia Liberman. 1972. *Curso moderno de Matemática para a Escola Elementar. Vol. 5*. [Modern Course of Mathematics for the Elementary School. Volume 5]. São Paulo: Companhia Editora Nacional.
São Paulo. Secretaria da Educação [Education Department of the state of São Paulo]1950a. *Programa para o ensino primário fundamental. 4° ano. Ato n° 5, de 9 de janeiro de 1950 [Program for elementary school 4th grade. Act 5, from January 9th, 1950]*. Rio de Janeiro/São Paulo/Belo Horizonte: Livraria Francisco Alves.
———. Secretaria de Estado dos Negócios da Educação [Education Department of the state of São Paulo]1950b. *Programa para o ensino primário fundamental. 5° ano. Ato n° 35, de 22 de abril de 1950 [Program for elementary school 5th grade. Act 35, from April 22th, 1950]*. São Paulo: Imprensa Oficial do Estado de São Paulo.
———. Secretaria da Educação. Departamento de Educação [Education Department of the state of São Paulo]. 1969. *Programa da escola primária do Estado de São Paulo—Nível 1 e 2* [São Paulo State Primary School Program - Level 1 and 2]. São Paulo.
———. Secretaria da Educação. Centro de Recursos Humanos e Pesquisas Educacionais "Prof. Laerte Ramos de Carvalho" [Center for Human Resources and Educational Research of the state of São Paulo]. 1973. *Guias Curriculares Propostos para as Matérias do Núcleo Comum do Ensino do 1° Grau* [Curriculum Guides for the Common Core of Elementary School]. São Paulo.
Sarrazy, Bernard. 2003. Le problème d'arithmétique dans l'enseignement des mathématiques à l'école primaire de 1887 à 1990 [The arithmetic problem in the teaching of mathematics in elementary school from 1887 to 1990]. *Carrefours de l'éducation* 15: 82–101.
Segundas noções, de gramática aritmética, história pátria, geografia: correspondentes ao programa oficial de admissão à primeira série ginasial [Second notions of grammar, arithmetic, national history, geography: corresponding to the official program for admission to secondary education]. 1955. Porto Alegre: Selbach.
Stacey, Kaye, and Ross Turner. 2015. The evolution and key concepts of the PISA mathematics frameworks. In *Assessing Mathematical Literacy. The PISA Experience*, ed. Kaye Stacey and Ross Turner, 5–33. Cham: Springer.
Swetz, Frank J. 1992. Fifteenth and sixteenth century arithmetic texts: What can we learn from them? *Science & Education* 1: 365–378.
Touyarot, Marie-Antoinette. 1968. Variations sur le thème des apllications linéaires à partir du cours moyen 2e année [Variations on the theme of linear applications from the fourth grade]. *Le Courrier de la Recherche Pédagogique* 33: 17–36.
Vandendriessche, Louis. 1968. Étude des fonctions et applications au CM 1 [Study of functions and applications in fourth grade]. *Le Courrier de la Recherche Pédagogique* 33: 37–56.
Walusinski, Gilbert. 1963. La règle de trois n'aura pas lieu [There will be no place for the rule of three]. *Bulletin de l'APMEP* 231: 386–387.
———. 1966. Mathématiques d'aujourd'hui pour hommes et femmes de demain [Today's mathematics for the people of tomorrow]. *Le Courrier de la Recherche Pédagogique* 27: 5–11.

In a Search for Innovative Approaches to Teaching Primary Arithmetic: David Eugene Smith (1860–1944) and His Early Works on Mathematics Education

Viktor Freiman and Alexei Volkov

Abstract The authors discuss the publications of David Eugene Smith (1860–1944) devoted to teaching of arithmetic in primary school; in particular, they focus on four papers bearing the common title "The evolution of method in arithmetic" published in 1895 and 1896. In these papers Smith provided a short but interesting analysis of several works on teaching arithmetic published in North America and Western Europe from the late 18th to the late 19th century. We argue that Smith stressed the importance of paying attention to the history of methods of teaching mathematics; we also suggest that their analysis offers an opportunity to explore the origins of Smith's ideas concerning mathematics education.

Keywords David Eugene Smith · History of mathematics education · Teaching of arithmetic · John Kelley Ellwood · Eduard Jänicke · Friedrich Unger · Johann Heinrich Pestalozzi · August Wilhelm Grube · Friedrich Adolph Wilhelm Diesterweg

1 Introduction

David Eugene Smith (1860–1944) is well known as the author and co-author of numerous publications on the history of mathematics and of mathematics education. In the mathematics education community he is recognized as one of the founders of The International Commission of Mathematics Instruction (ICMI) established in 1908, and The National Council of Teachers of Mathematics (NCTM) founded in 1920 (Donoghue 2003, 2006, 2008; Karp 2019a). He made a significant

V. Freiman (✉)
Université de Moncton, Moncton, NB, Canada
e-mail: viktor.freiman@umoncton.ca

A. Volkov
National Tsing-Hua University, Hsinchu City, Taiwan

A. Karp (ed.), *Advances In The History Of Mathematics Education*,
International Studies in the History of Mathematics and its Teaching,
https://doi.org/10.1007/978-3-030-95235-8_4

contribution to teacher education by developing a program of mathematics education at Teachers College, Columbia University and by writing a number of seminal papers on methods of teaching elementary mathematics. He was also a prolific (co-) author of numerous mathematics textbooks.

Being interested in studying didactical innovations in genetic-historical and cross-cultural perspective (Freiman and Volkov 2014), we identified the case of arithmetic teaching in the 19th century as one of the crucial elements for this period when several new didactical trends emerged; we suggested that these trends continue influencing mathematics classes in schools even now (Freiman and Volkov 2018, p. 6). In this respect, our investigation of Smith's contribution and of its connection to European (in particular, German) mathematical didactics furthers our quest for deeper analysis of the overall paradigm of primary and secondary mathematics education established in the late 19th and early 20th centuries.

In our paper we especially focus upon the early stages of Smith's academic career, when his views on teaching and learning mathematics started taking shape and defined his position in reform-oriented didactical debates, in particular, regarding the methods of teaching arithmetic in early grades. Several texts written by Smith in the 1890s and, particularly, his four very short papers (approximately six pages altogether) published under the common title "The Evolution of Method in Arithmetic" (1895; 1896a, b, c) summarize Smith's ideas concerning teaching arithmetic which were further developed in his publications of the first decade of the 20th century. We shall begin with a summary of his educational background.

2 D.E. Smith's Educational Background and His Work in the Field of Mathematics Education

Born in 1860, Smith received a solid classical education (with special focus on Classical Greek and Latin) and pursued his studies in philosophy and art history at Syracuse University (New York State). He obtained degrees of Bachelor of Philosophy (1881), Master of Philosophy (1884), and Ph.D. in the history of fine arts (1887). Smith also earned a L.L.D. (Doctor of Laws) diploma in 1905. However, despite his early interest in art history and his professional degree in law, he chose a career in mathematics and mathematics education.[1]

Surprisingly, Smith's mathematical background is not so well known. His biographies suggest that he took graduate courses in mathematics (along with history and modern languages) and that in 1884 he was invited to be a replacement mathematics teacher at Cortland Normal School (New York State). It is likely that this experience ignited his interest in mathematics and its teaching, so he became a staff member of the normal school where he worked for seven years until

[1] Elements of Smith's biography related to his educational background and his career can be found in Brasch (1944), Fite (1945), and Simons (1945).

obtaining, in 1891, the chair of mathematics at the Michigan State Normal School at Ypsilanti affiliated to Michigan University. He also served as principal at Brockport State Normal School, New York State, from 1898 to 1901. From 1901 until his retirement in 1926 he worked at Teachers College, Columbia University (Simons 1945, p. 40).

Smith started developing his approach to mathematics education for teachers when working in Michigan State Normal School where he designed a program for teacher training in the early 1890s (Donoghue 2006; Schubring 2020). Smith introduced new courses focused on the methods of teaching in which he emphasized the role of historical development of the subject. Also, he developed a new course on the history of mathematics which, according to Donoghue (2006), was unique in the US normal schools and a rarity in tertiary institutions of the country.

From the beginning of his career, he participated in didactical debates in which he sometimes took a rather critical position. For instance, in 1894 he published a critique of John Kelley Ellwood's (1858-1940) note on division published in the same year thus entering in debates on how to teach different aspects of arithmetic; we shall return to this note below. In the 1890s Smith started publishing his own mathematics textbooks. The earliest of them were co-authored with Wooster Woodruff Beman (1850–1922);[2] they included *Plane and Solid Geometry* (Beman and Smith 1895), *Higher Arithmetic* (Beman and Smith 1897), *New Plane and Solid Geometry* (Beman and Smith 1899), *Elements of Algebra* (Beman and Smith 1900), and *Academic Algebra* (Beman and Smith 1902). Beman and Smith together translated from German and published Felix Klein's (1849–1925) *Famous Problems of Elementary Geometry* (Klein 1897; originally published in German in 1895) and Karl Fink's (1851–1898) *A Brief History of Mathematics* (Fink 1900 originally published in German in 1890). More on cooperation of Beman and Smith can be found in Frick (1936).

D.E. Smith continued writing mathematics textbooks in the early 20th century, alone or in collaboration with other authors; in her bibliography of Smith's works, Bertha M. Frick (1894–1975) claimed that Smith authored or co-authored "some one hundred and fifty" textbooks and that about sixty percent of them were in the field of arithmetic (Frick 1936, p. 10); unfortunately, Frick did not provide references to the textbooks in her bibliography. It also remains unknown whether Frick counted subsequent editions of Smith's textbooks and their translations into foreign languages as different items. The list of Smith's works more recently published online by Wang Xiaoqin (汪晓勤) contained references to 24 textbooks authored by Smith and to 74 textbooks co-authored by him (Wang 2010, pp. 28–34).

Smith actively participated in the work of the International Commission on Mathematical Instruction since its creation in 1908 and served as its president in 1928–1932. For Smith's work in ICMI and his influence on this organization and its members see, for example, Furinghetti (2019), Giacardi (2019), Karp (2019b, esp. see pp. v-vi), Rogers (2019, esp. see pp. 101-103), and Schubring (2019). The

[2] For a short biography of Beman see Hinsdale (1906), p. 266; see also Anon. (1922), pp. 137–138.

book-length publications on mathematics education edited, authored or co-authored by Smith included *The Teaching of Elementary Mathematics* (1900), *Mathematics in the Elementary School* (Smith and McMurry1903) co-authored with Frank Morton McMurry (1862–1936) and its updated versions *The Teaching of Arithmetic* (Smith 1909a, 1909b) signed by Smith alone and reprinted with updated bibliography (Smith 1913),[3] *The Teaching of Geometry* (1911), *The Teaching of Junior High School Mathematics* (Smith and Reeve 1927a), and *The Teaching of High School Mathematics* (Smith and Reeve 1927b) (see also Reeve 1944). His papers on mathematics education are too numerous to be cited here; references to them can be found in Frick (1936) and Wang (2010).

3 First Steps: Smith's Polemics with Ellwood

John Kelley Ellwood (1858–1940) was a school principal in Pennsylvania; he published several mathematics textbooks (Ellwood 1892; Colaw and Ellwood 1900a, b) along with journal articles on mathematics and mathematics education (Ellwood 1894a, b, 1896). His short article "Remarks on Division" (1894a) questions the practice of introducing division of a "concrete" number by a "pure" number, for example, the division of $10 by 2 yielding $5. He sees it as problematic, since division is inverse to multiplication, which, in turn, is but a repeated addition. For example, one can repeatedly subtract $5 from $20, until "nothing remains" (p. 47). In this case the subtraction has to be repeated four times, and therefore, one can conclude that $20 divided by $5 equals to 4. But, Ellwood claims, it is impossible to divide $20 by 4, since "we cannot by mere subtraction [i.e., the repeated subtraction of 4 from $20. – V.F. & A.V.] find the multiplicand" ($5). So, "[t]o find one of the equal parts of a concrete number [i.e., perform the division of $20 by 4. – V.F. & A.V.] is more than division: it is a problem that involves the use of division, it is an 'application' of division just as 'profit and loss' is an application of percentage" (Ellwood 1894a, p. 48). He suggests that the problem should be solved according to the following steps:

(a) ¼ of $20 is as many dollars as there are 4's in 20.
(b) There are five 4's in 20.
(c) ∴ [therefore. – V. F. & A. V.], ¼ of $20 is $5. (Ellwood 1894a, p. 48).

In the next issue of the same 1894 volume of *The American Mathematical Monthly*, Smith ("reluctantly, feeling it is almost unnecessary", p. 74) takes on this article with his "Note on Mr. Ellwood's Remarks on Division" by pointing at the historical development of mathematics as "a history of the destruction of old definitions, old hobbies, old idols." To support his view, he provides four examples (unit fractions in Ancient Egypt; negative roots of equations in Arabic mathematics; the theory of exponents; the commutative law of multiplication) and claims that "the world is

[3] The mention of McMurry's co-authorship of the original work (1903) found in the Introduction of Smith (1909a, b) disappears in Smith (1913).

going to move along in spite of somebody's definition [of a mathematical object]." Smith quotes Ellwood as saying, "Multiplication is a mere process of adding" and states that this interpretation was changed (Smith makes this claim using the French expression "nous avons changé[4] tout cela" meaning "we have changed all that"). It is plausible to suggest that here Smith quoted Molière's (1622–1673) play *Le Médecin Malgré Lui* (The Physician in Spite of Himself) of 1666.[5] In the context of the remark about division, Smith suggests, this statement means that "the world of to-day says $10 ÷ 2 = $5, and a good part of the world expressly distinguishes between the two notions of division in the primary grades" (Smith 1894, p. 75).

By quoting Molière, Smith probably wanted to emphasize a need to adjust teaching according to "modern views" concerning mathematics, thus claiming that division of an amount of money by a natural number should no longer involve the outdated interpretation of division. Smith's "Note" provoked a prompt response from Ellwood in the next issue of the same journal; the latter pointed out some issues with Smith's epistemological position while labeling Smith as a self-proclaimed "exponent of progress" who was still "championing an old 'fogy notion'" (Ellwood 1894b). To the best of our knowledge, Smith did not reply to these remarks.

Other details of the debate fall beyond the scope of our chapter, but the discussed example provides evidence of Smith's engagement in didactical debates early in his career and his taking on the issue of perennity vs modernity of mathematical knowledge once it is transposed into didactical context. It also shows his epistemological position concerning mathematical knowledge as anchored in the past but reinterpreted and adapted to meet modern trends. This position was further developed in his series of articles on evolution of methods of teaching arithmetic published in the journal *Inland Educator* in 1895-1986 where he analyzed the historical development of didactical ideas based on the work of German scholars of the second half of the 19th century and offered a critical analysis of their innovative approaches that we shall discuss below.

4 Four Papers on Method in Arithmetic

In his four rather brief articles bearing the common title "The Evolution of Method in Arithmetic" (1895; 1896a, b, c) Smith laid out important connections between innovative methods of teaching arithmetic in the 19th century and his own

---

[4] In Smith's paper "changé" is misprinted as "change."

[5] In the play, the individual named Sganarelle pretending to be a medical doctor makes a mistake by stating that the heart is located on the right side of the human body while the liver is located on the left; when his interlocutor, Géronte, corrects him by mentioning their actual locations, Sganarelle replies: "Oui, cela était autrefois ainsi: mais nous avons changé tout cela, et nous faisons maintenant la médecine d'une méthode toute nouvelle," (Moliere 1845, pp. 631–632) that is, "Yes; this was so formerly; but we have changed all that, and we now-a-days practice the medical art on an entirely new system." (Molière. 1875, pp. 416–417).

didactical work. While more details about Smith's conception of arithmetic instruction were provided in his later publications (for example, in Chapters 1 to 5 of Smith 1900), these early papers were important contributions already expressing his essential ideas concerning importance of method in teaching arithmetic in elementary grades and of paying attention to historical roots not only of mathematics itself but also of methods of its teaching.

The journal *Inland Educator* in which these four papers appeared was founded by Francis Marion Stalker (1861–1930) and Charles Madison Curry (1869–1944) in 1895. F.M. Stalker worked at Indiana State Normal School from 1892 to 1929, while Ch. M. Curry was the author of numerous primary textbooks. The journal's subtitle *A Journal for the Progressive Teacher* suggests that its creators intended to place it within the movement of "progressive education" that arose in the United States in the 1890s. The focus of the *Journal*'s publications seems to be the methods of teaching, their role, and their development. For instance, the very first article in the Issue 1 of Volume 1, written by Howard Sandison (1850–1919), professor of "mental sciences and methods" and vice-president of the Indiana State Normal School, summarized his intuitions about method (Sandison 1895) which he defines as "a real activity according to, and in harmony with an ideal activity" (p. 17).

If we return to the *Inland Educator*, its last issue of 1895 (printed in January 1896) was introduced in its Editorial as follows: "The January number will be a strong one and among other things will contain: The first of a series of papers upon The Evolution of Method in Arithmetic by Dr. David Eugene Smith of the State Normal School, at Ypsilanti, Michigan, which will prove of much value to our readers" (Stalker and Curry 1895, p. 318).

The series of Smith's four papers indeed remains valuable even now, but for a different reason: it offers an opportunity to explore how exactly he tried to establish connections between the history of teaching arithmetic and the "innovative methods" catering to the interests of "progressive teachers."

4.1 Smith's Analysis of the Methods of Teaching Arithmetic Featured in European Publications of the 19th Century

While specifically focusing on methods of teaching arithmetic and their evolution, Smith (1895, p. 345) argues that "it is impossible to find much that is valuable" in the American publications on mathematics education and claims that the evolution of the methods "has chiefly taken place" in Germany and Switzerland. The leading role of the German and particularly the Prussian education systems during the transition from the 18th to the 19th century was highlighted, among others, by Ellis et al. (2014) who, for example, mentioned that Prussia was one the first in the world to try to introduce a compulsory primary education via so-called *Volksschule* (People's School). By applying Enlightenment principles to its educational organization (Schubring 2012) Prussia at the beginning of the 19th century became one the most progressive school systems, thus attracting attention of educators in other countries, such as Russia and United States (Ellis et al. 2014, pp. 8–9). While the emphasis was put on general education (*Allgemeinbildung*) with a neo-humanist conception and an organic view of human knowledge, it turned out that (perhaps under influence of French schools) mathematics started to play a key role in this system (Schubring 2012). While being rather unique among German states, Prussia succeeded, towards the second part of the 19th century, to extend its influence to establish a better coordination of economic, political, and educational systems with other parts of Germany (Schubring 2012).

A particular role in the reform-oriented development of methods of teaching arithmetic in early schooling during this transition period (from the late 18th to the early 19th century) was played by the Swiss educator Johann Heinrich Pestalozzi (1746–1827) and his German followers. As Kristín Bjarnadóttir (2014) pointed out, "Pestalozzi set out to find ways of helping children to understand number and not merely to develop speed and accuracy in the mechanical working of examples. Children were to discover for themselves the mathematical rules through activities based on sense-impressions" (p. 443). These "sense-impressions," along with observation and intuition were key elements of the methods based on *Anschauung* (Bjarnadóttir 2014, p. 443) that is, direct or immediate intuition of sense data. Other didactical innovations introduced by August Wilhelm Grube (1816–1884) and Friedrich Adolph Wilhelm Diesterweg (1790–1866) as well as genetic-historical development of methods of teaching arithmetic later became the objects of critical analysis by other German scholars and educators. Among those works, the publications authored by Eduard Jänicke (1829–1891), Matthäus Sterner (1844–1893), and Friedrich Unger were chosen by Smith for his brief analysis.

More specifically, as the main sources of his analysis, Smith selected the three following books, all focused on the historical development of arithmetic in connection to teaching: Eduard Jänicke's *Der Rechenunterricht in der Deutschen Volksschule* published in 1879, Matthäus Sterner's *Geschichte der Rechenkunst* of 1891, and Friedrich Unger's *Die Methodik der praktischen Arithmetik in historischer Entwicklung* published in 1888. Smith does not explain the reasons for choosing these three particular sources while other relevant works of the same authors, such as Jänicke's earlier publication on the subject (1877), were presumably available to him. Nevertheless, it is Unger's work which was apparently of particular interest for Smith, since he quoted it "at length" (according to his own words; see Smith 1895, p. 345). Its author, Friedrich Unger, worked as a head teacher in Realschule Leipzig-Reudnitz (Leipzig, Germany) in the 1880s. A part of his *Methodik* was already published in 1883 and its full version, published in 1888, helped him earn a Ph.D degree. Hartmann (1893) in his bibliographic notice on Unger's work praised it as a "rich treasure trove and a reliable guide for those who want to delve deeper into the history of our subject."

In the very beginning of the first of the four papers, Smith (1895) claims that "for those who have access to such works, these papers [i.e., the four papers of Smith under consideration. – V.F. & A.V.] will have no value; for others it is hoped that they may be of use." (p. 345). Furthermore, Smith adds to his analysis some indications of how the didactical innovations described by German scholars might have influenced the development of methods of teaching arithmetic in the United States during the 19th century and notes the need for critical analysis of them by further teachers.

4.2 Teaching Arithmetic Prior to the 19th Century: Smith's Reading of Unger

Smith (1895), following the structure of Unger's text, begins his analysis by depicting the general situation in mathematics education in Germany in the 17th and 18th century when mathematics was considered "merely a side issue, as are drawing and music today" (p. 346); it was taught by teachers who had to deal with difficult life conditions, were poorly paid, and wrote teaching materials while doing additional jobs (e.g., book binding). In Latin schools, practical arithmetic was completely neglected, while in German schools reading was taught before arithmetic as prerequisite ("all children who can read shall study arithmetic", Smith 1895, p. 346). The teaching consisted of giving students individual work and solving textbook problems, but the recitation of tables (e.g., multiplication table) was also prioritized. The teacher's role was mainly in helping students when needed.

Smith mentions that, as far as the tasks that students had to do are concerned, "[t]he work was always mechanical, the operations being performed by rule, with no effort at a clear understanding of the process" (p. 345). The contents included operations with integers and fractions in lower grades, with an increasing amount of practice and introduction of the Rule of Three in middle grades; in higher grades, decimal computations, and trigonometry were added.

Apparently, by the end of the 18th century the situation gradually changed in several aspects ("wherever arithmetic was taught, it was taught better than ever before," p. 346). More connections to other subjects and to real life situations were established, logical reasoning ("strict proofs") started to take over a "mere mechanism" (in terms of teaching approach), and a clearer distinction was made between the culture and utility of mathematics (cultural values were promoted instead of utilitarian values). Smith wrote, "The best books of the first part of the eighteenth century began to recognize this [cultural. – V.F. & A.V.] value and to set it forth as a prominent reason for teaching" (p. 346). There was also a recognition of the need to begin learning mathematics from the beginning of the primary grades, and, accordingly, the methods of teaching started to play an important role.

Smith emphasizes the role of the philanthropist movement to set up the "rules of the method" (Smith 1895, p. 346). This movement was related to *The Philanthropinum*, the school in Dessau, Germany, founded in 1774 by Johan Bernhard Basedow (1724–1790) and Christian Heinrich Wolke (1741–1825) to implement the educational ideas of Jean-Jacques Rousseau (1712–1778). When describing the philanthropists' "rules of the method," Smith mentions that when selecting subject matter more attention had to be paid to the pupil taking into

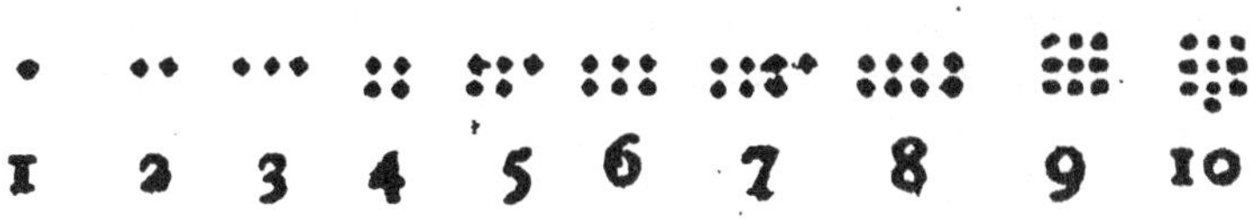

Fig. 1 Busse's diagrams for numbers from 1 to 10 (Busse 1786, p. 1)

consideration her or his age, position, and future occupation. Also, the traditional way of introducing numbers needed to be revised towards a gradual increase in the size of the quantities. Smith makes reference to Friedrich Gottlieb von Busse (1756–1835), a teacher and member of administration of the school since 1779, who advocated a three-step process of introducing numbers, first from 1 to 10, then from 1 to 100, and finally, from 1 to 1000 (Smith 1895, p. 347). The teaching of topics related to numbers was supposed to be supported by perception rather than memorization, for example, by using number figures "Zahlbilder" in which dots were used to replace concrete objects, such as apples, nuts, etc.; see Fig. 1.

When discussing these methods, Smith (1895), p. 347), as Unger (1888) before him, refers to Busse's *Gemeinverständliches Rechenbuch* (Common Arithmetic Book for Schools) (Busse 1786), in which "distinctness and understanding must occupy the foremost place; there must be accommodation of the matter to the thought and language of the child" (Smith 1895, p. 347). Smith's final remarks on Busse's point deal with the value of mental (oral) arithmetic which had to be emphasized.

At the end of the first of his four papers, Smith claimed that the 18th century innovators' work on bringing arithmetic to young children led to the development of the method (of teaching arithmetic) which was completed early in the 19th century by Pestalozzi: "This was a period of preparation for method as we now know it [...]. The time was ripe for a leader with a method, and this leader came in the person of Pestalozzi" (Smith 1895, p. 347). The method of Pestalozzi's was identified by Unger (1888) as one of "three noteworthy theories" on teaching arithmetic in early grades that the 19th century produced. Indeed, Smith (1896a), like Unger, begins his analysis with Pestalozzi (who considered perception to be the foundation of reckoning, p. 17) and his followers, then moving on to Grube (who suggested that counting and arithmetical operations should be treated simultaneously), and then to Tanck and Knilling (who claimed that the "foundation of all reckoning is counting"). We will discuss Smith's analysis below.

4.3 Pestalozzi's Method (Anschauung)

In Unger's *Methodik* (1888), a section devoted to the *Anschauung* method is structured around the work of Pestalozzi (part 1), that of his followers (part 2), and the work of Diesterweg (part 3). Following Unger (1888, p. 176) who called Pestalozzi the "father of modern method," Smith (1896a, p. 17) sees in his work a precursor of "the movement to adapt instruction at every stage to the thinking power of the child, as determined by psychological study" because he advocated valuing the child's work with concrete objects (not teaching the sign "until after the thing"), as well as fulfilling a need by "developing and strengthening the mental power." Without presenting Pestalozzi's arithmetic in detail, Smith seemed to focus on later critics of his method while still recognizing its great value for modern (of his time) teaching of arithmetic. The information on Pestalozzi provided below can help the reader get a sense of the method.

The educators who followed Pestalozzi's movement created materials to support a visual (perceptual) approach to arithmetic; these materials included "Pestalozzi's tables" described in his work *Anschauungslehre der Zahlverhältnisse* (Intuitive teaching about number) (Pestalozzi 1803). The "tables" were designed by Pestalozzi in collaboration with other instructors of his school, in particular, with his student and, later, active collaborator Hermann Krüsi (1775–1844).[6] Krüsi himself describes the creation of these tables as follows:

> At the suggestion of Pestalozzi, he [i.e., Krüsi. –V.F. & A.V.] prepared a course of Mental Arithmetic on the objective plan. He also prepared the necessary means of illustrating the tables of units and fractions, the latter based upon the division of a rectangle. The teaching of Arithmetic on this plan produced a rapidity and accuracy in the computation of numbers which appeared almost miraculous to those who had been accustomed only to the old system of ciphering according to given rules. Some of the pupils educated at Burgdorf, such as Ramsauer and Schmid, gave to this branch a more extended application, for which they obtained considerable celebrity, while Krüsi, who was the real originator of the system, modestly resigned to them his share of the honors.[7] (Krüsi 1875, p. 71)

Unger (1888) explains how the table shown in Fig. 2 ("Table of units") can be used to introduce, for example, multiplication:

> Die erste Übung bezweckt die Orientierung auf der Tabelle, das Kind zeigt und spricht: „1 mal 1, 2 mal 1 etc bis 10 mal 10."
>
> [The first exercise is aimed at orientation on the table; the child points at it and says, "1 times 1; 2 times 1; etc. till 10 times 10."]

The table was also used for other exercises.

Drawing on Pestalozzi's legacy, Smith acknowledges that critical voices about the former's approach (especially about its elements based on perception) were already present in his time, for instance, in his home country, Switzerland. The critics objected to insufficient emphasis on the decimal feature of the number system, to the lack of introduction of operations with the help of "special exercises" (except for multiplication), to the excessive emphasis on oral method that could not provide a sufficient place for "arithmetic drill," to the fact that notation was introduced too late, to the insufficient use of computations with concrete numbers, etc. The attacks on Pestalozzi became even more "serious" by the end of the 19th century (Smith 1896a, p. 17, with reference to Knilling 1884, 1886b).

[6] Krüsi (1875, p. 41) reports that he "worked out exercises in Mental Arithmetic and Language." On Krüsi and his collaboration with Pestalozzi see Krüsi 1830 and Schiess 1928.

[7] All translations from German and French are by the authors unless indicated otherwise.

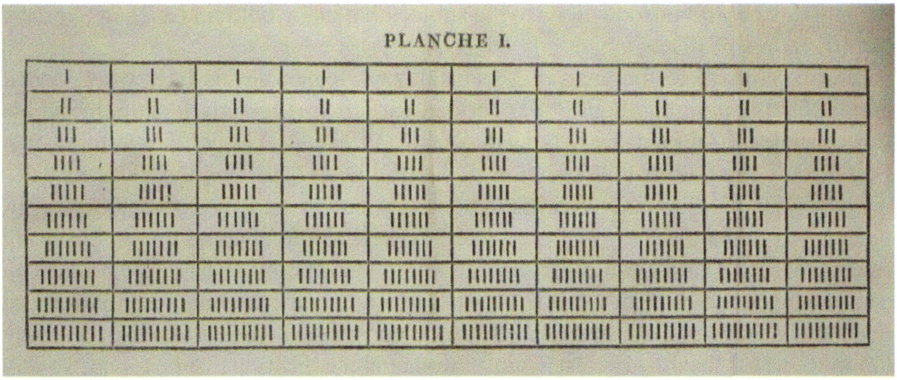
PLANCHE I.

I	I	I	I	I	I	I	I	I	I
II	II	II	II	II	II	II	II	II	II
III	III	III	III	III	III	III	III	III	III
IIII	IIII	IIII	IIII	IIII	IIII	IIII	IIII	IIII	IIII
IIIII	IIIII	IIIII	IIIII	IIIII	IIIII	IIIII	IIIII	IIIII	IIIII
IIIIII	IIIIII	IIIIII	IIIIII	IIIIII	IIIIII	IIIIII	IIIIII	IIIIII	IIIIII
IIIIIII	IIIIIII	IIIIIII	IIIIIII	IIIIIII	IIIIIII	IIIIIII	IIIIIII	IIIIIII	IIIIIII
IIIIIIII	IIIIIIII	IIIIIIII	IIIIIIII	IIIIIIII	IIIIIIII	IIIIIIII	IIIIIIII	IIIIIIII	IIIIIIII
IIIIIIIII	IIIIIIIII	IIIIIIIII	IIIIIIIII	IIIIIIIII	IIIIIIIII	IIIIIIIII	IIIIIIIII	IIIIIIIII	IIIIIIIII
IIIIIIIIII	IIIIIIIIII	IIIIIIIIII	IIIIIIIIII	IIIIIIIIII	IIIIIIIIII	IIIIIIIIII	IIIIIIIIII	IIIIIIIIII	IIIIIIIIII

Fig. 2 Pestalozzi's table (Chavannes 1805, Planche 1). This table, originally published in Pestalozzi 1803, vol. 1, was reproduced in works of other authors; see, for example, Diesterweg and Heuser 1829, p. 57.

Unger (1888, p. 181) mentions that "Pestalozzi himself recognized the imperfection of his arithmetic exercises in the later years of his life" leaving the task of improvement to his successors. Indeed, in the second part of the section on the *Anschauung* method, Unger mentions two types of followers. The first group was formed by those who, attracted by the reputation of Pestalozzi's school (he describes its reputation with the words "wie ein Lauffeuer verbreitete sich sein Ruf über ganz Deutschlandand", that is, "like a wildfire, it spread its reputation all over Germany") and taking advantage of its openness to all, started visiting the school, and, being excited about it, tried to implement the method in their own schools (Unger 1888, p. 181). Yet, according to Unger (1888), because of the lack of a clear understanding and critical reflection of how and why it could work, they did it in a way which was rather harmful to students, and, to some extent, to Pestalozzi's reputation.

Among the second group of Pestalozzi's students ("Jünger Pestalozzis") there were those who followed him "à la lettre" taking the accessories (tables, exercises, etc.) that he developed as models and implementing them. Others tried looking at "the method" rather critically, "grasping the spirit of the method, correct[ed] and expand[ed] what the master left imperfect; they alone [were] important for further development" (Unger 1888, p. 182).

Indeed, Smith (1896a, p. 17) recognizes that Pestalozzi's method did make an important impact ("the influence was remarkable") on the development of teaching arithmetic in the United States due to the work of Warren Colburn (1793–1833).[8] Smith also argues that Pestalozzi's influence was noticeable beyond arithmetic by making reference to Schmid's *Die Elemente der Zahl als Fundament der Algebra* (Elements of Number as Foundation of Algebra) (1810) which he praised as "one of

[8] On Colburn and his methods of teaching arithmetic see Kegley (1947) and Monroe (1912a, 1912b, 1912c, 1913a, 1913b).

the most successful introductions to the subject that one would find even in our day" (Smith 1896a, p. 17).

Smith returned to the analysis of Pestalozzi's legacy in his later work, and it seems that his opinion about it was formed earlier; it is already identifiable in his short articles discussed here. Indeed, Smith (1909a, p. 11), while claiming that Pestalozzi's method was "a creation of his followers," praises the ideas of early introduction of arithmetic into the school curriculum and the systematic use of objects "to make number relations clear" as well as the abandoning of "arbitrary rules," incessant drills "on abstract oral work," and the emphasis on the unit "by considering a number like 6 as '6 times 1'." Moreover, for the time when Pestalozzi lived and worked (that is, in the late 18th and early 19th century) all this, according to Smith, was a "healthy protest against the stagnant education that he found. To-day it has only an incidental lesson to the teacher, although Pestalozzi's spirit and several of his ideas may well command the admiration and respect of all who study the results of his great work" (Smith 1909a, p. 11).

Overall, Smith argues that Pestalozzi was "the greater of Knilling, Grube and Tanck, and all the host of method writers that abound in Germany today" (Smith 1896a, p. 17). In one of his later publications, Smith (1916) adds that Pestalozzi's method "was the first great external influence, one based on a mixture of child psychology and common sense, that caused any change in the sluggish course of American arithmetic, and it is one of the few influences that have been exerted on the subject which are really significant" (Smith 1916, p. 112).

4.4 Ernst Tillich: "Rechnened Denken and Denkend Rechnen" (Thinking by Computing; Computing by Thinking)

Ernst Tillich (1780–1807) was one of Pestalozzi's followers mentioned by Smith (1896a) who "devised the most coherent system on the basis of Pestalozzi" (Bullynck 2008, p. 581) while making the first attempt toward an "independent working out [of] the subject of primary arithmetic." By considering Pestalozzian method "too rigid and too empirical" (Bullynck 2008, p. 579), Tillich devised his own plan in which he put out some combinatorial ideas related to the concept of number. For instance, he considered 4 as combinations of additive components, such as 1 + 1 + 1 + 1, 1 + 2 + 1, etc. Earlier, Unger described Tillich's work as follows:

> Tillich justified his emphasis on the decimal system as opposed to going back again and again to the unit in Pestalozzian method by stating that with large numbers it is absolutely impossible to be particularly aware of all of their units, and that this could only be done in groups. The decimal structure of the number system itself provides the means for group-wise presentation of the numbers (according to ones, tens, hundreds, etc.). (Unger 1888, p. 183)

Unger thus sees Tillich's work as an extension of the "principle of naturalness" (*Naturgemässheit*) of the method which Pestalozzi carried out exclusively regarding the subject of the lesson (the child) and of the object of the teaching (the subject of arithmetic). The naturalness of the method with regard to the object (arithmetic) was no less necessary than that with regard to the subject (child) (Unger 1888, p. 183). This was also briefly noted by Smith (1896a, p. 17).

More precisely, Tillich (1806) in his *Anleitung zur Rechenkunst* (Introduction into the art of calculation) valued (as did Pestalozzi) mental arithmetic based on perception of numbers 1–10 as points, marks, etc., and paid special attention to the choice of a suitable material for children's perception while stressing the importance of the "decimal idea" (which represented an advancement if compared with Pestalozzi, according to Smith 1896a, p. 18). He was also putting oral practice (mental arithmetic) before written practice. Smith also noticed (*idem*) that to support "objective teaching," Tillich designed a special "reckoning box"; this box contained 100 sticks shaped as parallelepipeds. The box had the size 10×10 square inches at the bottom, and the depth of layers decreased from 10 inches to 1 inch. The first layer counted from the base contained $10 \times 10 = 100$ elementary cubes, the next layer $10 \times 9 = 90$ cubes, and so on. A cube with all dimensions equal to one inch represented unit, while prisms of 10 inches in height and a square base of 1×1 sq. inch represented the number ten (Fig. 3).

From a general perspective, Smith characterized Tillich's plan as a "methodical and systematic procedure, a slow, judicious, and gradual advance on the part of the pupil" (Smith 1896a, p. 17).

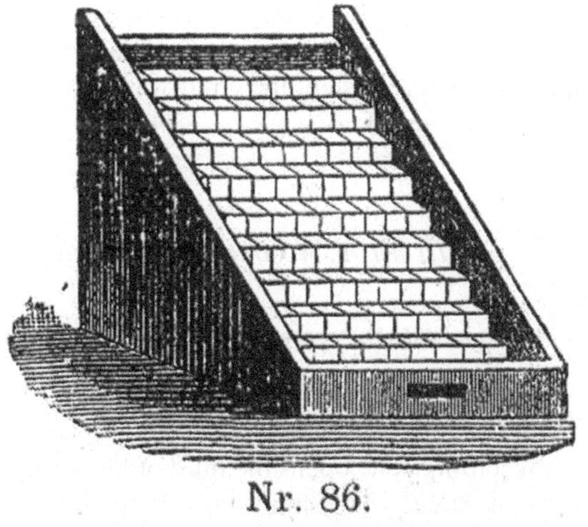

Fig. 3 Tillich's "reckoning box" (Rechenapparat oder Rechenkasten zur Veranschaulichung der Zahlen und Zahlenoperationen von 1–1000. http://www.schulmuseum.at/ausstellung/wechselausstellung/wa06/ausstellung_haenschen/04%20Rechnen/slides/01

4.5 Diesterweg: Two Methods, One Arithmetic

Following the interaction between mental and written arithmetic, Smith (1896a) draws readers' attention to the work of Friedrich Adolph Wilhelm Diesterweg (1790–1866) who was another follower of Pestalozzi. Modern historians suggest that he "advocated didactical ideas according to which pupils at elementary schools should reach a profound and reflected understanding of arithmetic and geometry, and not only be trained by drill and practice" (Gellert et al. 2019).

According to Smith (1896a), by distinguishing mental and written arithmetic (subjective vs objective method), Diesterweg and Heuser in their *Handbuch für Gesammt-Unterricht im Rechnen* (Handbook in Arithmetic for General Education) (1844) attempted to harmonize both methods: "There is only one arithmetic, the arithmetic of reason." An "heuristic method of teaching" connects two methods while the main principle would be to "teach that at every step the independence of the child is built up" (Smith 1896a, p. 18).

Diesterweg formulated several principles of such teaching that were supposed to be applied across all the school levels. These principles were summarized by Smith (1896a, p. 18). This summary listed the steps that he considered necessary for teaching numbers. It included three stages of development of the subject (i.e., arithmetic): clearness of concept, exercise, and application. Perception (external and internal) was supposed to support the correct understanding of a concept. Students were encouraged to discover a rule or a law on the basis of the clearly understood concept. All the new contents needed to be presented as such, visually, at first, but later they had to be put into relation with the materials learned earlier. Children were supposed to have enough time for exercise and application. Associating computations with abstract and concrete numbers was also important. Use of figures was valued over exercise with pure numbers. Mental arithmetic was seen at first as oral work without symbols of digits, and only later with written numbers. When solving applied problems, the learners needed to deal with applications to commonly known units (of money, weight, etc.). Complete verbal statements were also considered valuable. At every step, the child was encouraged to design appropriate problems for her- or himself.

4.6 Smith on A.W.Grube's Method

Smith concludes the second part of his series of articles with the following statement:

> Thus for nearly a half century Pestalozzianism, more or less modified, held complete sway. It was Grube who suggested the next radical change (Smith 1896a, p. 18).

Accordingly, he opens the third of his short papers with a brief discussion of the work of August Wilhelm Grube (1816–1884) (Smith 1896b, pp. 78–79). Grube's only work devoted to mathematics education was titled *Leitfaden für das Rechnen*

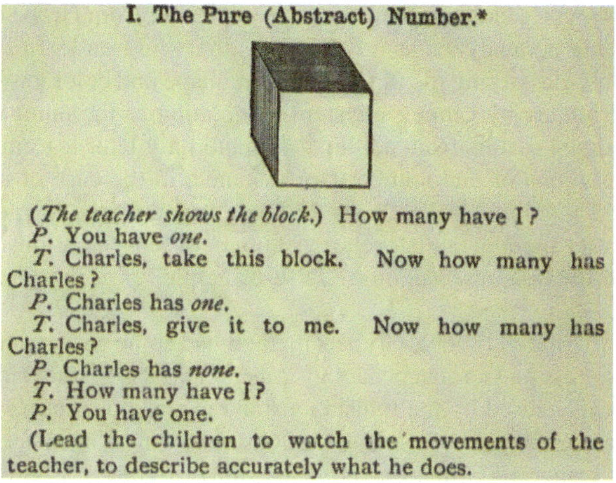

I. The Pure (Abstract) Number.*

(*The teacher shows the block.*) How many have I?
P. You have *one.*
T. Charles, take this block. Now how many has Charles?
P. Charles has *one.*
T. Charles, give it to me. Now how many has Charles?
P. Charles has *none.*
T. How many have I?
P. You have one.
(Lead the children to watch the movements of the teacher, to describe accurately what he does.

Fig. 4 Introduction of a number according to Grube (as interpreted by Seeley): The number one can be measured only by itself. The child thus has to learn the idea of unity. From Seeley 1891, p. 20

in der Elementarschule nach den Grundsätzen einer heuristique Méthode (Guide for Arithmetic in Elementary School, According to the Principles of Heuristic Method). It was originally published in 1842 and was reprinted several times with minor modifications. Smith's analysis of Grube, once again, follows the structure of Unger's work (1888) while focusing on another important method, that of considering a given number from different perspectives ("allseitigen Zahlbehandlung", lit. "all-round handling of numbers"). Smith begins with a description of Grube's method that the latter introduced in his monograph of 1842 (Smith does not mention the title of this book and provides only its date of publication, p. 78). It appears appropriate to begin our analysis of Smith's description with a short introduction to the work of Grube in the field of mathematics education.[9]

According to Smith (1909a), the method was "an adaptation of the concentric circles plan that has already been used" (p. 12). Grube was treating each number up to 100 as an "individual, exhausting all four operations before proceeding to the next number" (Smith 1896b, p. 79) and as "foundation of all number work." The numbers 1 to 10 were introduced one by one using the same scheme. Below is an example of introducing number 1 (Fig. 4) found in Levi Seeley's monograph (1888 [1891]) of teaching arithmetic according to Grube's method.

Numbers from 2 to 14 are introduced on the subsequent pages of Seeley's book with the help of pictures of two- and three-dimensional blocks reminiscent of the rectangular blocks used in the original edition of Diesterweg and Heuser's book (1844, p. 1). For numbers 15, ..., 20, Seeley uses only pictures of arrays of dots and vertical strokes. He concludes his presentation by passing to numbers 30, 50 (for an unknown reason forty is omitted), and 100.

[9] For an early critic of Grube's method see Badanes (1895).

We find in Unger's later publication of 1888 a quote from Grube (1842) stating that numbers can be analyzed as other objects that children learn through observational lessons, classifying the objects by size, shape and color, or when studying beginnings of botany by Linney's system. In relation to the numbers, instead of learning first that 4 results from adding 2 + 2 and only later learning that 4 – 2 = 2, all representations of one number (for instance, in the case of number 4 also including 2 × 2 = 4 and 4 ÷ 2 = 2) need to be grasped by the learners simultaneously so as to let them establish relationships between operations. This method was later identified as based upon a "false principle" (*falsches Princip*) (Unger 1888, p. 192).

According to Smith (1896b), Grube's method had some success in Germany but "made little impress on German pedagogy"; its popularity ("fame") in America was a "pure accident" caused by the translations and explanations provided by Soldan (1877, 1878) and Seeley (1888). Smith (1896b), following Unger's example (Unger 1888, p. 192), criticized Grube for the "false aim" of getting a "complete notion" (*Vorstellung*) of each number before moving to the next one: for mastering a number by the child most probably would happen "when he knows how to find from the totality of numbers one unknown number from certain data" (Smith 1896b, p. 79). For example, when children were required to decompose number 97 in different ways (6 × 16 + 1, 4 × 23 + 5; etc.), it proved to be a difficult task as the method was clearly above their capacity to get a correct representation despite the teacher's good intention of strengthening their reasoning power. Smith concludes his discussion of this method by saying that

> [I]t is doubtful if this method is followed in a single school [in America]. Certain it is that the essential feature [of this method] is quite generally not recognized. None of them [the notions such as concentric number circles, etc.] are Grube's; the one essential feature [of his method] is generally condemned by the great educators of the world. (Smith 1896b, p. 79)

The feature Smith mentions here is Grube's idea of making the learner "know" all the arithmetical relationships that exist for the consecutive natural numbers, that is, of making her or him able to tell that 97 is 6 × 16 + 1, 4 × 23 + 5, 3 × 31 + 4, etc. (Smith 1896b, p. 79).

4.7 Praising Ideas of German Educators at the End of the 19th Century: Das Zählprincip, Knilling's and Tanck's Counting Approach

The final part of Smith's analysis of 19th century innovations in the method of teaching arithmetic is devoted to the works of German educators of the 1870-80s who, based on the critiques of Pestalozzi and Grube, brought back Busse's idea that "foundation of all number work is counting" (Smith 1896c, p. 134). Unger (1888) mentions particularly the works of Tanck and Knilling published in 1884; he calls these authors "zwei Gegners of Grubischen Method", that is, "two opponents of Grube's Method" (p. 195).[10]

It seems relevant to provide at least minimal biographical information about these two authors.[11] Lars Friedrich Wilhelm August Tanck (1831–1928) became a schoolteacher in 1868. From 1875 to 1907 he was a co-founder and director of Fortbildungsschule (The School for Advanced Education) in Neumünster. In 1884, he published a book titled *Das Zählen und erste Rechnen* (Counting and First Arithmetic; second edition 1906) in which he paid special attention to counting methods.

Rudolf Knilling (1852–1923) was an elementary school teacher from Oberbayern who had "completely broken with the previous [that is, Grube's. – V. F & A. V.] method in arithmetic lessons" (Knilling 1886a[12]). Knilling explains his position as follows:

> But the reasons that made me a decided and passionate opponent of the Grube teaching method were less theoretical than practical, I already explained it in my lecture titled "Viewpoints and Principles concerning the practical reform of arithmetic instruction" that I read in 1884 on the occasion of a general meeting of the Bavarian Teachers' Association in Ansbach. […]
>
> I have learned from my own experience in the elementary class of no fewer than 103 boys about all the Sisyphean work we had to do; it consisted of the all-round treatment of numbers that teachers "pushed" on us. Our Method was like a heavy stone that we had to roll up the mountain day after day only to see that once we brought it at the top it inevitably rolled back to the abyss again. (Knilling 1884, p. 10)

According to Unger (1888), both Tanck and Knilling arrived at the counting approach on the basis of genetic-historical analysis of the concept of number (Entstehnung der Zahlbegriffe). According to this analysis, the concept of number did not develop on the basis of *Anschauung* (i.e., perception, direct observation, and intuition) but resulted from the *acts of counting*. For instance, knowledge of the exact number of objects in a given collection cannot be obtained through senses, but only through the act of counting. The process of counting can be accelerated by grouping the objects in small subsets (for instance, by 3, 4, or 5 objects); the counting can be made easier if the objects are arranged in rectangular arrays rather than circular ones (Fig. 5).

Furthermore, Unger (1888, p. 196) affirms that:

> [C]ounting itself is not a simple but rather a complicated act. First of all, it requires having a fixed series, the natural series of numbers, and being able to quickly dispose of it; secondly, such a control of our muscles that the movement of the muscles could safely and easily accompany the recitation of numerals, and thirdly, an insight into the connection between the reciting of numerals and the progression of things or processes brought about

[10] Some authors provide a reference to "Tanck and Knilling 1884" thus suggesting that this was a joint publication; yet we deal here with two separate works published in the same year, Tanck (1884) and Knilling (1884).

[11] The source of the biographical data on Tanck is the website https://www.wilhelm-tanck-schule.de/unsere-wts/wilhelm-tanck/ consulted on June 18, 2021.

[12] Originally published in 1885.

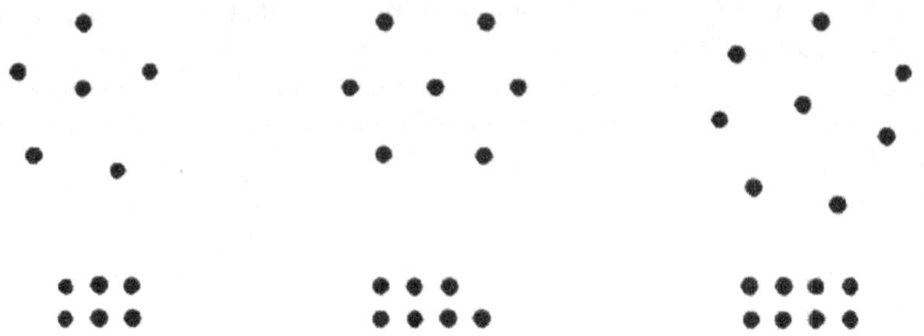

Fig. 5 Counting small sub-sets. Unger 1888, p. 196

by muscle movement (pushing coins, touching immovable bodies, nodding when the clock strikes, etc.).

Indeed, what follows from this complexity is that

> [W]e do not have specific and clear ideas and concepts of numbers. The person who calculates does not need them either, he can achieve his goal even without them. If he only knows that his result is the sum, or the difference, or the product, or the quotient of certain given numbers, he has solved his problem, and he does not need to worry about the other properties of the number found. It is not the nature of the number that matters, but the ways of finding it. What work would the brain of the accounting officer expect to do is to add up many columns [...]; he should think definitely and clearly, vividly and tangibly! Rather, the "calculator" [i.e., the person who calculates. - V.F. and A.V.] pushes back every thought and every concept that prevents him from exercising the law of numbers. (Unger 1888, p. 197)

According to Smith (whose summary is much shorter than Unger's), this method returns back to Busse's work stating that the ability to count could be drilled with the number series (Unger 1888, p. 195). Citing Unger (1888), Smith (1896c) also provided an outline taken from two German textbooks. The first of them is the textbook designed by Tanck for the first school year (1884) where only numbers 1 to 10 are introduced. The studied topics are as follows:

- Naming numbers;
- Counting objects;
- Writing numbers;
- Addition and subtraction within the first dozen (e.g., 5 + 3, 8 – 3);
- Multiplication presented as a special case of addition;
- Division;
- Review (using Grube's method).

The second textbook mentioned by Smith is Knilling's outline for year 1 discussing operations with the numbers from 1 to 100:[13]

- Counting and writing numbers;

[13] It is possible that there is a misprint here, as the teaching of year one was devoted mostly to the numbers 1 to 10.

- Easy exercises in counting by 1s, 2s, … 9s;
- Easy exercises in subtracting;
- Counting objects: fingers, marks, beads of the classroom abacus, etc.

Smith's overall evaluation of the work of Tanck and Knilling is favorable: according to him "their scheme has vitality and definiteness, and their aim is a high one" (Smith 1896c, p. 135). Interestingly, Smith's work was later cited by Knilling himself in a short review of contemporary American literature on teaching arithmetic (Knilling 1899). Along with the aforementioned quotation, Knilling quotes Smith when the latter identifies Knilling's and Tanck's call for returning to counting as the mainstream approach to introducing a number as "the whole movement" which "is a reaction as much against Pestalozzi as against Grube" (Knilling 1899, p. 134). Another quotation from Smith states that "Knilling labours with the fire and zeal of a crusader, and while his utterances are far from being calm and judicious, they are vigorous and striking" (Smith 1896c, p. 134, cited by Knilling 1899, p. 256).

In the same survey, Knilling also cited the work authored by Grace H. Ford who was, according to him, Smith's student ("Smith's Schülerin"). In 1896 Ford published a paper in the journal *Michigan School Moderator* about the development of the method of teaching arithmetic in Germany based on the works of Pestalozzi, Grube, Knilling and Tanck, clearly under influence of Smith's work. Smith introduced her article as a "brief resume of the development of methods of teaching arithmetic in this century" (Ford 1896, p. 108).

As conclusion of the series of his articles, Smith (1896c) mentions the work by McLellan and Dewey (1895) on the psychology of numbers published a short time before his own paper. He stresses that its novelty is a clear departure from Grube's "fixed unit system," praises it as "deserving of careful perusal by every teacher of arithmetic," and suggests that every work on mathematics education should be based upon psychology (Smith 1896c, p. 135). However, Smith thinks that more attention should be paid to German didactical heritage whose influence "upon American schools will doubtlessly be considerable" (Smith 1896a, b, c, d, p. 135).

5 Discussion and Conclusions

When reflecting on Smith's articles in the light of important changes in American mathematics education and its relation to the international community in the transitional decades from the 19th to the 20th century (1890–1910), one should keep in mind that the years when Smith joined mathematics education community (1890s) were pivotal in the way in which it was structured and shaped through "numerous associations, journals, and teacher preparation programs" (Karp 2019c, p. 194). American mathematics education was also a part of a "common movement with Europe—or at least an interconnected one" (Karp 2019c, p. 195). Karp makes an explicit reference to the books Smith (1900) and Young (1907) as the earliest American works specifically written as resources for teacher preparation which referred to the European experience and to European literature (Karp 2019c, p. 195). More particularly, the influence that German speaking countries had on the development of methods of teaching arithmetic at the elementary level since the 18th century can explain the decision of Smith to analyze the evolution of American methods based on German experience. The history of German mathematics education that Smith takes into consideration covers the time period beginning prior to 1700 and includes the educational theories and experience of the time when numerous elementary schools were founded (for example, in Prussia this process started in the early 18th century, see Bjarnadóttir 2014).

A genetic-historical approach adopted by German scholars at the end of the 19th century in their studies of the development of method of teaching arithmetic was particularly praised by Smith who began its popularization in the four papers that we reviewed in the present chapter. For example, while McLellan's and Dewey's *Psychology of the Number* (1895) was only briefly mentioned at the end of the fourth paper, Smith wrote, in 1896, a more detailed review of this book (Smith 1896d).[14] As a critical suggestion, he mentioned that the "work entirely ignores the historic growth of number-method, a thing that we have no right to demand, to be sure, but a feature that would have been of utmost value to teachers." He then argued that:

> A few pages on the evolution of method, say from Busse at Dessau, through the great Swiss teacher, through Tillich and Stephani and Diesterweg, through Grube and his small following, down to Tanck and Knilling and the writers of our time—such a chapter would have been very valuable. (Smith 1896d, p. 104)

Moreover, in his first book on teaching of elementary mathematics, Smith (1900) devotes a section to "How Arithmetic Has Been Taught" in his analysis of evolution of method which is clearly based on the four papers of 1895–96 discussed in the present chapter; in this section he provides even more details concerning the works of Unger, Jänicke, Knilling, among others. He also returns to the importance of method while discussing the contributions of Pestalozzi, Tillich, and Grube in his works on the teaching of arithmetic (1903; 1909a, b).[15]

Certainly, Smith was not the only one who studied the development of method in teaching arithmetic based on innovative work of the 19th century German educators. For instance, in Russia, Mihail I. Uspenskij (1866–1942) wrote a review (Uspenskij 1899) of Unger's *Methodik* that resembled that of Smith in many respects, while Knilling's work was discussed in Japan by Fujisawa Rikitarō 藤沢利喜太郎 (1861–1933) in his report (Rikitarō 1912) on teaching of mathematics in Japan; see also Karasawa 2014.

Nevertheless, the reform efforts were always a subject of debate, in Germany itself, as well as in Russia, in Japan, and in the United States. And apparently these

[14] For a review of the same work of McLellan and Dewey published in the same year see Badanes 1896.

[15] For a recent study of the contributions of Pestalozzi, Grube, and Dewey see, for example, Da Costa 2014, Mesquida et al. 2017.

debates (in particular, the debates on the method) are still alive even in the 21st century. Therefore, we find it meaningful to conduct an analysis of the didactical innovations of the past and of the trajectories along which they traveled across time and space. In this perspective, we consider our contribution on Smith's work in the field of didactics as only an initial stage of a forthcoming study.

References

Anon. 1922. Notes and News. *The American Mathematical Monthly* 29 (3): 136–142.

Badanes, Saul. 1895. *The Falsity of the Grube Method of Teaching Primary Arithmetic*. Thesis for the Doctorate in Psychology. New York.

———. 1896. The Psychology of Number [a review of McLellan and Dewey 1895]. *The Journal of Education* 43 (1) [1059]: 5–6.

Beman, Wooster Woodruff, and David Eugene Smith. 1895. *Plane and Solid Geometry*. Boston & London: Ginn & Co.

———. 1897. *Higher Arithmetic*. Boston & London: Ginn & Co.

———. 1899. *New Plane and Solid Geometry*. Boston & London: Ginn & Co.

———. 1900. *Elements of Algebra*. Boston: Ginn & Co.

———. 1902. *Academic Algebra*. Boston: Ginn & Co.

Bjarnadóttir, Kristín. 2014. History of Teaching Arithmetic. In *Handbook on the History of Mathematics Education*, ed. Alexander Karp and Gert Schubring, 431–458. New York: Springer.

Brasch, Frederick E. 1944. David Eugene Smith [obituary]. *Science [New Series]* 100 (2595): 257–259.

Bullynck, Maarten. 2008. The transmission of numeracy: Integrating reckoning in Protestant North-German elementary education (1770–1810). *Paedagogica Historica* 44 (5): 563–585. https://doi.org/10.1080/00309230802042714.

Busse, Friedrich Gottlieb [von]. 1786. *Gemeinverständliches Rechenbuch für Schulen, worin hauptsächlich nach Thalern, Groschen und Pfennigen gerechnet wird [A generally understandable arithmetic textbook for schools where calculations are mainly introduced with thalers, groschen and pfennigs]*. Leipzig: Crusius.

Chavannes, Daniel Alexandre. 1805. *Exposé de la méthode élémentaire de H. Pestalozzi, suivi d'une notice sur les travaux de cet homme célèbre, son institut et ses principaux collaborateurs*. [Presentation of the elementary method by H. Pestalozzi, followed by a notice on the work of this famous man, his institute and his main collaborators]. Paris: Levrault-Schoell.

Colaw, John M., and John Kelley Ellwood. 1900a. School arithmetic: Primary Book. Richmond (VA): B.F. Johnson Publishing Co.

———. 1900b. School arithmetic: Advanced Book. Richmond (VA): B.F. Johnson Publishing Co.

Da, Costa, and David Antonio. 2014. The conceptions and contributions of Pestalozzi, Grube, Parker and Dewey for teaching arithmetic in elementary level: The concept of number. *Revista História da Educação* 18 (42): 37–59.

Diesterweg, Friedrich Adolf Wilhelm, and Peter Heuser. 1829. *Methodisches Handbuch für den Gesammt-Unterricht im Rechnen: als Leitfaden beim Rechenunterrichte und zur Selbstbelehrung [Handbook of Methods for General Teaching in Arithmetic: Guidelines for Arithmetic Teaching and for Self-Learning]*. Vol. 1. Elberfeld: Büschler.

———. 1844. *Methodisches Handbuch für den Gesammt-Unterricht im Rechnen: als Leitfaden beim Rechenunterrichte und zur Selbstbelehrung* [Handbook of Methods for General Teaching in Arithmetic: Guidelines for Arithmetic Teaching and for Self-directed Learning]. Elberfeld: Büschler (In his publication Smith 1896a refers to the third edition of this work published in 1839).

Donoghue, Eileen F. 2003. The emergence of a profession: Mathematics education in the United States, 1890–1920. In *A History of School Mathematics*, ed. George Stanic and Jeremy Kilpatrick, 159–194. Reston, VA: National Council of Teachers of Mathematics.

———. 2006. The Education of Mathematics Teachers in the United States: David Eugene Smith. *Early Twentieth-Century Pioneer. Paedagogica Historica* 42 (4-5): 559–573.

———. 2008. David Eugene Smith and the founding of the International Commission on the Teaching of Mathematics. *International Journal for the History of Mathematics Education* 3 (2): 35–45.

Ellis, Arthur, Reinhard Golz, and Wolfgang Mayrhofer. 2014. The Education Systems of Germany and Other European Countries of the 19th Century in the View of American and Russian Classics: Horace Mann and Konstantin Ushinsky. *DE-Online Journal (International Dialogues on Education: Past and Present)* 1 (1): 7–28.

Ellwood, John Kelley. 1892. *Table book and test problems in mathematics*. New York etc.: American Book Company.

———. 1894a. Remarks on division. *The American Mathematical Monthly* 1 (2): 47–48.

———. 1894b. More Remarks on Division. *The American Mathematical Monthly* 1 (4): 116–118.

———. 1896. Number and Fractions. *The American Mathematical Monthly* 3 (11): 263–265.

Fink, Karl. 1900. *A Brief History of Mathematics*. Chicago: The Open Court.

Fite, W. Benjamin. 1945. David Eugene Smith [obituary]. *The American Mathematical Monthly* 52 (5): 237–238.

Ford, Grace H. 1896. Development of Method in Arithmetic in Germany in the Nineteenth Century. *The Moderator* 17 (4): 108–110.

Freiman, Viktor, and Alexei Volkov. 2014. What did the students learn from mathematics textbooks? The case of L. F. Magnitskiĭ's *Arithmetic*. In *Proceedings of the 38th Conference of the International Group for the Psychology of Mathematics Education and the 36th Conference of the North American Chapter of the Psychology of Mathematics Education*, eds. Oesterle, Susan, Peter Liljedahl, Cynthia Nicol, and Darien Allan, 3: 113-120. Vancouver.

———. 2018. History of computations, computing devices, and mathematics education from the teaching and learning perspective: Looking for new paths of investigation. In *Computations and Computing Devices in Mathematics Education before the Advent of Electronic Calculators*, ed. Alexei Volkov and Viktor Freiman, 3–21. Cham: Springer.

Frick, Bertha Margaret. 1936. Bibliography of the Critical, Historical and Pedagogical Writings of David Eugene Smith. *Osiris* 1: 9–78.

Furinghetti, Fulvia. 2019. Challenges, Hopes, Actions, and Tensions in the Early Years of the International Commission on the Teaching of Mathematics. In *National Subcommissions of ICMI and their Role in the Reform of Mathematics Education*, ed. Alexander Karp, 1–34. Cham: Springer.

Gellert, Uwe, Lisa Hefendehl-Hebeker, Hans Niels Jahnke, and Timo Leuders. 2019. Educational Research on Mathematics—A Short Survey of Its Development in German Speaking Countries. In *Traditions in German-Speaking Mathematics Education Research*, ed. Hans Niels Jahnke and Lisa Hefendehl-Hebeker, 1–23. ICME-13 Monographs. Cham: Springer. https://doi.org/10.1007/978-3-030-11069-7_1.

Giacardi, Livia. 2019. The Italian Subcommission of the International Commission on the Teaching of Mathematics (1908–1920): Organizational and Scientific Contributions. In *National Subcommissions of ICMI and their Role in the Reform of Mathematics Education*, ed. Alexander Karp, 119–147. Cham: Springer.

Grube, August Wilhelm. 1842. *Leitfaden für das Rechnen in der Elementarschule nach den Grundsätzen einer heuristischen Methode: Ein methodischer beitrag zum erziehenden Unterricht*. [Guide for Calculations in Elementary School upon Principles of Heuristic Method: One Methodical Contribution to Educational Teaching]. Berlin: T.C.F. Enslin (repr. 1881; 1883).

Hartmann, Berthold. 1893. *Der Rechenunterricht in der Deutſchen Volksſchule vom Standpunkte des Erziehenden Unterrichts [Teaching Arithmetic in German People's Schools from the*

Perspective of Educational Teaching]. Leipzig & Frankfurt: Kellelringsche Hofbuchhandlung. (E. v. Mayer) Verlag.

Hinsdale, Burke Aaron. 1906. *History of the University of Michigan, with biographical sketches of regents and members of the University Senate from 1837 to 1906*. Ann Arbor: University of Michigan Press.

Jänicke, Eduard. 1877. Geschichte des Rechenunterrichtes und der Raumlehre [History of Arithmetic and Space Geometry]. In *Geschichte der Methodik des deutschen Volksschulunterrichtes*, ed. C[arl] Kehr, vol. 1, 280–460. Gotha: Thienemanns Hofbuchhandlung.

———. 1879. *Der Rechenunterricht in der deutschen Volksschule. I. Teil: Grundzüge der Geschichte und Methodik des Rechenunterrichts [Arithmetic Teaching in German People's Schools. Part 1. Main Stages in the History and Methods of Teaching Arithmetic]*. Gotha: Thienemann.

Karasawa, Toshimitsu. 2014. A Composition of Content Based on Count Principle Method for Reeducation of Mathematics Teacher. *Journal of Modern Education Review* 4 (1): 50–56.

Karp, Alexander, ed. 2019a. *National Subcommissions of ICMI and their Role in the Reform of Mathematics Education*. Cham: Springer.

———. 2019b. In *Introduction. In National Subcommissions of ICMI and their Role in the Reform of Mathematics Education*, ed. Alexander Karp, v–viii. Cham: Springer.

———. 2019c. The American National Subcommission of ICMI. In *National Subcommissions of ICMI and their Role in the Reform of Mathematics Education*, ed. Alexander Karp, 193–234. Cham: Springer.

Kegley, T.M. 1947. Warren Colburn. *Peabody Journal of Education* 25 (1): 19–26. Retrieved on June 15, 2021, from http://www.jstor.org/stable/1489474.

Klein, Felix. 1895. *Vorträge über ausgewählte Fragen der Elementargeometrie [Lectures on selected questions of elementary geometry]*. Leipzig: Teubner.

———. 1897. *Famous Problems of Elementary Geometry. An authorized translation of Wooster Woodruff Beman and David Eugene Smith*. Boston etc: Ginn and Co.

Knilling, Rudolf. 1884. *Zur Reform des Rechenunterrichts in der Volksschulen. 1. Mein Programm. Kritik der Pestalozzi'schen, Grube'schen und gegenwärtigen Rechenmethode*. [On Reform of Arithmetic Teaching in People's Schools. 1 My Program. Critique of Pestalozzian, Grubean, and Modern Methods of Teaching Arithmetic]. München: Ackermann.

———. 1886a. Der Rechenunterricht im Leipziger (Volksschul-)Lehrerverein [Arithmetic Teaching In Leipzig's People's Schools-Teacher's Union]. *Zeitschrift für mathematischen und naturwissenschaftlichen Unterricht* 17: 298–299.

———. 1886b. *Zur Reform des Rechenunterrichtes in den Volksschulen. 2. Fortsetzung und Schluß des theoretischen Teiles. Beschreibung des Rechentisches, Teillineals und verbesserten Zweisatzes. Praktischer Teil*. [On Reform of Arithmetic Teaching in People's Schools. 2 Follow-up and Conclusion to Theoretical Parts. Description of Arithmetic Tables. Practical Part]. München: Ackermann.

———. 1899. Aus der amerikanischen Rechenlitteratur der Gegenwart [From the Modern American Arithmetic Literature]. *Neue Bahnen: Monatsschrift für Haus-, Schul- und Gesellschafts- Erziehung. Zugleich Organ der Freien Vereinigung für philosophische Paedagogik* 10: 256–259.

Krüsi, Hermann. 1830. Geschichte der Entstehung und Bearbeitung der Pestalozzi'schen Anschauungslehre der Zahlenverhältnisse [History of Origins and Treatment of Pestalozzian *Anschuungslehre* of Numerical Proportions]. *Allgemeine Monatschrift für Erziehung und Unterricht* 13 (5): 161–181.

———. 1875. *Pestalozzi: his life, work, and influence*. Cincinnati & New York: Wilson, Hinkle & Co.

McLellan, James Alexander, and John Dewey. 1895. *The Psychology of Number and Its Applications to Methods of Teaching Arithmetic*. New York: D. Appleton & Co.

Mesquida, Peri, Fabio Ignacio Pereira, and Mauricio Eduardo Bernz. 2017. The Pestalozzi Method: Mathematics as a Way to the Truth. *Creative Education* 8: 1088–1098.

Moliere. 1845. *Œuvres de Molière, avec de notes de tous les commentateurs [Works of Molière, with Critical Notes]*. Paris: Librairie de Firmin Didot Frères.

Molière. 1875. *The Doctor in Spite of Himself.* In *The dramatic works of Molière*, vol. 3, 389–433. Edinburgh: William Paterson.

Monroe, Walter S[cott]. 1912a. Warren Colburn on the Teaching of Arithmetic Together with an Analysis of His Arithmetic Texts: I. The Life of Warren Colburn. *The Elementary School Teacher* 12 (9): 421–426.

———. 1912b. Warren Colburn on the Teaching of Arithmetic Together with an Analysis of His Arithmetic Texts [Continued]. *The Elementary School Teacher* 12 (10): 463–480.

———. 1912c. A Chapter in the Development of Arithmetic Teaching in the United States. III. *The Elementary School Teacher* 13 (1): 17–24.

———. 1913a. Analysis of Colburn's Arithmetics. IV. *The Elementary School Teacher* 13 (5): 239–246.

———. 1913b. Analysis of Colburn's Arithmetics. V. *The Elementary School Teacher* 13 (6): 294–302.

Pestalozzi, Johann Heinrich. 1803. *Anschauungslehre der Zahlverhältnisse* [The Teaching of Arithmetic by Sense-perception, vols. 1–2]. In *Pestalozzis Elementar-Bücher*, vol. 4-5. Zurich & Bern: H. Gessner, and Tübingen: J.G. Cotta.

R[eeve], W[ilfred] D[ouglas]. 1944. David Eugene Smith [obituary]. *The Mathematics Teacher* 37 (6): 278–279.

Fujisawa Rikitarō. 1912. *Summary report on the teaching of mathematics in Japan*. Tokyo.

Rogers, Leo. 2019. The British National Subcommission of ICMI and the Mathematics Education Reform. In *National Subcommissions of ICMI and their Role in the Reform of Mathematics Education*, ed. Alexander Karp, 93–117. Cham: Springer.

Sandison, Howard. 1895. The Real Province of Method. *The Journal of Education* 42 (6): 116.

Schiess, Emil. 1928. Hermann Krüsi: Pestalozzis ältester Gehilfe und Mitarbeiter (Fortsetzung und Schluss) [Hermann Krüsi: Former Pestalozzian Helper and Collaborator]. *Appenzellische Jahrbücher* 55: 1–76.

Schmid, Joseph. 1810. *Die Elemente der Zahl als Fundament der Algebra nach Pestalozzischen Grundsätzen [Elements of Number as Foundation of Algebra upon Pestalozzian Principles]*. Heidelberg: Mohr und Zimmer.

Schubring, Gert. 2012. Antagonisms between German states regarding the status of mathematics teaching during the 19th century: processes of reconciling them. *ZDM Mathematics Education* 44: 525–535. https://doi.org/10.1007/s11858-012-0407.

———. 2019. The German IMUK Subcommission. In *National Subcommissions of ICMI and their Role in the Reform of Mathematics Education*, ed. Alexander Karp, 65–91. Cham: Springer.

———. 2020. David Eugene Smith: Cortland, New York 1860—New York, New York 1944, *History of the International Commission on Mathematical Instruction*. http://www.icmihistory.unito.it/portrait/smith.php, last update 27/03/2020. Accessed on June 9, 2021.

Seeley, Levi. 1888. [repr. 1891]. *Grubé's [sic. – V.F. & A.V.] method of teaching arithmetic. Explained and illustrated. Also the improvements upon the method made by the followers of Grubé in Germany*. New York and Chicago: E. L. Kellogg & Co.

Simons, Lao Genevra. 1945. David Eugene Smith—In Memoriam. *Bulletin of the AMS* 51: 40–50.

Smith, David Eugene. 1895. The evolution of method in arithmetic. I. *Inland Educator* 1 (1): 345–347.

———. 1896a. The evolution of method in arithmetic. Second paper. Pestalozzi. *Inland Educator* 2 (1): 16–18.

———. 1896b. The evolution of method in arithmetic. Third paper. *Inland Educator* 2 (2): 78–79.

———. 1896c. The evolution of method in arithmetic. Fourth paper. *Inland Educator* 2 (3): 134–135.

———. 1896d. Review of McLellan and Dewey 1895. *The School Review* 4 (2): 102–104.

———. 1900. *The teaching of elementary mathematics*. New York: The Macmillan Company & London: Macmillan & Co., Ltd.. [Repr. 1902].

———. 1903. *The Teaching of Elementary Mathematics*. New York and London: MacMillan.

———. 1909a. The Teaching of Arithmetic. *Teachers College Record* 10 (1): 1–100.

———. 1909b. *The Teaching of Arithmetic*. New York: Teachers College, Columbia University.

———. 1911. *The Teaching of Geometry*. Boston, etc: Ginn.

———. 1913. *The Teaching of Arithmetic*. Boston, etc: Ginn.

———. 1916. The Development of the American Arithmetic. *Educational Review* 52: 109–118.

Smith, David Eugene, and Frank M. McMurry. 1903. Mathematics in the Elementary School. *Teachers College Record* 4 (2): 1–70.

Smith, David Eugene, and William David Reeve. 1927a. *The Teaching of Junior High School Mathematics*. Boston: Ginn.

———. 1927b. *The Teaching of High School Mathematics*. Boston: Ginn.

Smith, David Euguene. 1894. Note on Mr. Ellwood's Remarks on Division. *The American Mathematical Monthly* 1 (2): 74–75.

Soldan, F[rank] L[ouis]. 1877. Grube's Method. Outline of A.W.Grube's *Guide for Primary Instruction in Arithmetic. New England. Journal of Education* 5 (7): 81–82.

Soldan, F[rank] Louis. 1878. *Grube's Method of Teaching Arithmetic Explained. With a Large Number of Practical Hints and Illustrations*. Chicago: The Interstate Publishing Company.

Stalker, Francis M., and Charles M. Curry. 1895. Editorial. *Inland Educator* 1 (5): 318–319.

Sterner, Matthäus. 1891. *Geschichte der Rechenkunst [History of Art of Calculation]*. Munchen und Leipzig: Oldenbourg.

Tanck, Wilhelm. 1884. *Das Rechnen auf der Unterstufe nebst Beitrag zur Frage nach der Entstehung der Zahlenbegriffe [Arithmetic at the Basic Level along with a Contribution to the Question of the Origin of Number Concepts]*. Meldorf: Bremer.

Tillich, Ernst. 1806. *Allgemeines Lehrbuch der Arithmetik oder Anleitung zur Rechenkunst für Jedermann. [General Textbook of Arithmetic or Introduction to the Art of Calculation for Everyone]*. Leipzig: Gräff.

Unger, Friedrich. 1888. *Die Methodik der praktischen Arithmetik in historischer Entwickelung: vom ausgange des Mittelalters bis auf die Gegenwart. [Methods of Practical Arithmetic in Historical Development: From the Beginning of Middle Ages to the Modern Times]*. Leipzig: B.G. Teubner.

Uspenskij, M[ihail Ivanovich]. 1899. Nemetskaya i russkaya metodika arifmetiki za tekuschee stoletie [German and Russian methodology of (teaching) arithmetic in the present century, in Russian]. *Russkaya shkola* 4: 145–170.

Wang Xiaoqin. 汪晓勤 2010. D. E. Smith 论著目录 [List of publications of D.E.Smith]. Online publication https://faculty.ecnu.edu.cn/picture/article/1284/ee/f7/f981fe394b67ab075ed727ab3097/a790be31-a76a-441f-9717-bea21bcf8b1c.pdf.x. Accessed on August 22, 2021.

Young, Jacob William Albert. 1907. *The teaching of mathematics in the higher schools of Prussia*. New York, London and Bombay: Longmans, Green, and Co.

Entrance Exams to Higher Educational Institutions in Russia Before the Revolution: Problems, Procedures, People

Alexander Karp

Abstract This paper is devoted to entrance exams to higher technical institutions in Russia prior to the revolution of 1917. These exams, almost unresearched both with respect to organizational and administrative procedures and forms, and with respect to their mathematical content, exerted an enormous influence on the development of Russian mathematics education. This pertains above all to changes in the attitude toward problem solving and to the place of problems in the process of studying mathematics in general. The paper examines both general issues in the organization of exams and exam preparation, as well as various problem books for students seeking admission to higher educational institutions, which formed a foundation for the industry of exam preparation, which became the main component of the extracurricular, "informal" study of mathematics in Russia. The paper relies on materials from Russian (mainly, St. Petersburg) archives.

Keywords Russia · Entrance examinations · Higher technical institutions · Problems · Collections of problems

1 Introduction

This article addresses the history of entrance mathematics examinations to colleges in Russia. Exams constitute one of the most important means of selection and social stratification. Sociologists have long noted that education (if it is genuine education, of course) facilitates not only the growth of equality in a society, but also, conversely, the formation of a kind of new aristocracy (Sorokin 1959). To be sure, ideally this new aristocracy is formed not on any hereditary basis, but on the basis of individual achievements. Competitive entrance exams to the École Polytechnique in

A. Karp (✉)
Teachers College, Columbia University, New York, NY, USA
e-mail: apk16@columbia.edu

A. Karp (ed.), *Advances In The History Of Mathematics Education*,
International Studies in the History of Mathematics and its Teaching,
https://doi.org/10.1007/978-3-030-95235-8_5

Paris were one of the main steps in the creation of this new aristocracy (Belhoste, & Chatzis 2007). But even here, matters are not always simple: selection to colleges is not always and everywhere based on exams, and exams themselves—even in such a subject as mathematics, which is considered rational and objective—are not always conducted rationally and objectively. The politics of conducting exams in mathematics and their social role are an important aspect of the educational system, which merits study (note, for example, the enormous number of publications about exams in Russia today, the so-called Uniform State Exam—or the "EGE," as it is abbreviated in Russian—see Karp 2020).

The importance of exams explains the influence they exert on the content of education. This role is not prescribed by any directive from any ministry, but in practice students planning to enroll in a college cannot avoid thinking about what that college will require of them on its entrance exams; and teachers who are concerned about their students, and who wish to teach them that which might be useful to them, are guided by entrance exams in one way or another. Meanwhile, entrance exams may be extremely varied—different colleges may have different requirements.

In addition, an enormous literature for those enrolling in colleges inevitably appears, courses and additional classes begin to be offered, along with everything that is sometimes designated as informal education and contrasted with ordinary schools, but which plays an extremely large role, comparable to the role played by ordinary schools. The study of the content of exams—that is, of the types of problems that have been given on exams—is no less important than the study of school curricula.

In this way, entrance exams dictate an unofficial curriculum and exert an enormous influence on how a subject taught in school is understood as a whole. Nonetheless, Russian entrance exams, as Zharova (2016) has rightly remarked, constitute a lacuna in the history of education (and not merely the university education referred to by Zharova, but also and especially higher technical education). The normative regulations and procedures governing how exams are administered—let alone how they are perceived by applicants or, more broadly, the general public—have been little studied or often not studied at all. This is all the more true with respect to the methodological aspects of the process: the problems and theoretical questions in mathematics that have been given on exams; the system for preparing for them, with the whole industry associated with this preparation. Meanwhile, to repeat, the influence and importance of these aspects for the whole development of mathematics education are obvious. The present paper will address entrance exams before the revolution of 1917, beginning with the last quarter of the nineteenth century.

2 About Entrance Exams before 1875

A few words should be said about the period preceding the one investigated in the present paper. Entrance exams existed at that time also. Alexander Herzen, an acute and perceptive chronicler of his age, even somewhat exaggerates their role:

> Before 1848, the organization of our universities was purely democratic. Their doors were open to anyone who could pass an exam and was neither a serf, nor a peasant, not released by his community [obschina] (Herzen 1956, p. 107)[1].

Without going into a discussion of how democratic the whole organization of universities and admission to them actually was at that time, we will note that exams were by no means always given. The university statute of 1804 triumphantly declares:

> No one may be admitted as a student to a University without possessing the requisite knowledge for attending courses taught at the University.

Further, it specifies that:

> Those wishing to enjoy them [the courses] must present the Directorship of the University with proof of their social estate and a certificate from the Director of the Gymnasium concerning their conduct, diligence, and achievements in the subjects taught there; the rest, who did not attend gymnasia, must be tested by a Committee appointed by the Rector in foreign languages and basic necessary sciences (Ministry of Education 1901, p. 184).

The statute of 1835 expresses itself somewhat more decisively in favor of exams: "All those who wish to enroll as students at the University must pass a preliminary trial"; but it is stipulated at once that proof of completion of a full gymnasium course gives applicants the right to stand trial before others or even be released from testing altogether" (Ministry of Education 1901, p. 183).

At that time, students acquired their education in the most varied ways: not a few members of the nobility were educated, like Herzen, at home, and they, just like, say, graduates from theological seminaries, had to take exams; but, to repeat, exams were by no means taken by everyone (note that Herzen does not describe how he took exams, although he mentions that he was initially not permitted to take them, as someone who was already officially employed). Zharova (2016) estimates the number of students admitted to the Universities of St. Petersburg and Moscow during the period 1855–1861 without exams and comes to the conclusions that their total varied from 30% to 70% in different years.

The rules for testing students seeking admission to the University in 1860 describe in detail the categories of people released from entrance exams altogether (these are mainly graduates from gymnasia who completed them sufficiently successfully) and permitted to take an abridged test—that is, only in particular subjects of the gymnasium course, for example, when applying to the physics-mathematics faculty, in mathematics, physics, natural history, the Russian language, and one modern language. Such abridged testing had to be undergone by individuals who had graduated from gymnasia and received certificates of graduation, but with insufficiently high marks. All the others, that is, those who had not attended gymnasia, or who had, but had not received graduation certificates from them, were subjected to a full trial—that is, in all subjects of the gymnasium course. As far as can be judged from the surviving documents, for example, during the years 1860–1861

[1] All translations from Russian are by the author.

at the University of St. Petersburg exams were administered in precise accordance with these rules (St. Petersburg University 1860–1861). Samples of exams with lists of subjects, grades, and the signatures of exam administrators have survived. But over the years, various changes appeared in the procedure.

The rules (Pravila 1860) describe in detail how the examination committee must be formed, but they leave to the committee a considerable degree of independence in choosing the form of the exam—even in determining whether it is to be oral or written (with certain stipulations, however—the exam in the Russian language must be written). The most important question, however, is not even the form of the exam, but its degree of seriousness. Zharova (2014), quite rightly in our view, discerns a certain dynamic here: citing the reminiscences of students who had taken the exam, she writes about the exams' increasing level of difficulty—in the early nineteenth century, the entrance exam was a formality and could almost take place over dinner, while after the statute of 1835, during the reign of Nicholas I, it became something far more serious. It is natural to connect this, as Zharova does, with Emperor Nicholas's desire to make admission to the University more difficult for people from the lowest classes, but we must also note the striving on the part of the leadership of the country to "establish order" and "tighten the screws." Zharova herself (Zharova 2016) notes that Alexander II was also in favor of a rigorous entrance exam, even though he was far more liberal than his father.

The content of exams in mathematics can be judged to some extent, for example, based on the published *Questions for Testing Young People Who Wish to Be Admitted as Students to Imperial Kharkov University* (Voprosy 1849). The questions in mathematics include questions in arithmetic, algebra, geometry (plane geometry and solid geometry), plane trigonometry, and the applications of algebra to geometry, and on the whole correspond to the gymnasium course. The questions include descriptions of various sections ("The Concept of Logarithms and the Advantages of Using Them in Numerical Expressions," (Voprosy 1849, p. 81)) as well as, sometimes, separate theorems or problems ("Through two sides of a given triangle, draw a straight line that is parallel to the third side and equal to a given straight line" (Voprosy 1849, p. 87)). This, however, is obviously not enough to draw any conclusions about what exactly the exam looked like and how students needed to prepare for it—if only because the exam was not administered identically everywhere and at all times. The cited rules of 1860, for example, explain that "testing takes place at the university building or wherever is more convenient," and at two different times—in May, together with the usual gymnasium exit exams, and also in August (abridged form) (p. 6). This distinction itself is already significant.

Consequently, the true history of entrance exams and what might be called the industry of exam preparation begins with a noticeable increase in applicants seeking admission to higher educational institutions and the appearance of a comparatively large number of such institutions. Colleges appeared in the country gradually over the course of the nineteenth century (one could even say, from the second half of the eighteenth century), often as military or semi-military academies. Without going into the history of the formation of higher technical educational institutions in

Russia here, we will choose the 1870s as a nominal date from which to begin our discussion, when the number of such colleges was already sufficiently large, and when a relatively large number of people were thinking about enrolling in them, which in accordance with the established rules required them to pass exams (graduation certificates were not enough). Admission to such colleges will be the subject of the next section. Universities at that time admitted graduates from gymnasia without exams, while for the comparatively small group of applicants to whom an exam did have to be given (for example, graduates of theological seminaries), this exam was given not at the university (Posse 2018).

3 About Entrance Exams During the Years 1875–1917

Choosing 1875 as the nominal date inaugurating the period under investigation, we will not attempt to trace all of the relatively minor changes that occurred during the ensuing years, but will consider the subsequent period of approximately forty years as a unified whole (although this, of course, is an oversimplification). The transition to this period from the period that preceded it did not, of course, happen all at once. Nonetheless, it is characteristic that the change in the role of exams was recognized at that time by such a competent government agency as the political police.

A memorandum composed by one of its agents in 1875 addresses rising discontent among the youth.

> Since 15 July, young people have begun arriving in the capital from different parts of Russia, who have completed the gymnasium course with a view to enrolling in higher educational institutions. The University, as the only [institution] whose admission requirements do not present young people who have a graduation certificate with the difficulties of entrance exams, attracts them least of all. On the contrary, the majority aspire to enroll in the Technological, Mining, and Transport Engineers' Institutes. But since the establishment of the new examination rules, admission to these institutions has been associated with extraordinary difficulties, forcing each of the young people to work quite assiduously (Agenturnaya zapiska 1875).

The author of the memorandum poses the question: how is this possible, given that those seeking admission to colleges had only recently graduated from gymnasia, in which the exact sciences were taught in accordance with the same programs on which the exams were based? The memorandum makes the assertion that the exact sciences are not being taught sufficiently well, thus giving rise to the aforementioned difficulties, which force students to spend almost a year on additional studies and needlessly give rise to social tensions.

The existence of entrance exams did not, as we know, lead to any particular social upheavals or perturbations, while the issue of the unpreparedness for entrance exams of students who had graduated from schools continued to be discussed for many years. Nonetheless, let us repeat, it is characteristic that the change in the situation did not pass unnoticed, including by parties who were far from the teaching of mathematics. In what follows, our main interest will be in mathematical-methodological issues, including the analysis of exam requirements, the system of preparation for them and the associated publications, instructors, and organization, but first we should briefly characterize the general principles of the organization of entrance exams and the procedure of college admission in general.

3.1 General Observations about Admissions to Higher Educational Institutions

At a specific time of year, announcements about admissions to institutes appeared in the country's leading newspapers (the information contained there was usually already known to applicants long before this from numerous publications). Let us quote from one such announcement by the St. Petersburg Technological Institute from 1902:

It is hereby announced by the Emperor Nicholas I Technological Institute that admissions to the Institute will be conducted in the second half of August in accordance with the following rules.

1. The Institute will admit young people who have completed courses of study at higher educational institutions, gymnasia, or real schools with an additional class from the Ministry of National Education, or other secondary educational institutions equivalent to these, with graduation certificate grades attesting to excellent conduct.
2. Those wishing to enroll at the Institute must petition the Director of the Institute, between 15 July and 10 August, while persons of the Jewish persuasion who have completed a course of study at a higher educational institution must apply no later than 1 August. Petitions received after the indicated dates will not be accepted.

   The petition, which should indicate the department in which the applicant wishes to enroll, must include the following documents along with copies on plain paper.

   (a) a graduation certificate or proof of education acquired; (b) a duly authenticated birth certificate; (c) a document indicating the social estate to which the applicant belongs by birth, as well as a St. Petersburg residence permit.

   For members of the tax-paying estate, a leave warrant from their Community [for instance, peasant village community] may serve as a document indicating social estate, wherein the Community must express its consent to release the individual enrolling in the Institute, with the Community's' assumption of all payments due by the released party for his entire term of study at the Institute; (d) proof of assignment to an assigned precinct for compulsory military service; and (e) three photographic (ordinary album) shots, each personally signed by the applicant.

   Independently of the documents listed above, those wishing to enroll in the Institute must present proof of their impeccable conduct from the local police. Only those individuals are released from presenting such a proof who, in June of the current year, have completed a course of study at a secondary educational institution that grants them the right to enter the Institute.
3. Individuals who have completed a course of study at higher educational institutions will be admitted without competitive tests in the subjects of the gymnasium course.
4. All others who wish to be admitted as first-year students must undergo competitive testing in the Russian language, Algebra, Geometry, Trigonometry, and Physics, within the bounds of the

> gymnasium course on August 16, 17, 19, 20 and 21, while those who have been granted permission to enroll in higher courses must undergo testing in subjects from prior courses in the days following August 21, in accordance with the schedule posted in the building of the institute. (Tekhnologicheskii institut 1902).

These rules are rather typical. Let us note at once that passing entrance exams and documents pertaining to the completion of secondary schools were far from the only requirements for enrollment. There were constraints of a political nature. Confirmations of sound character were by no means handed out automatically. For example, here is how Viktorin Fyodorov, the son of a priest living in Uralsk, had to petition the authorities as late as 1912:

> Acknowledging the full worthlessness and inadequacy of my former political convictions, I sincerely repent my revolutionary actions and consider them merely the result of childish fantasy and the influence of the environment; I acknowledge them to have been unquestionably harmful and pernicious. At the present time, I reject all anti-government parties, do not involve myself in any politics, and presume most respectfully to solicit YOUR EXCELLENCY to excuse my revolutionary past and, if it proves possible, not to refuse to send to the St. Petersburg Konstantinovskaya Artillery Academy, the St. Petersburg Nikolayevskaya Engineering Academy, and the St. Petersburg Imperial Institute of Transport Engineers attestations of my political reliability, which are required for me to complete my education, so that in the future I might at least in part atone for my offense against the State and be a useful servant to the TSAR and the Fatherland. (Institut Inzhenerov putey soobscheniya 1917, p. 15).

As it happened, those in charge turned out to be favorably inclined, and while noting that during the previous year (1909) the said Viktorin had been "unreliable," nonetheless noted that he mended his ways since then.

Restrictions on persons of the Jewish persuasions are evident: each year, a determination was made at the Technological Institute as to how many such individuals could be admitted as students. In this connection, technical questions arose, for example, how to calculate the percentage (3%) of persons of the Jewish persuasion that could be admitted, and it would be specifically explained to the directors of the Technological Institute by their higher-ranking superiors that the figure was to be calculated with respect to the total number of students being admitted, and not for each department separately (Tekhnologicheskii institut 1902, p. 4).

We should note that at other institutes, the situation could be even worse: only Russian subjects of the Christian persuasion were admitted to the Electrotechnical Institute, for example. Petitions for admission by a certain Swiss citizen have survived, who was told that in order to be admitted, he had to become a Russian subject (Elektrotekhnicheskii institut 1909a, p. 1); as for Russian subjects of non-Christian persuasions, they were simply not permitted to take the exams.

The rules just cited (and those similar to them) were followed quite rigidly even after the February Revolution of 1917, which overthrew the tsar. An appeal by a certain Borukh Rapoport to the Provisional Government's Minister of Railways N. V. Nekrasov has survived (in this letter, the author expresses his hope that the addressee will respond to his appeal not only in his capacity as minister, but also as an "esteemed electee of the people" who is "sick at heart for the humiliated and the insulted"):

> I am a foreign polytechnic student. Obviously, I studied outside Russia not of my own free will, but due to circumstances established under the old regime, when as a Jew I was denied the possibility of obtaining an education in my homeland. I will not expand on this topic, since I would not be able to tell you, who are a leading fighter for the ideas of Russian civil society, anything new, while what is old is already sufficiently well known to you. I will merely say that at the University of Liège, I, like everyone else, had to undergo mandatory competitive entrance examinations in mathematics (arithmetic, algebra, geometry, linear and spherical trigonometry, analytic and descriptive geometry), drawing, mechanical drawing, as well as the Russian and French languages, history, and geography. I passed all of these subjects and was admitted as a valid student. Unfortunately, the war forced me to interrupt my studies during my second year, and I, who had accidentally happened to be in Austria at the beginning of the war, with great difficulty fled from there through Germany to Belgium and hurried to my homeland in the hopes of making myself as useful to her as is within my powers. (Institut inzhenerov putey soobscheniya 1917, p. 72).

He requested to be permitted to take entrance exams to the Transport Engineers' Institute, since he had obtained a secondary education at home and not at an official institution (while indicating that he was prepared subsequently to take all necessary exams for the secondary school course). His petition was denied because it failed to conform to the rules.

We should note, too, that members of the tax-paying estate (that is, first and foremost, peasants) evidently found themselves in a special position, if only because they were required to present the consent of the Community to which they belonged. There was, finally, yet another consideration, which the same rules for enrollment in the Technological Institute addressed as follows:

> On August 26, the results of the competitive tests for students seeking admission to the first year of studies will be announced. Those who, based on these results, will receive the right to enroll in the first year of studies at the Institute must pay a fee for their education (50 rubles) no later than September 1. Should they fail to do so, their vacancies will be filled by those who have passed the test and wish to enroll in the Institute. On September 2, the results of those seeking admission to the higher-level courses will be announced, and those of the exam takers who will be accepted are required to pay a fee for their education (50 rubles) no later than on September 5, after which they will be enrolled as students at the Institute. (Tekhnologicheskii institut 1902).

The existence of a fee for education was another filter for those wishing to enroll at a higher educational institution. All this does not mean, however, that the role of entrance tests was not great.

3.2 *One Case Study of Entrance Exams*

Before going on to the investigation of entrance exams as a whole, we will cite certain excerpts from people's personal recollections of how exams were perceived and taken. To this end, we will turn to the little known memoirs of Vladimir

Bonch-Bruevich (who at one time was effectively Lenin's secretary) concerning his education (Bonch-Bruevich 1902).

His recollections begin with a short chapter titled, precisely, "The Entrance Exam," which in its turn begins with the following words: "Long before the beginning of fall 1883, one of the main topics of conversation in our family was my upcoming 'competitive exam' to Konstantinovsky Land Surveying Institute" (p. 270). Konstantinovsky Land Surveying Institute, which was located in Moscow and prepared geodetic engineers, as we would now call them, had by that time already ceased being a military institution, but of course differed from the more liberal University.

Before being admitted, Bonch-Bruevich studied at a preparatory boarding school maintained by a certain Simov, who served as senior educator at the Institute, and was therefore preferred by Bonch-Bruevich's parents to self-preparation. Bonch-Bruevich relates how they scared him with exams—"Just you wait—you'll fail, then you'll find out!"—and how around him stories circulated about exam administrators—strict "fathers," a nitpicky 'Frenchman," a diminutive "gopher" mathematics teacher—who found a special form of pleasure in failing students on exams" (p. 271), and about a "frightening cramming of subjects."

But now at last the fateful day of August 23, 1883 arrived, when, taken leave of with kisses and the sign of the cross, Bonch-Bruevich entered the enormous and gloomy hall in which the exams were to take place. There were 50 vacancies and 100 applicants; consequently, one half were destined to fail (Bonch-Bruevich does not miss the opportunity to remark how children's souls were corrupted by such competition, which spurred them to hope that someone else might fail). Finally, 9 o'clock arrived, the bell rang, some boys were crying and clinging to their mothers (who, let us note, like the fathers, remained in the hall). "The pale faces expressed an indescribable fear." The roll was called. Then the directors and exam administrators entered, about 25 persons in all. A prayer service was held; the exam takers prayed with all their hearts. Then they were directed to their seats, given paper and writing implements, and the dictation commenced. From time to time, harsh cries would be heard: "Don't peek at your neighbor's exam!" "No talking!" "I'll send you away from the exam right now!"

"All these words," the sensitive Bonch-Bruevich remarks, "hit us like a cudgel over the head, since almost all of us had sinned and were guilty of this crime" (p. 276). At last, the dictation ended and the notebooks were taken away.

> The exam administrators distributed themselves at different desks. We rose from our desks and crowded in the hallway. We were divided into groups. Some were taken to "arithmetic," others to "religious instruction," others to "the Frenchman and the German," and so on. After having taken an exam with one, we passed on to the next, to the third, and so on, presenting all of our learning with greater or lesser success. (p. 276).

The teachers posed questions, and the students answered them until the teachers said "enough," after which they would be examined in another subject. This, Bonch-Bruevich remarks, continued all day. "Good fortune would sometimes smile, sometimes slip away. The sense of alarm now waned, now waxed. By 4 p.m. I had successfully passed all the exams and took off for home at breakneck speed" (p. 277).

Despite all the horror conjured up by the author (here we cannot fail to remember that during the first years after the revolution, when Lenin, and Bonch-Bruevich along with him, were in power, all exams were canceled), the picture painted is rather idyllic, if only because all of the exams fit into one day. In the future, this was usually not the case.

3.3 General Observations about Entrance Exams

As the Director of the Technological Institute reported (Tekhnologicheskii institut 1901, pp. 18–19), during the 1901–1902 academic year 250 openings were announced for the first-year course of the Institute. The number of requests received from students wishing to be permitted to take competitive entrance tests was 1143. The number of people who showed up was 1007. Scores in the Russian language, Algebra, Geometry, Trigonometry, and Physics were given on a five-point system, but fractional scores were also allowed. What total scores were received and by what number of test-takers may be seen from the following table (the rest scored lower) (Table 1).

These 237 people were admitted as students. Another 13 people were admitted to the first-year of studies from among those who had completed a course of study at other higher educational institutions and who had decided to obtain this education as well (for example, university graduates).

In addition to this, the Institute petitioned to be allowed to admit those who had received the full score of 5 in four of the specialized subjects, but reached a total score of only 23 ¼ because they had received only 3¼ in the Russian language. There were 25 such people (an increase in the number of admitted students by 10% was considered in principle possible). And on top of this, the Institute submitted a special petition to be permitted to admit the nephew of the priest of the church of the Institute non-competitively, who had received a total score of only 22 ½; and without any special petition at all the heads of the institute were informed of a request by Grand Duke Konstantin Konstantinovich that a certain Zhukov, who had received a total score of 23, be admitted non-competitively (p. 35).

For the sake of comparison, we can note that in 1904, the Institute reported that the number of students who had submitted requests to be admitted to exams was 1182 for the mechanics department and 217 for the chemistry department (Tekhnologicheskii institut 1904). Subsequently, 1031 people went through competitive examinations at the mechanics department, and 186 at the chemistry

Table 1 Scores received by the best test-takers

Total score	25	24 ¾	24 ½	24 ¼	24	23 ¾	23 ½
Number of test-takers	11	10	20	22	62	46	66

department. The number of applicants who had successfully completed their exams was 627 for the mechanics department and 116 for the chemistry department. There were 175 openings at the mechanics department and 55 at the chemistry department. As a result, all those who had received a total score of 23¼ or higher, 178 people in all, were admitted to the mechanics department; and all those who had received a total score of 22 or higher, 57 people in all, were admitted to the chemistry department (in reality, this was not quite the end of the procedure—some of those admitted failed to pay the fee and their places were given to previously determined "candidates," as they were called, so that it still took a relatively long time for the list of enrolled students to become finalized).

Also, for the sake of comparison, we can note that in 1909, at the Electrotechnical Institute, according to the Institute's report to the Minister, matters stood as follows (Elektrotekhnicheskii institut 1909b, p. 15): There were 80 places available. 576 applications were submitted; 535 people were admitted to take exams; 446 actually started taking the exams; 200 fully completed the exams. The passing score was set at 16¾ (three students with lower scores were also admitted). Based on the total scores, students who had reached such a score were distributed as follows (it should be kept in mind that four exam grades were given) (Table 2).

The distribution by score of the others who had fully completed the exams was as follows (p. 21) (Table 3).

It should be noted that while at the Technological Institute, those who had received an unsatisfactory grade were not allowed to take the next exams, at the

Table 2 Scores received by the best test-takers

Total score	20	19¾	19½	19¼	19	18¾	18½	18¼	18	17¾	17½	17¼	17	16¾
Number of test-takers	2	1	1	1	3	1	5	8	8	5	12	15	8	9

Table 3 The distribution by score of the others

Total score	16½	16¼	16	15¾	15½	15¼	15	14 ¾	14½	14¼	14	13¾	13½	13¼	13
Number of test-takers	10	7	13	6	7	7	6	6	4	4	11	10	4	3	1
Total score	12 ¾	12½	12	11¾	11½	11 ¼	10¾	10½	10	8½					
Number of test-takers	5	4	2	2	4	1	1	1	1	1					

Electrotechnical Institute, they could continue to take exams. As we see, competition to be admitted to the Institutes was intense and the grading was quite exacting.

G. K. Shmulevich's (1915) handbook offers the following estimate:

> The total number of people who take exams to be admitted to St. Petersburg's institutes attains 1900–2000 every year, rarely more than that; 1000–1100 are admitted, and 800–1000 are left overboard, that is 40–50% of those who took the exams (p. 6).

At first glance, the estimate seems not to accord with the numbers that we cited above for only two institutes, while the total number of institutes who admitted students based on exams in St. Petersburg at that time was six—in addition to the Technological and the Electrotechnical, there were also the Mining, Transport Engineers', Forestry, and Civil Engineers' Institutes. The explanation for this is that many students were applying to several Institutes simultaneously (which in part also explains the long time it took for the list of enrolled students to be finalized—a person could be admitted to two places at once; not enroll in one of them, obviously; and his place would become vacant again, and so on). Shmulevich in fact writes that "it would not be difficult at all to combine together the exams to two-three institutes" (p. 14). Still, to see the competition "as a whole" as involving two people for each available place, as Shmulevich does, is hardly justifiable (he himself decisively remarks that to consider the scores based on each higher educational institution individually is "not statistics, but simply a joke" (p. 5)).

About the institutes in Moscow, Shmulevich (1911, 1915) writes with the pride of a St. Petersburg resident that they are much easier to get admitted to than even not very demanding institutes in St. Petersburg:

> The exams are conducted in a completely unrigorous manner, there is usually a decent number of openings, and only 800-900 people take the exams (for a total number of 600-700 openings at four Institutes); in this way, only a small fraction of those who take the exams are left overboard (Shmulevich 1911, p. 64).

Competitive examinations were also conducted at colleges in other cities.

The actual examination procedure was described at the Technological Institute as follows (Tekhnologicheskii institut 1902). Those being examined were divided into three cohorts, and the exams began at 9 a.m. and continued until they ended without a break. Grades were announced daily after the end of each day's exam. All of the exams were oral, except the Russian language exam, for which therefore exam takers were specifically told that they needed to bring writing implements. For the oral exam, exam takers were specifically told that:

> Those taking the exam will be called on to present answers in the same order in which their names appear on the list, and they must, presenting the examiner with their entrance card, immediately come forward for an answer; those failing to do so the third time they are called will not be examined. ...
>
> Those who appear for the exam will be seated in the corresponding classrooms, which they are not prohibited from leaving and returning to in the event of necessity, but they must make no noise in doing so.
>
> Silence must be observed in the classrooms. Those taking the exam must approach the desk and the blackboard only when they are called (p. 6).

At the Electrotechnical institute, exams were conducted somewhat differently. There were three written exams: in geometry with trigonometry, in algebra, and in the Russian language (composition). In addition, there were three oral exams: in geometry with trigonometry, in algebra with theoretical arithmetic, and in physics. One grade was given in algebra with arithmetic (averaging the results of the oral and written exams), one grade was given in geometry with trigonometry (again averaging the oral and written exams), and one each in physics and in the Russian language (Elektrotekhnicheskii Institut 1914–1915).

Schukin's (1915) handbook offers examples of "cards," that is, problem sets that were assigned on oral exams at the Institute of Transport Engineers. One of these is a card with the following four problems (p. 15):

1. Derive the formulas. sin $(a + b)$, cos $(a + b)$. Give generalizations.
2. Solve the equation $1 - \cos x = \sin x \cdot (1 - \cos x)$.
3. Find the period of the function $y = 3\sin\frac{x}{3} - \cos 5x$.
4. Solve a triangle based on two sides and the angle between them (with numerical givens).

The same handbook explains that the exam taker solves problems on paper, and once he has completed them, approaches the examiner, who checks his work and concurrently poses simple (as Schukin stresses) questions. This, of course, was not always and everywhere the case.

Among those invited to administer the exams were, on the one hand, institute employees, and on the other hand, teachers from secondary education institutions. Thus, for example, among those administering the exams at the Technological Institute in 1904 we see Boris Koyalovich (Tekhnologicheskii institut 1904), a famous professor who subsequently became a member of the Russian subcommission of the International Commission on Mathematics Education (Karp 2019). Those invited to serve as exam administrators were usually sent invitations to this effect, to which they also responded in writing, for example, as follows:

> I have the honor of informing you that I gratefully accept your invitation to be an examiner in trigonometry for the entrance tests to the Mining Institute this year. (Gornyi institut 1914–1915, p. 10)

The number of exam administrators oscillated—for example, in 1904 at the Technological Institute, the exam in algebra was administered by four people, while the exam in geometry was administered by five.

Naturally, the examiners were recompensed for their work. For example, in 1912 at the Electrotechnical Institute, 200 rubles were paid for checking 500 exams in geometry with trigonometry, 180 rubles were paid for checking 450 exams in algebra, and 210 rubles were paid for checking 350 exams in the Russian language (it may be supposed that it was assumed that not everyone would make it to the exam in the Russian language). For monitoring the written exam in geometry with trigonometry, algebra, and the Russian language, 90 rubles, 70 rubles, and 60 rubles were apportioned, respectively. For each oral exam, a payment of 300 rubles were apportioned (Elektrotekhnicheskii Institut 1912, p. 47).

It is not always easy to calculate the exact number of people who administered the exams and how much they worked, but sometimes the amounts received by each examiner are directly indicated. For example, the invitations sent out by the Mining Institute immediately indicated that the examiner in algebra and arithmetic would be recompensed 130 rubles for his work (Gornyi institut 1914–1915, p. 23). The amount could also be greater or smaller. For example, the director of the Technological Institute informed his superiors that "To conduct competitive testing this year at the Institute of which I have been placed in charge, State Counsellor Kon, a teacher from St. Petersburg's Tenth Gymnasium, has been invited [to administer] exams in the Russian language, and Titular Counsellor Adamovich and Collegiate Secretary Ivanov have been invited to assist the Institute's instructors to conduct exams in Geometry." The director went on to request permission to disburse a payment of 250 rubles to Kon, and of 50 rubles each to Adamovich and Ivanov (Tekhnologicheskii institut 1901, p. 37).

Without attempting to assess the purchasing power of the ruble at that time, we will merely note, for the sake of comparison, that in the same year of 1901, Andrey Kiselev, the famous author of textbooks, retired from teaching with an annual pension of 743 rubles and 43 kopeks (Karp 2002). Consequently, we may suppose that teachers were quite interested in receiving, for example, 130 rubles for a few days of exams, however difficult they might be.

We can only guess how long the exam in mathematics lasted for each applicant and how thoroughly each written test was checked—naturally, this could by no means have been the same everywhere and at all times. However, comparing the number of exam days (for example, at the Technological Institute, there were five exams days for all subjects with three cohorts of students) with the number of exam takers and exam administrators, we can suppose that the time for direct contact between the exam administrator and the exam taker (without counting preparation for an answer) was brief. Shmulevich (1915, p. 14), who has already been quoted above, explained that an oral exam takes up one to two hours, while they continue until 5–6 p.m. (although in some institutes only until 1–2 p.m.); from this, he concluded that it was possible to make it to several places in one day. He also writes about cases in which examiners checked problems by looking only at the answers, without counting correct solutions because they themselves had made a mistake (p. 7), which shows that examiners by no means always checked the solutions fully. But here we must discuss the question of the difficulty of the exams in mathematics at different places and the preparation for them.

3.4 On the Distinctive Characteristics of Entrance Exams and on Preparing for Them

The number of those enrolling in higher educational institutions across the country, as we have already said, was in the thousands, not even in the tens of thousands; nonetheless, this was no longer a situation in which everything could be handled on a case-by-case basis: general problems arose and so did general recommendations. The same Shmulevich (1915) praises exam administrators, on the whole, writing that they strive to select the best, even if there are also among them those who make no effort or are excessively harsh, and so on; but the exams themselves differ from institute to institute, and not even so much in terms of their form—oral exams everywhere involved both answering a theoretical question and solving given problems—as in terms of their difficulty and the importance of their parts. For example, he writes about the Mining Institute that nowhere else are things as serious and rigorous as there. At the same time,

> The requirements in theory do not differ from the requirements at other institutes, except only perhaps more attention than anywhere else is paid to the theory of trigonometry. The problems that are given are quite difficult, sometimes requiring considerable acumen, or else extensive experience, but the main thing is that usually a successful answer in theory receives a score of 3 points, which is in fact quite enough, since in recent years at the Mining Institute there has been effectively no competition, and everyone who did not receive a single score of 2 points is accepted (p. 17).

The theoretical issues are assessed there quite rigorously (p. 18).

At the Transport Engineers' Institute, according to Shmulevich, the problems are very difficult—so difficult that "most of them will not be solved by anyone who has not prepared specifically for the Transport Engineers' Institute and thus has not solved problems of this kind. And the inability to solve a problem often leads to a grade of 2" (p. 19). The problems at the Electrotechnical Institute supposedly also have the defect of being excessively difficult, and furthermore their level of difficulty is not uniform in all cases—the exam taker might happen to pick a quite simple question or an extremely difficult one. Exam takers there are given three problems to complete in three hours on their written exams, and for a grade of 3 it is usually enough to solve two of them. At the Technological Institute, in Shmulevich's opinion, exams in mathematics are not very difficult, and few grades of "2" are given, but in order to be admitted, the grades exam takers receive on their Russian language exams turn out to be important. At the Forestry Institute and the Institute of Civil Engineers, according to Shmulevich, the questions are more rigorous than at the Technological Institute, but the role of the exam in mathematics is not that important; at the Institute of Civil Engineers, for example, after the first exam—in drawing—not many more applicants are left than there are open places.

Each Institute had many other distinctive characteristics, including not only the distribution of levels of difficulty, but also the types of problems. At some Institutes, special attention was paid to the course in arithmetic (today, it would probably be more accurate to call this course elementary number theory); in others, there was a particular focus on, say, the applications of trigonometry to geometry. Speaking of all these distinctive characteristics (note that there was also a relatively large number of Institutes around the country that Shmulevich does not mention, each with its own distinctive characteristics), Shmulevich pushes the reader toward the idea that special preparation was required for each Institute in particular (even though the programs of the exams were formally practically identical).

There were quite many exam preparation courses of all different kinds at the time—various publications from such institutions have survived (Kursy inzhenera Dmitrieva 1913; Schukin 1914). The business was quite profitable: for example, the book Almoev 1915b includes a printed advertisement about Almoev's courses, from which we learn that these courses cost 200 rubles per person. Among those who taught the courses were many who presented themselves as engineers (that is, as people who had themselves at one time made it through entrance exams): engineer Schukin's courses, engineer Dmitriev's course, and so on. Petr Shmulevich, whose brother was quoted above, and who will be discussed below, as perhaps the most prominent figure in the exam preparation industry, was also an engineer. But ordinary teachers and people with a university education, such as Grigory Shmulevich, labored in this industry as well, of course. Grigory Shmulevich himself writes about the usefulness of lessons with a tutor quite diplomatically, beginning with the acknowledgement that students can also prepare quite well on their own, but immediately pointing out that according to his observations (how they were obtained is not quite clear), out of 100 who prepare on their own, 30–40 end up being accepted, compared to 60–80 of those who prepare with a teacher, and in rare cases 90–96 of those who prepare with an especially good teacher (p. 10). This naturally gives him an occasion to bring dishonest teachers out into the open:

> It should also be taken into account that preparation for competitive exams is often offered by persons who have no relation whatsoever to mathematics and its teaching. (Naturally, not every engineer, let alone every student, knows mathematics and is able to teach it, even though he did pass competitive exams at one time.) Such individuals have given competitive exam preparation the sad reputation of "drilling and cramming," since having learned only those questions which are offered on exams more frequently than others, they exercise their students in them almost exclusively. At times their pupils perform quite successfully, but more often if the type of problems given or the character of the theoretical questions posed is even slightly changed, their pupils fail in large numbers (Shmulevich 1915, p. 10).

Shmulevich even tells about a certain cunning student who had announced that he would take money (200 rubles) only from those who are accepted. Twelve hapless applicants fell for this bait. Three of them were accepted, which gave the enterprising student 600 rubles, quite a sizable sum.

Shmulevich also reports that certain undiscriminating applicants enroll in the university, which can be done without exams, in the hope that mathematics classes there will prepare them for enrolling at the engineering institute of their choice in the following year. In reality, in Shmulevich's opinion, the year is lost uselessly, since the study of higher mathematics does not prepare students to pass exams in school mathematics.

This leads him to a natural transition to the remark that there are also (even though they are a minority) conscientious teachers, who aim at mathematical development, rather than drilling and cramming. It may be supposed that the author has in mind his brother and himself.

3.5 Petr Shmulevich, the Biography of a Teacher

Before the revolution of 1917, Petr Shmulevich was probably the most famous teacher who prepared students for competitive exams. His problem books enjoyed wide success, as did his course. His name became proverbial, and already much later, at the height of the anti-Semitic "anti-cosmopolite campaign," writers wishing to smear Mark Vygodsky, a well-known mathematician and historian of mathematics, and a prominent author of a handbook for college applicants that had come out at the time, compared him with Shmulevich, likening his book to the "notorious pre-revolutionary collections by Shmulevich, who turned the publication of competitive problems into a commerical enterprise" (Karp 2018).

In a questionnaire from the Soviet period (Shmulevich 1925–1926) Shmulevich indicated that he had been born on December 26, 1873 in the city of Cherkassy in Kiev Gubernia in the family of Samuel Shmulevich, a pharmacist. The future author of handbooks began to work at the age of 15 (giving lessons). He graduated from a gymnasium in Samara in 1891, and already before this, on November 9, 1890, he had taken an important step: he had been baptized, acquiring a new patronymic in honor of his godparent, a teacher at the Samara classical gymnasium by the name of Kronid Fedorovich (note that this same teacher would later facilitate the baptism of Petr's brothers, Grigory and Mark, each of whom also became Kronidovich).

Baptism, of course, made the possibilities of finding a place in the world considerably easier, but did not in itself automatically bring any money. In 1893, Shmulevich enrolled in the St. Petersburg Institute of Transport Engineers, but he had nothing to live on—during his second year, he was expelled for nonpayment. Several years later, he submitted a request to be reinstated at the Institute, having saved up money, as he explained in his petition. The request was granted and Shmulevich was reinstated at the Institute and graduated from it. In the same autobiography, he wrote as follows: "in 1895, published a problem book in mathematics, having collected materials for it from problems that had been offered on entrance exams in various institutes, and after that rapidly began to prosper and to earn a great deal of money."

Shmulevich's enterprise was set up on a large scale. The sessions with students attending "The Courses of Engineer P. Shmulevich" were taught not only (and maybe not mainly) by Petr Kronidovich himself, but by other individuals, first and foremost his brother Grigory, who had successfully graduated from St. Petersburg University. In addition to classes in St. Petersburg, there was also the option of classes at a boarding house outside the city—Shmulevich had acquired a property in the vicinity of the Siverskaya railway station and had built there, as we would now say, a mathematics camp, which included buildings for classes, a dining hall, a kitchen, and residential buildings.

It may be supposed, however, that most of the revenue still came from the publishing house that he founded. P. K. Shmuleivch annually published reference manuals for college applicants, and in addition his collections of problems were regularly reprinted—in arithmetic, algebra, trigonometry, and geometry. Nor were nonmathematical subjects forgotten, for example, manuals preparing students to write compositions on given topics were published (Karp 2018).

In the same autobiography (Shmulevich 1925–1926) he writes: "This went on until 1912, when, without interrupting my prior activities, I became director of the Tsementon Paint Factory, where I had been invited to order and organize production."

With the start of the war, things took a sharp turn for the worse—demand for services in preparing students for exams obviously dropped, and the paint business probably did not fare too well either, if only because of the expensiveness of the raw materials. Shmulevich began to look for other work, remembering his education in engineering. After the revolution of 1917, things obviously got much worse still—without listing all of his attempts to adapt to the new life, let us merely note that he died in Leningrad (St. Petersburg) during the blockade in 1943. His books, however, clearly continued to live—sometimes new editions of them appeared, revised in one way or another, but above all, the authors of new books and exam administrators clearly drew on his books for problems (Karp 2018).

3.6 Critical Remarks About Competitive Exams

Below, certain books by Petr Shmulevich will be analyzed both from the point of view of their methodological construction and from the point of view of the selection of problems in them, which reflected what appeared on exams (at least in St. Petersburg). But first we will turn to a criticism of his work, examining a book by Yakimov (1901) entitled *The Thoughts of a Provincial Pedagogue About Competitive Exams in Mathematics at Higher Educational Institutions,* which bears the subtitle *On Two Collections by P. Shmulevich Containing Problems for Such Exams.*

The position of Yakimov (a provincial teacher, apparently, from Kazan) is simple—on the very first page, he poses the question: why do many young people fail competitive exams? And he offers two possible answers: either their schools do not provide sufficient knowledge, or the competitive exams are no good. As one can immediately guess, he himself considers the second answer to be correct. After complaining about the difficulties of working at a school, he immediately turns his attention to the differences between the curricula of entrance exams and secondary schools, which are expressed first and foremost in the fact that exams require enormous skill at problem solving, which schools cannot possibly provide. In addition, not infrequently exam problems refer to topics that are not studied or practically not studied in schools (appearing in textbooks in fine print). He lists (as examples!)

twenty such topics, beginning with various numeral systems and certain elements of number theory, including solving various higher degree equations, the theory of complex expressions (studied, according to him, only in real schools, but not in gymnasia), solving trigonometric equations (which even in real schools are covered only minimally), and much else, ending with inverse trigonometric functions.

Then he goes on to discuss the fact that the problems offered on competitive exams "in terms of their level of difficulty and the artificiality of the techniques required to solve them would be absolutely unthinkable in secondary schools!" (Yakimov 1901, p. 8). In support of his point, the author reports that examiners and the authors of problems themselves make mistakes (here, however, it is not entirely clear whether he is really talking about examiners or about the authors of problem books for exam preparation). To support this thesis, Yakimov cites a problem that requires students to prove that the number $p^2 - 1$, where p is a prime and $p > 3$, is divisible by 12 (actually, this number is divisible by 24, which, however, does not negate its divisibility by 12).

From general considerations, the author passes to Shmulevich's handbooks, noting that these are the most widespread handbooks, which sell out so quickly, despite their price of four rubles, that only a year after the first edition a second edition has appeared, while the handbook on trigonometry has already come out (by the time Yakimov's critical remarks were published) in a third edition. The author quotes Shmulevich himself, who claims that the handbooks contain genuine problems from exams and, what is especially important, problems that continue to appear on exams, so that of the multitude of problems given on exams to 24 applicants who had studied under Shmulevich, only one was unfamiliar to them. From this, the critic draws the conclusion that students buy up the handbooks specifically in order to learn the solutions that appear in them. Following this, he moves on to what in fact is the main content of his book: the analysis of the solutions that appear in the handbooks, as well as of the problems themselves. His conclusions may be summarized as follows.

1. Relatively many problems are solved incorrectly. Yakimov presents examples that show that Shmulevich had indeed rushed and did not always think about what he was writing. Thus, in solving the equation $\sec x + \operatorname{cosec} x = 0.5$ he remarks: "clearly, the problem is impossible. The sum of two values, each of which is >1. Cannot be <1" (p. 19). But it is evident that both the secant and the cosecant can be negative, and the equation does, of course, have solutions.
2. In a relatively large number of cases, the problems are solved in a way that is not the best way to solve them, while other ways exist. For example, the following example is cited: define the number k such that the equation $3x^2 - kx + 18 = 0$ has 1 as a root. Shmulevich makes use of the properties of roots: their product is equal to 6, and their sum must be equal to $\frac{k}{3}$. Meanwhile, one can simply plug the number 1 into the equation and find the desired number from the equation $3 - k + 18 = 0$ (p. 13).

3. Most interesting of all, however, are not attacks against Shmulevich, but general remarks about the problems in the collection (that is, the problems given on exams). Yakimov writes as follows:

> Some of the problems are so easy they can be solved in one's head without making use of written computations; others are, conversely, so difficult that they would probably exceed not only my strength, but also the strength of riper minds than mine (p. 26).

In support, Yakimov cites several problems that are indeed capable of causing difficulties. We will confine ourselves to two examples. The first example: Solve the equations (today we would say, solve the system of equations) $x^2 - yz = a, y^2 - xz = b, z^2 - xy = c$. (p. 27). Decades later this problem could be found in Krechmar's famous problem book (Krechmar 1972, p. 48), devoted to difficult problems. The second example: Solve the equations $x^7 - y^7 = 2186, x - y = 2$ (Yakimov 1901, p. 27). The same problem in a simplified form—with a fifth-degree rather than a seventh-degree equation—is offered in a problem book for pedagogical colleges (Lyapin et al. 1973, p. 141), likewise published many years later.

The book concludes with the following impassioned paragraph:

> That time must still be far when the inalienable right of a young man to a higher education will not be bought at a price too dear for him; that hour will not soon strike when everything that is not included in the secondary school curriculum will be struck from the curriculum of competitive tests, and when funeral hymns will be sung over all problems whose solutions require extremely artificial techniques. (Yakimov 1901, p. 29).

3.7 *On Petr Shmulevich's Problem Books*

Shmulevich published many different handbooks, and even those with the same title came out in different editions that were not fully identical. Below, we will focus on three publications.

3.7.1 Problem Book in Trigonometry (Shmulevich 1897)

Shmulevich published a problem book in trigonometry first (in 1895). In the preface to the next edition, which is the one we will be discussing, Shmulevich noted that "despite its major shortcomings and extreme costliness," one thousand copies were sold in eight months (quite a large quantity for the time). In the new edition, he added many new problems, thanking in the same preface those who had sent them from such places as Moscow and Kharkov.

The book consists of the following sections. Theoretical information comes first: the author discusses the introduction of the radian measure of an angle, general considerations about angles whose sinuses or tangents are equal, and so forth. Then come two sections devoted to problems: the first is about trigonometric relations and equations, while the second is about the solution of triangles. The book concludes with a third section: the solutions to the problems given.

It should be remarked at once that the terminology as well as the system of explanation used by the author often differ from those accepted today. Below, we will ignore this, as far as possible.

Also, let us note at once the high level of skill requirements in computation and algebraic transformation in general presupposed by the problems in the book. A typical (and comparatively easy) problem involving transformations looks as follows (№ 111): "Prove that $\frac{\sin^2 2\alpha - 4\sin^2\alpha}{\sin^2 2\alpha - 4 + 4\sin^2\alpha} = tg^4\alpha$ " (p. 32); but there are also problems such as # 128: "Prove that $\cos\frac{\pi}{5} - \cos\frac{2\pi}{5} = \frac{1}{2}$ " (p. 32), which in subsequent decades would appear in numerous handbooks as a "challenge" problem, and which indeed usually involves using an artificial technique (one can also compute the respective cosines, of course, but that hardly makes the task any easier—in a practically analogous problem in Almoev's handbook (1915b), which will be discussed below, the respective values of the trigonometric functions are simply given, apparently on the assumption that students already know them from somewhere). Different types of equations are represented: there are equations with inverse trigonometric functions, equations that combine solving exponential and trigonometric equations. For example, in problem № 315, the following equation is given: $\arcsin x + \arcsin\frac{x}{\sqrt{3}} = \frac{\pi}{2}$ (p. 40). There are, however, also problems that are much easier, for example, the section devoted to equations opens with № 250 $\sin x = 3\cos x$ (p. 38).

In the section devoted to the solution of triangles, quite varied problems are offered, which require students to find certain facts about a triangle based on certain other given facts about it (today, the problem of the solution of triangles is understood in a more narrow fashion). As an example, let us cite one of the first problems from this section (№ 471), in which, knowing that the angle at vertex A is a right angle, and that the sum of the length of the side opposite vertex A and the height of the triangle drawn from vertex A is equal to m, and also knowing the value of angle C, students are asked to find the area of the triangle (p. 47). Unquestionably more difficult is problem № 605, which is among those that conclude the section: in this problem, given all three heights of a triangle, students are asked to find the radius of the circle circumscribed about it. Without going into a discussion of the usefulness of such problems (as far as can be judged, they have practically universally fallen out of use), we should recognize that their solution presupposes quite solid knowledge and confident command of the school techniques of computation and transformation.

3.7.2 Problem Book in Algebra and Arithmetic (Shmulevich 1900)

The problem book begins with an introduction which informs readers that it is based on materials from competitive exams at St. Petersburg's institutes, that it contains 12 sections and over 1200 problems. Then the author explains that it is decidedly impossible to do all of these problems, and therefore in each section he highlights the most typical problems (15–25% of the total number). Then comes a clarification:

> Undoubtedly, in light of the particular typicality and difficulty of these problems, the variety of their content and the greater or lesser artificiality of the ways of solving them, a closer acquaintance with them can provide fundamental skills and develop the ability to use different types of artificial techniques, which is indeed the purpose of my book (pp. I–II).

Readers are then told that, although the author had planned to open the book with an exposition of those theoretical topics which are included in the curricula of entrance exams and not included in the curricula of secondary schools, because after the Institutes' new curricula came out there turned out to be too many such topics, he did not do so, planning instead a separate edition of additional sections in algebra.

In conclusion, the author criticizes other similar problem books for either being photographic reproductions of his books, or for offering problems that have no relation to competitive exams. Also, Shmulevich doesn't miss to inform readers that at a certain institute, examiners draw problems directly from his problem book.

The first section of this book is somewhat unexpectedly devoted not to arithmetic (the latter is given only in the third section), but to the roots of equations, usually quadratic ones. This section contains various problems on the relations between the roots of an equation and its coefficients, on finding certain parameters when the root is known (such a problem was cited above), and so on. Then follow problems about arithmetic and geometric progressions; sufficiently typical is problem № 105, in which students are asked to find four numbers that constitute a geometric progression, given their sum and the sum of their squares (p. 12).

The section on arithmetic includes problems on numeral systems, on divisibility, and similar topics. As examples, we can cite № 17, which asks students to formulate a rule for divisibility by 37 in a numeral system with base 1000 (p. 17), and № 222: "Prove that the product of four consecutive numbers increased by one is always a perfect square" (p. 20).

It should also be noted that the problems turn out to be easier to solve sequentially than out of context: they are organized in groups based on similarities between them or the techniques involving in solving them. Initially, relatively easy problems are given. For example, Section IV, devoted to transformations, opens with the following problem: Eliminate irrationality in the denominator $\frac{10}{\sqrt{3}+\sqrt{2}}$ (p. 21). And only after this come problems which also require students to eliminate irrationality in the denominator, but in which there are no longer just two square roots, but three or four, or in which the denominator contains both a square and a cube root, and so on.

Since it is not possible for us to discuss the content of all of the sections here, we will merely say that the author achieves a genuine breadth in Sections V and VI, which are devoted to equations and systems of equations (we cannot refrain from the hypothesis that such problems were convenient for oral exams, if only because they are easy to check based on their answers). These sections contain problems of the most varied kinds: with irrational and rational equations, equations in which a new variable must be introduced, and problems in which a polynomial must be broken up into factors, or rational roots must be found. The very first equation in the section (№ 281) is the following: $x^5 + 4x = 5x^3$, and it can indeed be solved "without using written calculations," as Shmulevich's critic Yakimov averred, except that doing so also requires a considerable level of mathematical culture.

Then follow sections devoted to exponential and logarithmic equations and systems of equations and transformations of expressions with exponents and logarithms. In our view, these sections contain substantially fewer technically difficult and artificial problems than the preceding ones. As an example of a problem from these sections, let us cite № 791, which asks students to solve the following system of equations: $2^x \cdot 3^y = 12$, $2^y \cdot 3^x = 18$, (p. 47). The equations can be divided by one another to obtain the equation $\left(\frac{2}{3}\right)^{x-y} = \frac{2}{3}$, from which it is already not difficult to express x in terms of y and so on.

Then follow problems on combinations, the binomial theorem, and continued fractions, and then a section devoted to proofs. The proofs in this section are of the most varied kinds: there suddenly appears a proof (№992) of the inequality between the arithmetic and geometric means of two positive numbers (one of the very few problems about inequalities in the book); but there are also such classic problems involving the transformation of expressions as, say, № 972: Prove that $\sqrt[3]{20+14\sqrt{2}} + \sqrt[3]{20-14\sqrt{2}} = 4$ (p. 58).

The very last section in the book contains problems which the author could not place in other sections, including problems from more recent exams, which he had no time to insert anywhere else. Here, we can see assignments that can be easily imagined as questions on oral exams, but which it is difficult to call "problems" (for example, №1195: What is a principal square root? (p. 73)). Solutions to all problems in all sections are provided.

3.7.3 Problem Book in Geometry (Shmulevich 1903)

Shmulevich writes in his introduction: "I must call students' attention to the fact that in all of the Institutes, the most important role on the exam in Geometry is played by knowledge of theory, which thus deserves especial attention" (p. IV). Consequently, the role of problems is necessarily smaller, and the problems themselves from today's point of view possess a special character—enormous attention is devoted to straightedge and compass construction problems.

They are the subject of a theoretical introduction in which various methods for solving such problems are discussed. Then comes Chapter I, devoted to straightedge and compass construction problems involving proofs that are given at the Transport Engineers' Institute. Fifty problems are given in all (some of them contain separate parts, so in reality the number of problems is greater), with the majority of them being straightedge and compass construction problems, solutions to which are immediately provided, which contain four parts: analysis, construction, proof, and investigation. The very first problem is quite substantive: students are asked to construct a triangle given a side, the adjacent angle, and the sum of the other two sides (pp. 11–12). In order to do so, they must first construct a triangle one of whose sides is equal to the given sum of the sides of the desired triangle. This problem is followed by several others whose solutions require making use of similar ideas (the author calls the method applied in these cases the "straightening method"). Then come a group of problems that make use of a different idea: the construction of "a part of the desired object"—for example, problem № 7 asks students to construct a triangle given its three medians (p. 20). The author indicates each time in a subheading what methods are used. The problems continue in a similar fashion, and as the cited examples already make clear, the problems given are quite challenging (although, to be sure, many of them are classical problems, that is, widely known).

We will proceed at once to the last problem in the chapter—one of the very few involving a proof (without a construction). In this problem, students are asked to prove that if the three sides of one triangle are equal to the three medians of another triangle, then the area of the first triangle stands in a relation of ¾ to the area of the second triangle (p. 70). Here, as is usual in such problems, the students must extend a median, obtaining a parallelogram, and analyze the areas of the resulting triangles. The problem can hardly be considered unsolvably difficult, but it too unquestionably presupposes a high level of preparation.

The next chapter contains straightedge and compass construction problems given at the Technological and Electrotechnical Institutes. Here, the problems are different, and the author's approach is different as well. The chapter's first problem is as follows: Divide a given segment in half. And the author simply indicates the relevant passage in Kiselev's standard textbook (p. 78). He goes on to proceed in a similar fashion, in relatively rare cases giving his own solution, and also sometimes confining himself to the remark that the method of construction is obvious (for example, №№ 6–8, p. 78). Indeed, the chapter for the most part contains problems studied in school; thus, the issue here is so much not problem solving as theoretical knowledge.

Chapter III is devoted to computation problems given at various institutes. The overwhelming majority of these problems turn out to be far less sophisticated than the preceding ones. For example, students are asked, given three sides of a triangle, to determine whether it is acute, to find its area, its heights or medians, the radii of circles inscribed in it or circumscribed about it, and so on. In other words, they are given problems in which they are indeed asked to compute something, but the algorithms for doing it are clearly not unknown to them.

This section includes not only problems in plane geometry, but also problems in solid geometry. These problems, however, are also fully in line with standard problems from the school textbook: for example, problem № 77 asks students to find the volume and lateral surface of a regular hexagonal pyramid given its height and the angle between its height and apothem (p. 94). Problem № 82 asks students to find the distance from the apex of a pyramid to a plane that is parallel to its base and divides the volume of the pyramid in half (p. 94). In problem № 92, the lateral surface of a cone and its height are given, and students are asked to find the lateral surface of a cylinder that has the same base and height as the cone (p. 95).

Finally, the book provides answers and solutions to the problems in Chapter III and contains a supplement on constructing the roots of a quadratic equation.

3.7.4 Some Conclusions About Shmulevich's Problem Books

As has already been said, the problem books examined above make it possible for us to analyze problems given on exams, but this will be discussed below, when we will examine several other similar sources. Here, we will merely comment on how the books are written, without reproaching the author for conscientiously reflecting the demands that existed at the time, even if we do not like them (as is done in large measure by Yakimov, cited above).

As we have seen, Shmulevich is quite exact in reproducing problems that were given at various Institutes and subdividing them into groups according to the requirements of various higher educational institutions. Moreover, the problems are organized methodologically—their subject matter and the methods used for solving them are taken into account, as is the problems' level of difficulty—and we can observe an adherence, even if it is not always absolutely rigorously maintained, to the principle of moving from the simple to the complex. For us, it is not always clear why sections appear in one or another order; why certain sections are missing (for example, why there are so few inequalities); sometimes the terminology or the approach to the presentation of the material are not especially convenient; but it should be borne in mind that the problem book reflected the style and content of education customary at that time.

In any case, it should be acknowledged that Shmulevich's problem books reflected the high level of working with problems that had evolved in Russia, which also produced other famous problem books, which were better preserved in the country's memory because they continued to be used for many years after the revolution of 1917 (such as Rybkin's problem book (1903), which remained officially in use even into the 1970s, and continued to be used unofficially far later still). The paper Karp (2015) discusses what might be called a methodological revolution that took place in Russia during the period between 1880 and 1900, which led to a sharp growth in the attention paid to problem solving in schools. It may be supposed that one of its causes was the importance of problems on competitive exams. Shmulevich's problem books were undoubtedly a part of the observed changes.

All of this, of course, does not negate any critical remarks about various solutions offered in the books. Indeed, Shmulevich himself writes explicitly about the shortcomings of various of his books (without omitting to emphasize, however, that his competitors' books are far worse). From today's point of view, many characteristics of the organization of the books can be criticized as well: today, what is most typical and what is most important would be highlighted more explicitly; but, to repeat, the books should be judged by the standards of the time—it is natural that, over the years, the principles and traditions laid then as foundations should have borne such fruit as was unthinkable back then.

3.8 On Certain Other Problem Books for Applicants to Higher Educational Institutions

Petr Shmulevich, as has already been said, clearly played a special role in the development of the exam preparation industry, but he was, of course, no monopolist. We can mention, for example, the problem book by Korotkevich (1900), oriented toward the same Transport Engineers' Institute and Technological Institute as Shmulevich's books, or Agapov's (1901) handbook for Orenburg. There were many other problem books for applicants, often with their own local color: in a large city, handbooks were published for applicants to the higher educational institutions in that city. Below, we will briefly discuss some of these handbooks, aiming on the one hand to represent different cities of publication, and on the other hand to demonstrate possible differences among the publications.

3.8.1 On Almoev's Problem Books (Almoev 1915a, b)

Almoev's problem books were designed for the institutes in Moscow. The books were published in different issues (using materials from exams from different years and different institutes). Below, we will discuss two issues from 1915. In discussing the first issue, we will confine ourselves to problems in geometry; in discussing the second issue, we will focus on problems in algebra and trigonometry.

The first issue was designed for the Moscow Transport Engineers' Institute and the Agricultural Institute. The second issue was oriented toward the Land Surveying Institute and the Technical Academy. For each of these institutes, problems are given that appeared on exams—usually, they are grouped by topic, and the solutions to these problems follow. We must probably agree with Grigory Shmulevich's verdict, cited above, that Moscow's institutes were easier to get into than St. Petersburg's: such intricate problems as were given at the latter appear far less frequently in Almoev's collection.

At the same time, the very first subsection of problems in algebra from the Moscow Technical Academy opens with equations that cannot be considered altogether simple. The first of them is the following: $8x^3 - 17x = 9$; to solve it, students are advised to add 8 to the left side of the equation and subtract the same number, and then to employ factorization (Almoev 1915b, pp. 145 and 155).

The concluding equation in the section is as follows: $x(x + 1)(x + 2)(x + 3) = 9$. In this case, a transformation must be carried out, rewrite the equation in the following manner: $(x^2 + 3x)(x^2 + 3x + 2) = 9$ and introduce a new variable, relative to which a quadratic equation will be obtained (pp. 145 and 159).

Also represented in the section are systems of equations, progressions, and series (including infinite ones), logarithms, exponential and logarithmic equations, combinations and the binomial theorem, the polynomial remainder theorem, irrational and complex numbers, and the investigation of quadratic equations. All of this, to repeat, was at a level that presupposed a sound preparation, but given such a preparation, did not usually require any special inventiveness (for example, № 533, in which students were asked to eliminate irrationality in the denominator of the fraction $\frac{1}{\sqrt{a}+\sqrt{b}+\sqrt{c}}$ (p. 154))—a lot had to be written down, but a schoolchild who had assimilated the standard material knew well what he had to write.

The same can be said about the next subsection, trigonometry. What we have here are mainly standard "school" problems, which were later repeated in textbooks for decades. To be sure, there were certain exceptions: certain details did change, for example, knowledge of the value of $\sin 54^\circ$, which the author makes use of (p. 203), would subsequently no longer be used to overburden the students. We should also note the presence of equations with inverse trigonometric functions.

The problems from the Land Surveying Institute in algebra and trigonometry are probably more simple on the whole (if only in terms of their technical demands) than the problems from the Technical Academy, although they resemble them in their subject matter and the techniques used for solving them. We will confine ourselves to one example: № 713 asks students to solve the system of equations $x^2 + 3xy = 18$, $xy + 4y^2 = 7$ (p. 231). Here, the equations may be added to obtain a linear equation for the given variables.

The problem book contains no geometry problems for either the Technical Academy or the Land Surveying Institute. The first issue for the Transport Engineers' Institute contains no separate section devoted to geometry problems, but it does have a section on solving triangles in the section devoted to trigonometry. Here, for example, we encounter the following well-known problem—№250 (Almoev 1915a): Prove that if the sum of the squares of the cosines of the angles of a triangle is equal to 1, the triangle is a right triangle (this is proved using trigonometric transformations).

A section on Geometry is provided for the Agricultural Institute. The problems for 1914 are broken into groups (as far as can be judged, corresponding to groups of exam takers). The problems include computation problems and problems in solid geometry. In one problem, for example, students must find the side of a regular tetrahedron circumscribed about four tangent spheres with radius r (№ 315). There are computation problems in plane geometry, for example, one in which students must find the sides of a triangle given the radii of the escribed circles (№314). There are also straightedge and compass construction problems or problems involving loci, for example, one in which three points M, A, and B lying on the same line are given, and students must find the locus of the midpoints of the tangents drawn from point M to circles passing through points A and B (№ 316) (p. 85). None of these problems are easy, but they might be called "theoretical" after a fashion—in the sense that ideas similar to the ones needed in order to solve them were probably discussed, if not in gymnasia, then in preparatory courses.

3.8.2 On Lyandau's Problem Book (Lyandau 1904)

Iosif Lyandau, a student at the Warsaw Polytechnic Institute, wrote and largely prepared this book for publication independently: the text is handwritten and reproduced. Problems offered during various years, beginning with 1898, are given in sequence, and a solution is provided after each problem. The problem selection is somewhat perplexing. The book contains many problems involving progressions. For example, the very first problem is as follows: Consider an arithmetic progression with an odd number of terms, such that the sum of the terms in even-numbered positions is equal to S_1, while the sum of the terms in odd-numbered positions is equal to S_2. Find the number of terms in the progression and its middle term (p. 5). In other words, we have before us a problem that requires students accurately to write down what is given and to formulate the necessary equalities.

There are also classic word problems about motion. There is a so-called composite problem, which effectively consists of several problems, whose results are used in the final solution. There is a system of four equations on which several assignments are based (pp. 40-41), and so on. Algebra problems are relatively few in number (fewer than 15). One may suppose that these are problems that became known to a student who decided to earn a little money by tutoring.

The same can be said about the problems in geometry using trigonometry. For 1898, the author gives four problems. The first of them is as follows: A plane is drawn through the edge of a cube which divides its volume into two parts that stand in a 2:3 relation. Find the angles into which it divides the dihedral angle at this edge (p, 51). For each of the subsequent years, until 1903, one problem is given. We will confine ourselves to another typical example (problem for 1901):

> The lateral edges of a triangular pyramid have the same length, equal to l meter. Of the three face-edge-face angles formed by these edges at the apex of the pyramid, two are equal to α, and the third is equal to β. Find the formula that expresses the volume of the pyramid and calculate its value for l=1.3275, $\alpha = 44^0 27' 14''$, $\beta = 31^0 29' 18''$ (p. 77).

3.8.3 On Andreev's Problem Book (Andreev 1909)

This is also a lithographed edition. In the preface (which, by contrast with the rest of the text, is printed, not handwritten), the author, who presents himself as an engineering technologist, informs readers that he undertook his labor "wishing to bring all possible help in the difficult task of preparing for competitive exams." Further, he notes that in its content the problem book "encompasses the minimum entrance exam curriculum and in a mere two or three weeks can provide [students] with quite solid skills for solving difficult and artificial problems." The problem book was published in Kharkov, but does not anywhere mention that it is based specifically on materials from Kharkov. On the contrary, the author speaks about higher technical institutes in general.

This collection is unquestionably written in a more thorough manner than the preceding one. The first section contains 16 problems on sequences; the second section contains six problems on summation. Then follow eight problems on eliminating irrationality. Then come equations and systems (with the usual artificial techniques, third-degree and higher equations, and so forth), exponential and logarithmic equations, logarithms, and finally a general section, in which the author includes everything that he could not include in other sections. Then follow detailed solutions. What the author's role consisted in—other than rewriting the collected problems in beautiful handwriting and possibly solving them—one can only guess. It may be supposed that the pamphlet was written for applicants who studied with him, who were required to buy it.

3.8.4 On the Problem Book by the Driubins and Ryss (Driubin et al. 1904)

The authors clearly formulate their goal in the preface: to help applicants to the Kiev Polytechnical Institute. The following elucidation immediately follows:

> As is well known, secondary schools, while giving their pupils sufficient knowledge of theory, do little to develop their ability to think quickly and grasp the essence of a question without going into protracted theoretical reasoning. Unfortunately, none of the existing problem books satisfies this long overdue need.

Further, the authors note that in recent years exams have changed dramatically; therefore, all of the existing problem books are unsuitable for preparation.

The problem book opens with problems in physics, followed by problems in algebra. There are no divisions into separate sections with their own subheadings here, but the material (407 problems in all) is clearly subdivided by topic. First come problems about the roots of quadratic equations and their connection with their coefficients, then follow problems involving transformations, including factorization (quite simple), expressions involving radicals, conditional equalities, problems involving the simplification of expressions and proofs of identities. All of these problems, in our view, are quite standard (in the sense that similar problems may be found in textbooks for gymnasia and other problem books for applicants) and of moderate difficulty. We cannot find any innovations, which would support the authors note that other manuals could not be used. As an example, let us cite one of the unquestionably difficult problems, which demands exactitude and a presupposes a high level of algebraic culture, but which hardly requires any special quick reasoning ability, and is rather oriented toward a solid knowledge of typical techniques, even if they are not simple ones:

> If $x + y + z = 0$, then x^2y^3 can be represented in the form $ayx^4 + bxy^4 + czx^4 + dxz^4 + ezy^4 + fyz^4$. Find a, b, c, d, e, f (p. 57).

The authors' idea, in the solution they propose, is to express two variables in terms of the third: $x = pz$ and $y = qz$ (from the statement of the problem, it is clear that $p + q + 1 = 0$); and to substitute these expressions into the equality $x^2y^3 = ayx^4 + bxy^4 + czx^4 + dxz^4 + ezy^4 + fyz^4$ in order to compare the corresponding coefficients, after performing some transformations.

Without listing all of the topic sections (which are not designated by individual subheadings, as we noted above), we will merely say that all of basic topics of the course are represented, including, of course, an abundance of equations and systems of equations, about which we can say what we have already said above: they require a sound preparation, but hardly anything more than that. We will confine ourselves to one more example, № 137: $\sqrt{x-2}-\sqrt{6x-1}+\sqrt{x+3}=0$ (p. 60).

One feature of this problem book appears to be a pronounced attention to inequalities: there are noticeably more such problems than in the other problem books that have been discussed so far. At the same time, the inequality problems are for the most part not difficult (for example, No. 303: solve the inequality $\frac{2x-3}{3x-2}>0$, p. 68), although sometimes somewhat unexpectedly far more difficult assignments are offered, for example, prove the inequality $a^3 + b^3 + c^3 > 3abc$ for three given positive numbers (No. 320, p. 69). Note that in this case, first, the stipulation that they are not all equal is omitted, and second, students are asked to carry out the factorization of the polynomial, from which the proof follows, only later, in No. 360. In the solution, however, the authors point out that for equal numbers, the strict inequality does not hold; and for a solution, they propose expressing two variables in terms of the third and some additional terms, after which, it is promised, the inequality will be easily obtained. The authors themselves do not provide any calculations (p. 118) and, indeed, it is not that easy.

Two other sections in the book are devoted to geometry and trigonometry, respectively. Without going into any detailed discussion of them, we will note that there are no subheadings here either (except the subheadings "Plane Geometry" and "Solid Geometry" in the geometry section). In geometry, straightedge and compass problems and problems involving proofs absolutely predominate, and they are not, in our view, very carefully sequenced or organized—next to a rather difficult problem there may be an altogether easy one, which, moreover, is not particularly connected with its neighbors. The trigonometry problems are not significantly different from those found in other problem books, except perhaps that they appear to be somewhat easier.

3.9 Several Remarks About Women's and Military Higher Educational Institutions

In this paper, it is obviously impossible to cover all the details of admission to all higher educational institutions. The whole question of women's education remains outside the scope of our consideration. We should say, however, that although the state of women's education changed during the period under study, there were practically never any entrance examinations at women's educational institutions. The reference book "All Higher Educational Institutions for Women in Russia" (Vse vysshie 1916) usually either only lists the documents that must be submitted for admission, or (for a few institutions, for example, for the St. Petersburg Bestuzhevsky courses) stipulates that if the number of applicants exceeds the number of open places, then a competition among graduation certificates will be held in accordance with the rules described in the same reference book. Exams are mentioned only in exceptional cases; for example, the Moscow construction courses for women stipulated that "for admission, individuals who have received an education at home are required to pass the courses' verifying tests in history and mathematics covering the equivalent of seven grades at a women's gymnasium" (Vse vysshie 1916, p. 29). However, these are by no means full-fledged competitive entrance tests.

Military educational institutions formed their own closed world, which will also not be examined in detail here. We should say, however, that entrance exams did take place at these institutions and were often quite serious. Let us quote General Alexei Ignatiev (1986), who graduated from the General Staff Academy in 1902 and described its entrance exam, which he had taken, as follows:

> Contrary to my expectations, the exam in mathematics turned out to be dangerous. Chevalier guard Goryainov, my regimental comrade, failed this subject, despite his university pin, and since this occurred on the eve of my own exam, one can imagine my trepidation.
>
> Two old men in ancient black general staff frock coats with silver aiguillettes and generals' shoulder marks that had faded with time, who had the appearance of revenants from the underworld, were sitting behind a long table [...].
>
> I first went to Sharngorst [one of the examiners]. Not satisfied with a problem I had solved, which involved extracting a cube root, he went on tormenting me with questions from number theory, which I answered more by guesswork, than by knowledge. I understood that, for this little man, the significance of curricula is secondary.
>
> "Go on to geometry. What do you have there? A circle? Excellent." And instead of theorems from Semashko's textbook, so well known to me, which were the foundation of all teaching of geometry in the corps, the little general ordered me to draw simply a circle; then another, slightly larger; and then asked me to find the centers of all third circles tangent to the first two.
>
> We had never solved such straightedge and compass construction problems in the corps, and they were not mentioned in the curricula. Sharngorst couldn't care less about that, and

> he forced me to suffer at the blackboard for a good two hours. I had to erase numerous chords and perpendiculars over and over again.

Then the second examiner proposed moving on to trigonometry, in which Ignatyev got the highest possible grade, which compensated for his setbacks in geometry.

By the time Ignatyev wrote his memoirs, he had probably partly forgotten various school subtleties, which is why he decided that he had been given a straightedge and compass construction problem; it is another matter that conic sections (relevant to the problem) might indeed not have been covered at the Page Corps, from which Ignatyev had graduated. As can be seen from the cited passage, even people with a higher education could enroll in the Academy.

However, Adamovich's problem book (Adamovich 1902), which contains problems given on competitive exams to the Academy, does not seem particularly difficult in the context of the problem books examined above, perhaps even the opposite, it contains fewer technically cumbersome problems, and more problems which it is natural to imagine on oral exams. Among them, for example, is № 120 (p. 15), in which students are asked to determine the first and last digit of the number 777^{777} (perhaps it is such a problem that Ignatyev had in mind when he talked about problems in number theory). Many word problems are included, as well as inequalities, problems in arithmetic; on the other hand, practically no attention is devoted to exponential equations and the like. It is noteworthy that, in addition to problems, the book includes so-called questions, which were probably posed on exams. These include the following, for example: № 35—Can the center of a circle inscribed in a triangle lie not in the plane of the triangle; № 42—How great is the greatest possible number of points of mutual intersection of three circles?; № 51—How can one find $\sin 3^{\circ}$ without relying on trigonometric tables? (pp. 29–31).

Traditionally, military-naval educational institutions were the centers of mathematics education. Therefore, let us say a few more words about the St. Petersburg Military-Naval Academy. Quite a large number of materials from entrance exams to this institution have survived, including both administrative materials—such as schedules of exams—and actual problems and checked tests. For example, the problems for the written exam in geometry held on September 5, 1894 were as follows (Nikolaevskaya Morskaya Akademiya 1894, p. 5):

1. Find the volume and lateral surface of a regular hexagonal pyramid, given that its height is $H = 63$, and the radius of a circle circumscribed about its base is $R = 17$.
2. The edge of a cube is equal to a. Find the surface and volume of a polyhedron whose vertices are the centers of the cube.
3. Find the height of a spherical segment, given that the relation of its spherical surface to the area of the base is equal to n, assuming the radius of the sphere = 1.

It should be noted that the tests were checked rather carelessly: there are practically no comments on them; the examiners usually prefer brevity in explanations—for example, if an acute angle must be found, and the exam taker gives the general formula first, it is crossed out. Moreover, there are missed mistakes: for example, after transformations of the expression $2\sqrt{3+\sqrt{5+\sqrt{13+\sqrt{48}}}}$, one exam taker obtained a number in front of which he wrote the sign ±; another wrote the same number, but without any signs—both received the same grade. (Nikolaevskaya Morskaya Akademiya 1908–1910, pp. 66 and 75).

The grades that were given were quite strict, however; the following grades, for example, were given on an exam in trigonometry (based on a 12-point system): 12, 11, 11, 10, 9, 8, 7, 7, 6, 5, 5, 4, 2 (p. 34).

Let us repeat, however, that military educational institutes merit a separate study.

4 Discussion and Conclusion

Probably the first thing that makes an impression on the contemporary reader who becomes acquainted with the problems from pre-revolutionary exams is how long they remained in use. The current USE—the Uniform State Exam—combines the functions of an exit and entrance exam. Consequently, the range of problems which students graduating from high school are preparing to solve has changed substantially and narrowed. Examining problems and problem books from previous decades, however, beginning with the problem book by Skanavi (1969), which became a classic and went through numerous editions, we see how much was taken over, although naturally not always directly, from pre-revolutionary collections.

Differences do exist, of course. Problems with excessively cumbersome computations have fallen away—problems involving the computation of cubic roots, obviously unnecessary in the age of electronic technology, had already lost their attraction for examiners long before this technology was introduced on a mass scale. Algebraic transformations endured for decades, and indeed to some extent endure to this day as a crucial part of the course, but such intricacy as was found in certain problems in Shmulevich's books was usually no longer desired. Certain sections lost their former importance—the same binomial theorem, for example. In geometry, straightedge and compass construction problems practically disappeared. Moreover, the ability to use various artificial techniques, which the authors of the introductions to the problem books examined above explicitly called virtually the most important skill of all, indispensable for the exam taker, ceased to be understood in this way. Nonetheless, let us repeat, very many problems continued to be used—above, only a few references to later books in which they appeared were cited, but the number of such references can be multiplied many times.

One can ask what caused the changes that have taken place, and here, in addition to the aforementioned development of technology, one must likely cite changes in the understanding of mathematics as well, one of whose manifestations was, for example, the international reform movement of the early twentieth century, which slowly penetrated into the schools and transformed them. There is another question, however, which is no less interesting: when were the problems that were subsequently assigned for over a century originally invented?

With respect to this question, we can hardly hope to find a first use for each problem (or even type of problem). We can, however, point out that the invention and selection of difficult problems in Russia unlikely occurred independently of the rest of the world. It would have been natural for a Russian examiner to look at exam problems given, say, at the École Polytechnique. And foreign problem books usually contain the same ideas as Russian ones as well, although one can also discern certain distinctive features (let us cite the problem book of de Aubert & Papelier (1941), translated and published in Russia only in 1941, but published in France much earlier and also not without precedents). It is natural to think that certain problems aimed at using classic ideas were invented by Russian examiners themselves. It would be interesting to conduct a detailed comparison of exam problems from different countries, investigating their origins, although, to repeat, a complete success in this matter can unlikely be hoped for.

It has already been said that exam problems played an obvious role in the change in the attitude toward problems. Above, we have repeatedly cited remarks by the authors of problem books to the effect that gymnasium graduates know theory well enough, but what they can't do so well is to think and invent. Exams effectively created a new curriculum, which was implemented initially outside the schools, but with time inevitably began to exert an influence on schools. This pertains not only to the introduction of additional sections, which were indeed many—let us agree with Ignatyev: examiners indeed often assumed that secondary school curricula were secondary in importance. But another circumstance is no less important: the emphases in teaching gradually began to be distributed in a somewhat different fashion—the understanding of what in particular was more important than anything else continued to change somewhat. The need to select the strongest college students exerted an influence on secondary education.

Here, however, we must be sufficiently careful. First of all, the understanding of what it means to "think" was often clearly rather limited: to "invent" meant inventing artificial techniques, tricks of sorts, in other words, was quite archaic. Second of all, exams were still theoretical in nature, and not only because, as has been said above, it was possible to receive a satisfactory grade and enroll even without solving the problems. Problem solving itself also possessed a theoretical character, in the sense that a great deal had to be known and sometimes simply reproduced (as with problems taken directly from the textbook). In other cases, students had to know techniques and be able to use them. It is not entirely clear how often the brainteasing problems quoted by Yakimov (1901) were given. They do indeed appear in problem books for applicants, but this does not mean that they were regularly given on exams. Later Soviet critics (see Karp 2018) to a certain extent reiterated Yakimov's criticism that problem books for applicants oriented them toward rote memorization of solutions and their reproduction—and this is probably accurate—Shmulevich and other writers of problem books themselves boast that students taking the exam knew all the given problems beforehand. However, the ability to reproduce a complex solution or to construct a solution on the basis of a model, in the image and likeness of what has been learned, is by no means the same thing as simple memorization by rote.

Naturally, it would be profoundly ahistorical to imagine the exam taker of 1901, say, as being similar to a contemporary American schoolchild, and to marvel at how he could figure out how to solve any equation with inverse trigonometric functions. He could figure it out because he had been taught to do it. At the same time, it is important and obvious that schools taught (or even more precisely, began to teach during these years) a very great number of ideas, techniques, and algorithms, and not merely as a handful of examples memorized by rote, but striving to teach students how to apply them in somewhat new situations and in combinations not necessarily known in advance. And schools succeeded in teaching many students how to do this.

Here it must be noted that, of course, not all Institutes required such skills to an equal degree. The materials cited above confirm the natural supposition (although it would be desirable to possess a greater number of facts) that Institutes in Moscow and St. Petersburg differed from the provincial ones, and that even between the former there were noticeable differences in requirements. As the cited materials make clear, at that time, too, students prepared not for mathematics exams in general, but for a specific higher educational institution with its requirements.

To this end, alongside the "official" system of education, another, "informal" system took shape. It possessed a frankly commercial character—the same Shmulevich did not in the least consider it necessary to conceal the fact that he worked for the sake of money and was very interested in questions of profit. We see attributes of the commercial world in the self-advertisements, in the desire to draw the client away from competitors, and in all manner of commercial techniques (for example, groups that signed up for courses received a discount—Almoev (1915b)). Moreover, as in any commercial enterprise, not all merchandise was equally good. Shmulevich himself wrote about the enormous shortcomings of his first problem book, the very same that radically changed his financial position, and his subsequent books have mistakes and omissions as well. While in the case of Shmulevich we might suppose that these came from hastiness, which itself arose from a desire to do more and to earn more, we cannot make the same supposition with equal confidence in all cases—since shortcomings might also be explained by an author's incompetence. We have seen problem books of obviously poor quality, which confirm the critical remarks that Petr Shmulevich and his brother Grigory poured on their rivals.

Nonetheless, one cannot fail to see that the commercial period in the development of Russian mathematics education brought it a great deal. Indeed, the legendary Kiselev (Karp 2002) himself was a part of it—he did not conceal even during Soviet years that he wanted to earn more. It is clear that even long before that, those involved in mathematics education were usually absolutely not selfless, but while in 1811 the academician and cavalier Semyon Guryev offered his composition "The Foundations of Differential Calculus" to the tsar (through the minister of education) so that money might be apportioned for its publication and the author might be given a diamond ring in gratitude (On Academician Guryev's Request 1811), by the end of the century the educational literature market had grown to such a size that it was possible to buy houses in St. Petersburg with proceeds (as Kiselev did) without disturbing the tsar.

The orientation toward the mass scale (even if it was very limited from today's point of view) spurred the search for better methodological and pedagogical approaches in order to help students to solve the problems facing them—including the problem of preparing for admission to higher educational institutions, as part of whose solution, for example, the first books of difficult problems came into being (to repeat once again: many Russian olympiad problems can be found in Shmulevich's collections—although, to be sure, those who suggested them for olympiads possibly did not remember or even know about this), as did the earliest summer math camps. As a businessman, Shmulevich was forced to think about his students and about those who used his books to study, and this bore its own fruits.

We must not forget, too, that the market was incomparably smaller than it is today, and its participants were not enormous corporations, but individuals or small "firms." From their activity, far fewer documents have survived than from agencies in one way or another connected with the Ministry of Education, including even private gymnasia, but this does not mean that their role was small.

Finally, we should mention another side of the process under consideration, which has been only marginally touched on above: criticism of the examination system. It would not be mistaken to assert that what was demanded of the applicant was first and foremost a high level of computational and algebraic skills, firm and broad knowledge, and industriousness. These qualities are undoubtedly useful, but they do not exhaust the list of what is needed by the future engineer. To what extent exams measured the ability to think can be debated, as indeed can the extent to which they measured other sides of the students' preparedness, too. Did Institutes select students well, even leaving aside all of the restrictions they introduced based on political, class, religious, national, financial, and other considerations? This discussion, which went on for decades, began in the period examined in this article, and many critical remarks were voiced already then.

Yakimov (1901), as we have seen, was convinced that exams were in principle something harmful, and that anyone wishing to study at a higher educational institution should have the opportunity to do so. Bonch-Bruevich (1902) painted heart-rending pictures of exams corrupting children morally and having an obviously harmful effect on them physically. These writers were by no means alone, although one cannot fail to see that on the whole the system functioned and prepared a considerable number of highly qualified experts. Posse (2018), noting that for the selection of future students to the Institutes, not only exams are used, but also graduation certificate competitions, went on in a melancholy tone: "The advantages, or better, the comparative shortcomings of such systems have been discussed extensively before and continue to be discussed to this day; however, nothing better has thus far been invented" (pp. 117–118).

A vast literature exists about the shortcomings of entrance exams in general and in mathematics in particular. Let us repeat that this literature appeared and developed long before the revolution of 1917. Let us remind readers that during the post-revolutionary years, this anti-exam viewpoint prevailed, albeit for a short time. But this already pertains to the next period in the history of exams, which the author hopes to investigate in the future.

References

Adamovich, S.V. 1902. *Sbornik voprosov i zadach po arifmetike, algebre, geometrii i trigonometrii, predlagavshikhsya na konkursnykh ekzamenakh pri postuplenii v Nikolaevskuiu Akademiiu General'nogo shtaba*. [Collection of Questions and Problems in Arithmetic, Algebra, Geometry, and Trigonometry Given on Competitive Exams for Entering the Nikolayevskaya General Staff Academy]. St. Petersburg: Tipografiya N.N.Klobukova.

Agapov, Dmitry V. 1901. *Sbornik tipichnykh i naibolee trudnykh zadach po Arifmetike, Algebre, Geometrii, Trigonometrii i Fisike, predlagavshikhsya na konkursnykh ekzamenakh vo vsekh vysshikh uchebnykh zavedeniyakh*. [Collection of Typical and Especially Difficult Problems in Arithmetic, Algebra, Geometry, Trigonometry, and Physics Given on Competitive Exams at All Higher Educational Institutions]. Orenburg: Tipo-litografiya B.A.Breslina.

Agenturnaya zapiska o nedostatkakh klassicheskoy sistemy obrazovaniya v gimnaziyakh i o bol'shikh probelakh znanii v oblasti tochnykh nauk u lits, sdavavshikh priemnye ekzameny v vysshie uchebnye zavedeniya s tekhnicheskim uklonom [Agents' memorandum on the defects of the classical system of education in gymnasia and on large gaps in knowledge in the sphere of the exact sciences among persons taking entrance exams to institutions of higher learning with a technical orientation]. (1875). State archive of Russian Federation, f. 109, op. 3a, d. 1035.

Almoev, Amayak K. 1915a. *Sbornik zadach s resheniyami predlozhennykh na konkursnykh ekasamenakh. Vypusk 1 [Collection of Problems with Solutions Given on Competitive Exams, Issue 1]*. Moscow: Izdanie Almoeva.

———. 1915b. *Sbornik zadach s resheniyami, predlozhennykh na konkursnykh ekasamenakh. Vypusk 2 [Collection of Problems with Solutions Given on Competitive Exams, Issue 2]*. Moscow: Izdanie Almoeva.

Andreev, A. 1909. *Sbornik zadach, predlagavshikhsya na konkursnykh ekzamenakh v vysshie tekhnicheskie zavedeniya: Algebra* [Collection of Problems Given on Competitive Exams to Higher Technical Institutions: Algebra]. Kharkov.

Belhoste, Bruno, and Konstantinos Chatzis. 2007. From Technical Corps to Technocratic Power: French State Engineers and their Professional and Cultural Universe in the First Half of the 19th Century. *History and Technology* 23 (3): 209–225.

Bonch-Bruevich, Vladimir D. 1902. Shest' let v zakrytom uchebnom zavedenii [Six Years at a Boarding School]. *Zhizn'*: 270–292.

de Aubert, P., and G. Papelier (Ober P., G. Papel'e). 1941. *Uprazhneniya po elementarnoy algebre [Exercises in Elementary Algebra]*. Moscow: Uchpedgiz.

Driubin, E., L. Driubin, and C. Ryss. 1904. *Sbornik zadach s resheniyami predlagavshikhsya na konkursnykh ispytaniyakh v Kievskom Politekhnicheskom institute po fizike, algebre, geometrii i trigonometrii*. [Book of Problems with Solutions Given at Competitive Tests to the Kiev Polytechnical Institute in Physics, Algebra, Geometry, and Trigonometry]. Kiev: Tipografiya D.N. Trakhtenberg.

Elektrotekhnicheskii institut. 1909a. *O prinyatii v Institut v chislo studentov lits po osobym khodataistvam* [On Admission to the Student Body of the Institute by Special Petition]. Central state historical archive of St. Petersburg, f. 990, op. 1, d. 1076.

———. 1909b. *O proizvodstve priemnykh ispytanii v 1909 godu i o zachislenii v chislo studentov litz, vyderzhavshikh takovaye ispytaniya*. [On the Implementation of Entrance Tests in 1909 and on the Enrollment of Persons Who Passed Such Tests Within the Student Body]. Central state historical archive of St. Petersburg, f. 990, op. 1, d. 1050.

———. 1912. *O proizvodstve v 1912 godu priemnykh ispytanii i o zachislenii v chislo studentov litz, vyderzhavshikh takovaye ispytaniya*. [On the Implementation of Entrance Tests in 1912 and on the Enrollment of Persons Who Passed Such Tests Within the Student Body]. Central state historical archive of St. Petersburg, f. 990, op. 1, d. 1282.

———. 1914–1915. *O proizvodstve v 1914 godu priemnykh ispytanii i o zachislenii v chislo studentov litz, vyderzhavshikh takovaye ispytaniya.* [On the Implementation of Entrance Tests in 1914 and on the Enrollment of Persons Who Passed Such Tests Within the Student Body]. Central state historical archive of St. Petersburg, f. 990, op. 1, d. 1398.

Gornyi institut. 1914–1915. *O priemnykh ekzamenakh s 1 yanvarya 1914 goda po sentyabr' 1915* [On Entrance Exams from 1 January 1914 until September 1915]. Central state historical archive of St. Petersburg, f. 963, op. 1, d. 13929.

Herzen, Alexander I. 1956. *Byloe i dumy, Chast'1. [My past and thoughts. Part 1]. Writings. v.4.* Moscow: Gosudarstvennoe izdatel'stvo khudozhestvennoy literatury.

Ignatiev, Alexey A. 1986. *Piat'desiat let v stroiu [Fifty Years in Service].* Moscow: Voenizdat.

Institut inzhenerov putey soobscheniya. 1917. *O prieme v Institut. Chast' I.* [On Admission to the Institute. Part I]. Central state historical archive of St. Petersburg, f. 381, op. 10, d. 20.

Karp, Alexander. 2002. *Klassik real'nogo obrazovania. [Classic of Genuine Education]. St.* Petersburg: SMIO-PRESS.

———. 2015. Problems in old Russian textbooks: How they were selected. In *"Dig where you stand" 3*, ed. Kristin Bjarnadottir et al., 203–218. Uppsala: Uppsala University.

———. 2018. Neromanticheskii geroy matematicheskogo obrazovaniya (O P.K. Shmuleviche) [Non-romantic hero of mathematics education (On P.K. Shmulevich)]. *Matematika v shkole* 5: 68–75.

———, ed. 2019. *National Subcommissions of ICMI and their Role in the Reform of Mathematics Education*. Cham: Springer.

———. 2020. Russian Mathematics Education After 1991. In *Eastern European mathematics education in the decades of change*, ed. Alexander Karp, 173–228. Cham: Springer.

Korotkevich, M.A. 1900. *Sbornik zadach po Fizike i Geometrii, predlagavshikhsya na konkursnykh ispytaniyakh v Instituty inzhenerov putey soobscheniya, Tekhnologicheskom, i dr.*. [Collection of Problems in Physics and Geometry Given on Competitive Tests to the Institute of Transport Engineers, the Technological Institute, and Others]. St. Petersburg: Izdanie V.L. Lebedeva.

Krechmar, Vasily A. 1972. *Zadachnik po algebre [Problem Book in Algebra].* Moscow: Nauka.

Kursy inzhenera Dmitrieva. 1913. *Sbornik zadach i tem sochinenii, predlozhennykh na konkursnykh ekazamnanakh v Moskovskikh spetsial'nykh vysshikh uchebnykh zavedeniyakh.* [Collection of Problems and Composition Topics Given on Competitive Exams at Moscow's Special Higher Educational Institutions]. Moscow: Tipografiya B. Rikhter.

Lyandau, Iosif. 1904. *Sbornik zadach, predlagavshikhsya na konkursnykh ekzamenakh v Varshavskom Politekhnicheskom Institute. Algebra i gometriya [Collection of Problems Given on Competitive Exams at the Warsaw Polytechnic Institute. Algbera and Geometry].* Waesaw: Tipografiya S. Orgel'branda i synovey.

Lyapin, Sergey E., Irina V. Baranova Irina, and Zoya G. Borchugova. 1973. *Sbornik zadach po elementarnoy algebre [Collection of Problems in Elementary Algebra].* Moscow: Proveschenie.

Ministry of Education. 1901. *Sravnitel'naya tablitsa ustavov universitetov [Comparative Table of University Charters].* St. Petersburg: Tipografiya Sankt-Peterburgskoy tiur'my.

Nikolaevskaya Morskaya Akademiya. 1894. *Pis'mennye raboty na priemnykh ekzamenakh po geometrii, trigonometrii, fizike* [Written Assignments on Entrance Exams in Geometry, Trigonometry, Physics]. Russian state archive of navy, f. 433, op. 1, d. 77.

———. 1908–1910. *Pis'mennye raboty na priemnykh ekzamenakh po algebra, geometrii, trigonometrii* [Written Assignments on Entrance Exams in Algebra, Geometry, Trigonometry]. Russian state archive of navy, f. 433, op. 1, d. 243.

Po prosheniiu akademika Gur'eva o podnesenii Ego Velichestvu sochineniya ego pod nazvaniem "Osnovaniya differentsial'nogo ischisleniya i o pozhalovanii avtoru brilliantovogo perstnya [On Academician Guryev's Request Concerning the Offering to His Majesty of His Composition Entitled "The Foundations of Differential Calculus" and on the Award of a Diamond Ring to the Author]. 1811. Russian state historical archive, f. 733, op. 118, d. 196.

Posse, Konstantin. 2018. Doklad o prepodavanii matematiki v universitetakh, vysshikh tekhnicheskikh uchebnykh zavedeniyakh i nekotorykh voenno-uchebnykh zavedeniyakh Rossii [Report on Mathematics Instruction in Universities, Higher Technical Education Institutions, and Certain Military-Academic Institutions in Russia]. In *Otchety Rossiiskoy natsional'noy podkomissii Mezhdunarodnoy komissii po matematicheskomu obrazovaniu*, ed. Alexander Karp, 55–144. St. Petersburg: Lema.

Pravila dlya ispytanii postupaiuschikh v studenty universiteta [Rules for Test for Applicants to the University]. 1860. St. Petersburg: Tipografiya Imperatorskoy Akademii Nauk.

Rybkin, Nikolay A. 1903. *Sbornik geometricheskikh zadach na vychislenie [Collection of Geometric Computation Problems]*. Moscow: Sotrudnik shkol.

Schukin, M.V. 1914. *Sbornik i resheniya zadach po planimetii, predlagavshikhsya v proshlom godu na konkursnykh ekazamenakh v institute inzhenerov putey soobscheniya*. [Collection of Problems in Plane Geometry, with Solutions, Given Last Year on Competitive Exams to the Institute of Railway Engineers]. St. Petersburg: Chernyshevskie podgotovitel'nye kursy.

———. 1915. *Institut inzhenerov putey soobscheniya. Konkursnye ekzameny 1915 goda* [Transport Engineers' Institute. Competitive Exams for 1915]. Petrograd: Izdanie Kursov Schukina.

Shmulevich Grigorii K. 1911. *Kak gotovit'sya k konkursnym ekzamenam v Vvysshie tekhnicheskie uchebnye zavedeniya Peterburgskie i Moskovskie* [How to Prepare for Competitive Exams to Higher Technical Educational Institutions in Petersburg and Moscow]. St. Petersburg.

Shmulevich Grigorii K. 1915. *Kak gotovit'sya k konkursnym ekzamenam v Vvysshie tekhnicheskie uchebnye zavedeniya Petrogradskie i Moskovskie* [How to Prepare for Competitive Exams to Higher Technical Educational Institutions in Petrograd and Moscow]. Petrograd.

Shmulevich, Petr K. 1897. *Sbornik zadach, predlagavshikhsya na konkursnykh ekzamenakh v Institutakh Inzhenerov putey soobscheniya, Grazhdanskikh inzhenerov, Technologicheskom i drugikh. Trigonometriya*. [Collection of Problems Given on Competitive Exams at the Transport Engineers' Institute, the Institute of Civil Engineers, the Technological Institute, and Others. Trigonometry]. St. Petersburg: Tipografiya A.E. Vineke.

———. 1900. *Sbornik zadach, predlagavshikhsya na konkursnykh ekzamenakh v Institutakh inzhenerov putey soobscheniya, Gornom, Technologicheskom i drugikh. Algebra i arifmetika*. [Collection of Problems Given on Competitive Exams at the Transport Engineers' Institute, the Mining Institute, the Technological Institute, and Others. Algebra and Arithmetic]. St. Petersburg: Tipografiya I.M.Komelova.

———. 1903. *Sbornik zadach, predlagavshikhsya na konkursnykh ekzamenakh v spetsial'nye vysschie uchebnye zavedeniya. Geometriya [Collection of Problems Given on Competitive Exams to Special Higher Educational Institutions. Geometry]*. St. Petersburg: Sklad izdaniya u avtora.

———. 1925–1926. *Delo Shmulevicha P.K.* [The Case of P. K. Shmulevich]. Central state archive of St. Petersburg Санкт-Петербурга, f. 1244, op. 18, d. 43.

Skanavi, Mark I., ed. 1969. *Sbornik zadach po matematike dly konkursnykh ekzamenov vo VTUZy [Collection of Problems in Mathematics for Competitive Exams to Higher Technical Educational Institutions]*. Moscow: Vysshaya shkola.

Sorokin, Pitirim. 1959. *Social and cultural mobility*. Glencoe, IL: Free Press.

St. Petersburg University. 1860–1861. *Priemnye ekzameny v 1860-1861 uchebnom godu. Chast' I*. [Entrance Exams for the 1860-1861 School Year. Part I]. Central state historical archive of St. Petersburg, f. 14, op. 2, d. 489.

Tekhnologicheskii institut. 1901. *O rezul'tatakh priemnykh sostyazatel'nykh ispytanii* [On the Outcomes of Competitive Entrance Tests]. Central state historical archive of St. Petersburg, f. 492, op. 2, d. 6889.

———. 1902. *O rezul'tatakh priemnykh sostyazatel'nykh ispytanii* [On the Outcomes of Competitive Entrance Tests]. Central state historical archive of St. Petersburg, f. 492, op. 2, d. 7239.

———. 1904. *O rezul'tatakh priemnykh sostyazatel'nykh ispytanii* [On the Outcomes of Competitive Entrance Tests]. Central state historical archive of St. Petersburg, f. 492, op. 2, d. 7979.

Voprosy dlya ispytanii molodykh liudey, zhelaiuschikh postupit' v chislo studentov Imperatorskogo Khar'kovskogo universiteta [Questions for Testing Young People Who Wish to Be Admitted as Students to Imperial Kharkov University]. 1849. Khar'kov: Khar'kovskii universitet.

Vse vysshie zhenskie uchebnye zavedeniya Rossii. Neobkhodimyi spravochnik dlya zhenschin, stremyaschikhsya k vysshemu obrazovaniiu [All Higher Educational Institutions for Women in Russia. Indispensable Handbook for Women Aspiring to Higher Education]. 1916. Moscow: Izdatel'stvo gruppy studentov.

Yakimov, V.P. 1901. *Dumy provintsial'nogo pedagoga o konkursnykh ispytaniyakh po matematike v vysshikh uchebnykh zavedeniyakh i dvukh sbornikakh P. Shmulevicha, soderzhaschikh v sebe zadachi dlya nazvannykh ispytanii.* [The Thoughts of a Provincial Pedagogue About Competitive Exams in Mathematics at Institutions of Higher Education and About Two Collections by P. Shmulevich Containing Problems for the Said Tests]. Kazan': Tipo-litografiya Imperatorskogo universiteta.

Zharova, Ekaterina Yu. 2014. Ekzameny v universitetakh Rossiiskoy imperii v pervoy polovine XIX veka [Exams at the Universities of the Russian Empire During the First Half of the Nineteenth Century]. *Voprosy obrazovaniya* 4: 246–259.

———. 2016. Ekzameny v universitetakh Rossiiskoy imperii vo vyoroy polovine XIX—nachale XX veka [Exams at the Universities of the Russian Empire During the Second Half of the Nineteenth Century and the Beginning of the Twentieth Century]. *Vestnik Permskogo universiteta* 5: 32–41.

Gnomonics in Secondary School Education in the Territories of Poland in the 17th–20th Centuries

Karolina Karpińska

Abstract The aim of the paper is to discuss the evolution of teaching gnomonics in secondary schools in the Polish territories in the 17th–20th centuries. The construction of sundials was taught as an example of the application of geometry in everyday life. Until the beginning of the 18th century, it was discussed only in selected schools as a consequence of the decisions of individual teachers and usually during "mathematics" and "astronomy" lessons. In 1711 gnomonics was introduced as an obligatory element of mathematics teaching in Polish Jesuit colleges. A hundred years later it became a compulsory topic in Polish secular secondary schools ending with matriculation examinations and was taught during "physics and chemistry" lessons. In the 1830s, ministerial ordinances were introduced, under which the construction of sundials began to be taught as part of "descriptive geometry" lessons in the technical department of gymnasia in the Kingdom of Poland. The paper will compare curricula implemented in selected schools, gnomonical issues included in textbooks, and general ordinances related to secondary school education in the Polish territories. Geometrical constructions, basics of astronomy and measuring instruments which were used for the construction, and setting of the sundials will be discussed.

Keywords Mathematics education · Gnomonics · Sundials · Sundials construction · Geometry · Descriptive geometry · Mathematics education in Poland · Secondary schools

K. Karpińska (✉)
L. & A. Birkenmajer Institute for the History of Science, Polish Academy of Sciences, Warsaw, Poland
e-mail: karolinakarpinska@ihnpan.pl

A. Karp (ed.), *Advances In The History Of Mathematics Education*,
International Studies in the History of Mathematics and its Teaching,
https://doi.org/10.1007/978-3-030-95235-8_6

1 Introduction

Looking at the façades of palaces, churches and town halls, or walking in the park, we can often see instruments used to measure time—sundials. Some of them were built several centuries ago, but there are also modern ones. Currently, sundials are treated primarily as decorations in public spaces. They operate on the basis of the daily change in the position of the Sun (the consequence of the Earth's rotation), so they belong to the class of natural clocks (next to sand, fire, water, etc.). Sundials indicate the solar time. Thus, in times when people function according to official time (depending on individual time zones and the introduction of summer and winter time), the time indicated by the sundial often differs from that shown by mechanical and quartz watches, etc. Nevertheless, using an appropriate time equation, official time can be calculated from the solar time.

Each sundial consists of a sundial face with an hour scale (sometimes this scale is more precise—a quarter of hours scale), and an indicator called a gnomon, which is attached to this face. The method of preparing a sundial face depends on the latitude of the place where the sundial is located, the position of the dial with respect to the horizon and directions of the world, and the shape of the face (flat, spherical, cylindrical, etc., or irregular). The gnomon is exposed to weather conditions, so in order to avoid its displacement, it is often supported with a cantilever, or it is shaped like a plate with a pointed end. The way it is positioned and tilted to the face depends on the type of sundial and the latitude.

The sundial shows the time only when it is illuminated by the Sun's rays. It works as follows: when a sunlit gnomon casts a shadow on the face of a sundial, then this shadow indicates the current solar time; if the shadow falls exactly on one of the lines (lines of hours or lines of quarters of hours) on the sundial face, then the time is indicated precisely, if it is not, the distance of the shadow from two successive lines indicates an approximation of solar time (Figs. 1 and 2).

Sundials that we can see on the walls of buildings, or those placed on monuments, etc., are stationary ones, but in the past, portable sundials and even pocket sundials were also used (they were the most popular in the 17th and 18th centuries). The correct operation of all these sundials was limited to the latitude for which they were prepared. Thus, the sundial prepared for the latitude of Warsaw would not work properly in Oxford, for instance.

The science dealing with sundials, describing the principles of their operation, their use and methods of construction, is called gnomonics. Sundials, in addition to determining the current solar hour, were used, for example, to determine the height of the Sun or determine the length of individual days and nights. Sometimes the lines of the zodiac were also placed on the faces of sundials, making them calendars and allowing for the determination of, for example, the time of the spring and autumn equinoxes and the dates of movable holidays.

The construction of sundials as a scientific study began to develop in ancient China and Babylon, and later it reached the Greeks, Italians and Arabs. However, the development of time-measuring methods did not intensify until the Middle

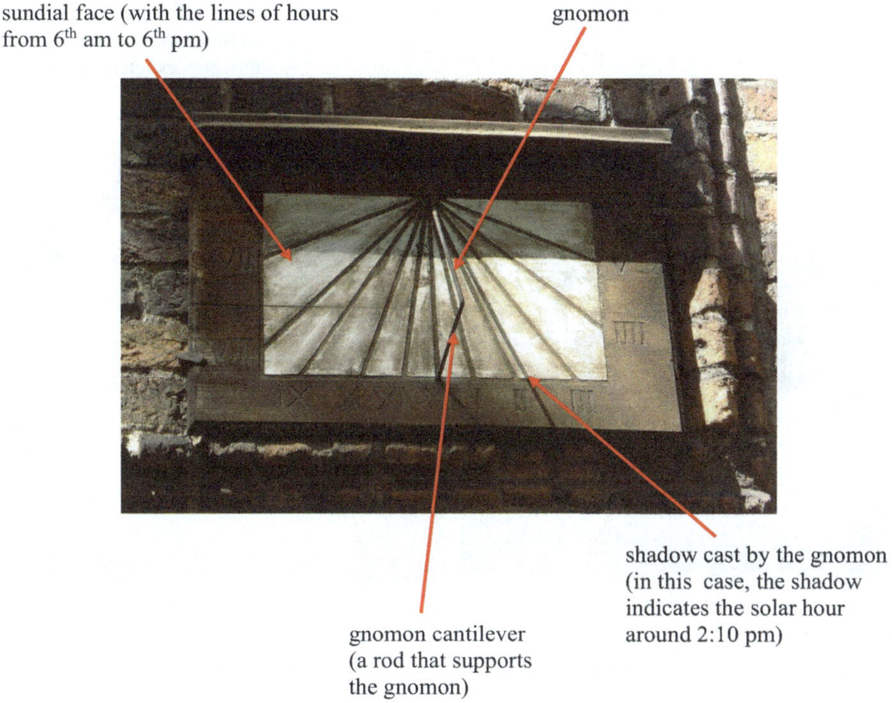

Fig. 1 Sundial on the wall of the church in Brodnica (latitude 53°15'). The sundial is built on a vertical wall facing south

Ages. People noticed the role of time in their lives more and more and considered it a factor necessary to achieve other goals; therefore, they tried to use it rationally and to a greater extent. In Europe, great interest in gnomonics arose in the 7th century, when Pope Sabinian ordered a sundial to be placed on the wall of each church (Przypkowski 1968, p. 13)—these sundials were supposed to indicate the time of the commencement of services to the faithful. However, it is only from the second half of the 15th to the end of the 18th century that we can speak about the European scientific movement, which brought the development of gnomonics to a level close to that of the 21st century. During those centuries, various types of sundials were constructed—those on flat surfaces (including horizontal, vertical and inclination sundials), and those on curved ones (including cylindrical, spherical and ring sundials). Gnomonics was taught at universities and in selected secondary schools, and one of the most important Polish discoveries from this period was the introduction by Nicolaus Copernicus (1473–1543) of the reflection method in the construction of sundials. Although mechanical clocks were already known at that time, they were difficult to access and therefore were not of a public nature. The situation was similar with pendulum clocks (with spindle and anchor escapements), chronometers, etc., which were invented in the 17th century. Only the formation of a modern

Fig. 2 Sundial at the Royal Lazienki Museum in Warsaw (latitude 52°13'). The face of this sundial is parallel to the horizon

industrial society in the 18th century and the introduction of public transport (e.g. stagecoach mail, mechanical rail) led to greater concerns about timing and resulted in its unification in large territories. Mean solar time (and later universal time) was established, which meant that from the beginning of the 19th century, most inhabitants of Europe and America possessed mechanical clocks.

Teaching carried out by specific nations often has individual, national characteristics. There are issues in the curricula that are not discussed in other countries. A characteristic feature of Polish teaching was the discussion of gnomonics. The construction of sundials was introduced to select Polish secondary schools from the beginning of the 17th century. In 1711, by ordinance, gnomonics was introduced to all Polish Jesuit colleges—gnomonics was a science particularly valued and developed by the Church, and priests and monks were considered experts in measuring time. In 1812, the construction of sundials was introduced to secular Polish secondary schools preparing students for matriculation examinations, which on the one hand was a continuation of the tradition started by the Jesuits, and on the other a consequence of French influence—the Ministry of Education of the Duchy of Warsaw asked the Warsaw teacher, Jan Bystrzycki, to recommend a textbook for astronomy, and he recommended *Traité élémentaire d'astronomie physique* by Jean

Baptiste Biot (Kukulski 1931, p. 471, 479), with a chapter on gnomonics (ed. 1811). The construction of sundials was an important part of Polish education until World War II.

Gnomonics was valued by Poles mainly because it showed the application of knowledge acquired at school in everyday life. It was an interdisciplinary science, combining skills in the fields of geography, mathematics, and astronomy; therefore, discussing it was an example of "Science, Technology, Engineering, Mathematics" (STEM) type of teaching—teaching through experience, learning about the world in practice and acquiring knowledge with a broad perspective.

At a time, the second half of the 18th century, when the world was experiencing a watchmaking revolution, Poland was in the most difficult moment of its history—internal conflicts and an unstable political situation as well as the collapse of culture, science, and technology led to three partitions of Poland—in 1772, 1793, and 1795. As a result of the Third Partition, all Polish lands were occupied by Austria, Russia, and Prussia, and the country itself disappeared from the map of Europe for 123 years. After the First Partition, Poles began to mobilize their forces to prevent further plunder of Polish lands. One of the actions taken at that time was the creation of the Commission of National Education—the first ministry of education in the world (Pardała 2010, p. 330). The aim of the Commission was to reform schools, introduce the Polish language as the language of instruction (previously it was Latin), and to develop new curricula, adapted to the needs of Polish society. These curricula were primarily focused on showing applications of school learning in everyday life. The Commission of National Education attached great importance to educating Poles because its members believed that only educated young people could lift the country from its decline. The actions of the Commission did not bring about the desired results. Just before the Third Partition of Poland, it was liquidated. Its work, however, remained alive. Until 1918, despite the occupation of Austria, Russia and Prussia, Polish schools operated in selected periods with Polish curricula focused on showing applications. The Duchy of Warsaw (1807–1815), the Kingdom of Poland (1815–1832), and Galicia (from 1868) although under the protectorate of France, Russia and Austria, respectively, had somewhat independent curricula. It turns out that in the 19th century there was a renaissance in teaching gnomonics in Polish schools.

Given that the construction of sundials involved mathematics, astronomy, and geography, through the ages gnomonics was treated as a field of mathematics[1], but also as a field of astronomy and geography. In Polish secondary schools gnomonics was usually taught during mathematics, astronomy, mathematical geography, physics, geography, and descriptive geometry lessons. Currently, gnomonics is not considered to be a part of mathematics, but in the history of mathematics teaching it is important also to "keep track of changes taking place in other subjects, which today we do not associate directly with mathematics" (Karp and Furinghetti 2016, p. 9).

[1] See e.g. *Acervus selelecius* [Kórnik Library, sign. BK 01025].

There are three types of curricula that should be considered when analyzing the history of teaching: intended, implemented, and attained curricula. Intended curricula are those established by the ministry of education, enriched with the content included in school textbooks. Implemented curricula can be seen in the curricula found in school reports[2], teachers' reports, and student notebooks. Attained curricula are learning outcomes that can be determined by analyzing written student essays, end-of-term and school-year exams, and matriculation exams. The article concerns itself with the intended and implemented curricula.

The sources for the research were primarily annual school reports (they included curricula, textbook lists, teachers' research, and teaching works), material contained in student as well as teacher editions of school textbooks, written teachers' reports for school management on the material to be implemented, regulations regarding curricula recommended for selected types of schools, orders of the ministry of education, and reports on school visits. The subject of this study's analysis was the teaching of gnomonics in several dozen schools operating in Poland.

Up to this point, no analysis has been made of the teaching of gnomonics in Poland. In the literature related to teaching, only rudimentary information on the subject is mentioned, e.g. the titles of the textbooks from which the subject was taught in selected schools[3]. This article is the first attempt to systematically analyze the teaching of gnomonics in secondary schools in the Polish territories.

2 Types of Sundials Discussed in the Polish Secondary Schools

The day and night was divided into 24 h, but the methods of counting and measuring individual solar hours were different. As a consequence, in school education in the Polish territories, four types of sundials were introduced: astronomical, Italian, Babylonian, and ancient. The face of the astronomical sundial is divided into two parts: the twelve hours from noon to midnight (at noon is the 12th hour, just after it is the 1st hour, etc.) and the next twelve hours from midnight to noon. On the Italian sundial face, hours are counted from the 1st to the 24th—sunset is the 24th hour, and just after it comes 1st hour, etc. Babylonian sundials operate according to the same principle, but here the 24th hour is located at the sunrise. The ancient sundials

---

[2] Directors and chancellors of selected secondary schools published annual brochures, usually called invitations to public examinations or school reports. Such brochures were published, for example, by Jesuit colleges. In 1788, matriculation examinations were established in Prussia, which later spread to other countries. The introduction of matriculation examinations resulted in the ministries of education of individual countries gradually obligating all secondary schools that prepare for these exams to issue such brochures. In Poland, the first matriculation exam was carried out in 1789. In 1812, Polish secondary schools preparing for matriculation examinations were required to issue annual information brochures.

[3] See: Kubik (1972, p. 125).

indicate twelve equal hours from sunrise to sunset and another twelve equal hours from sunset to sunrise—the length between the sunrise and sunset changes during the year, so in the summer days, hours are longer than in the winter, and at the same time, the night hours are shorter (Gąsiorowski 1871, pp. III–V). The constructions of all these types of sundials differed from each other.

Out of all 24 hour lines, only those that could be lit by the Sun were usually placed on the dials—the sunrise and sunset as well as the location of the sundial were taken into account (Fig. 3).

In school education within the Polish territories, attention was paid primarily to astronomical sundials with faces placed on flat surfaces. This type of sundial has been discussed with the greatest meticulousness, which is why in this article it will be analyzed in detail.

When we look at the sky on a cloudless night, then we can get an impression that all stars and other celestial bodies are placed on a large, moving sphere in the center of which the Earth is located. This is just an illusion, because all these objects are so far away that the "naked-eye" observation is not enough to assess their distance from the Earth. This imaginary sphere surrounding the Earth is called the celestial sphere (with its center in the center of the Earth) (Fig. 4).

"The celestial sphere [...] contains the Earth; and since the latter is only a point in relation to the immensity of the first, then any place on the surface of the Earth may be considered the center of the celestial sphere [...]. Therefore, each point of the Earth's surface as the center of the enormous celestial sphere is correctly used" (Kowalski 1868, pp. 10–12)—this statement became the basis for the explanation of how sundials work. The Earth, being the center of the celestial sphere, was treated as a point, then the relationships between the objects on the celestial sphere explained the structure and operating principle of all flat astronomical sundials.

Fig. 3 A sundial located on the south-eastern wall of the King Jan III's Palace at Wilanów (Warsaw; latitude 52°13'). In fact, it contains three sundials—in the center there is an astronomical sundial with the lines of hours and and the lines of the zodiac, on the right side there is a Babylonian sundial (the hour lines are largely obliterated), and on the left side there is an Italian sundial. Each of these sundials has a separate gnomon

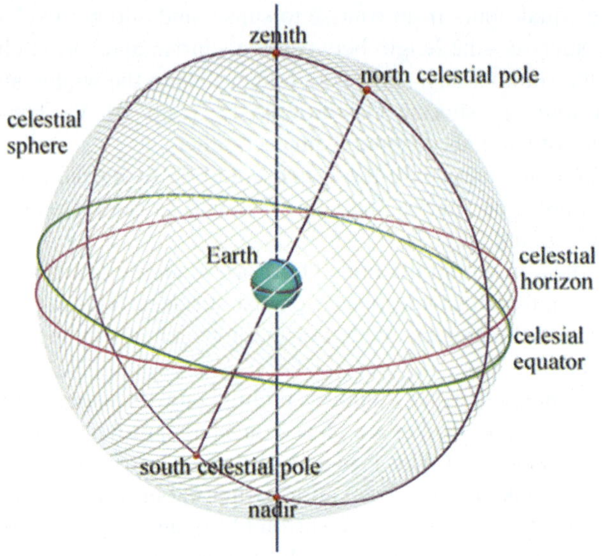

Fig. 4 Celestial sphere

Let us intersect the celestial sphere with an arbitrary plane. This plane will be the face of the sundial. How are the lines of the individual hours formed on this face? The general rule is as follows:

(1) The Sun makes its daily apparent motion around the Earth's axis; this motion is uniform and takes 24 equal hours.
(2) We draw 12 circles on the celestial sphere passing through the celestial poles, the first of which is the celestial meridian, such that they divide the sphere into 24 equal parts; then these circles are circles of the 12 individual hours. It follows from the movement of the Sun and the moment of its culmination that the celestial meridian is the circle of 12th hour (at noon and at midnight); the next circle, in the direction of the Sun's movement, is the 1st hour circle (pm and am), etc. (Fig. 5).
(3) The intersections of the planes on which these circles are located with the face of the sundial will designate the lines of corresponding hours—the 12th hour line (at noon and at midnight), 1st hour line (pm and am), etc. (Fig. 6).

The shadow cast by the sunlit Earth's axis will indicate the actual solar time on the sundial face (in this case, the Earth's axis is a gnomon).

Depending on the location of the face relative to the celestial sphere, there are different types of flat astronomical sundials.

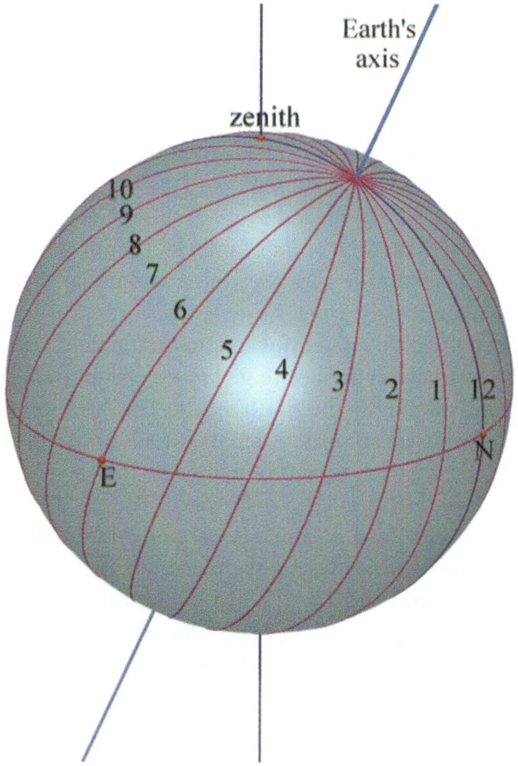

Fig. 5 Circles of hours on the celestial sphere

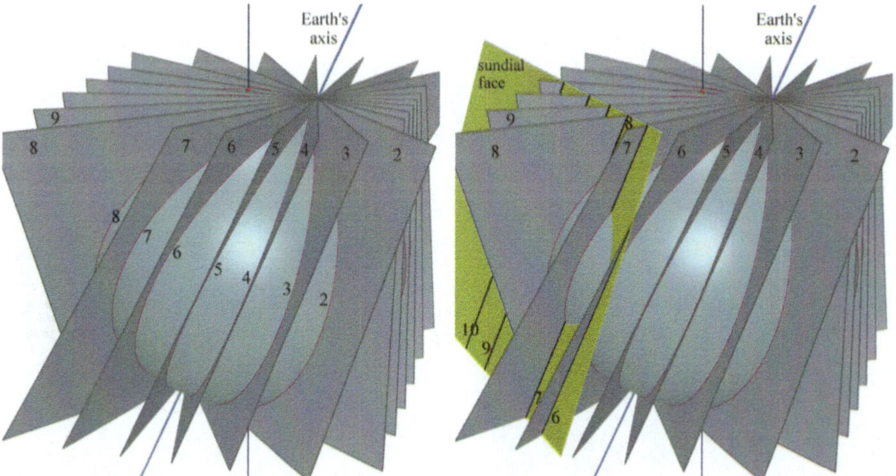

Fig. 6 Determining the lines of hous on the sundial face (the case of the celestial sphere)

2.1 Equatorial Sundial

The equatorial sundial is defined by a face that is the celestial equator plane (Fig. 7).

The celestial equator is perpendicular to the Earth's axis, so the planes dividing the celestial sphere into 24 equal parts also divide the celestial equator plane into 24 equal parts, and since 360° ÷ 24 = 15°, the lines of hours on the equatorial sundial face are evenly spaced every 15°. The gnomon is attached to the face of the equatorial sundial at the point of intersection of hour lines; the gnomon is the Earth's axis.

Considerations for the construcion of an equatorial sundial in the celestial sphere provide the idea that an equatorial sundial is prepared and set in the field. When preparing the equatorial sundial in the field, the celestial equator should be replaced by a small flat surface (e.g. a marble slab) parallel to the equator, the gnomon is then a rod parallel to the Earth's axis. The hour lines are located on both sides of the flat surface—in summer, when the Sun is above the equator, the sundial shows the hours on the upper side, and in winter on the lower (facing the ground).

Equatorial sundial in the field: The dial of the equatorial sundial is inclined to the local horizon at an angle equal to the inclination of the equator, so at an angle equal to: 90° – the latitude of the place where sundial is set. The lines of hours on the dial are spaced every 15°. The 12th hour line (noon and midnight) is located on the plane of the local meridian. The gnomon is attached to the sundial's face at the point of intersection of all lines of hours and is parallel to the Earth's axis (so it is perpendicular to the equatorial sundial face) (Figs. 8 and 9).

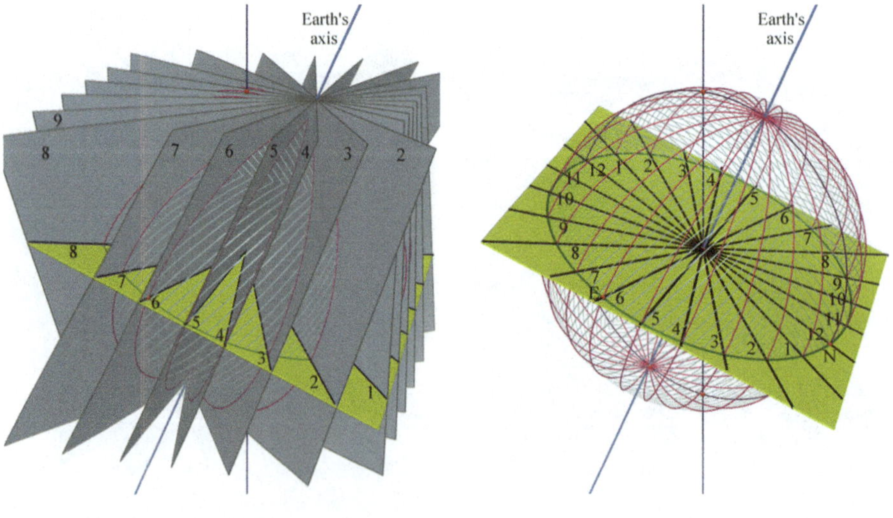

Fig. 7 Equatorial sundial in the celestial sphere

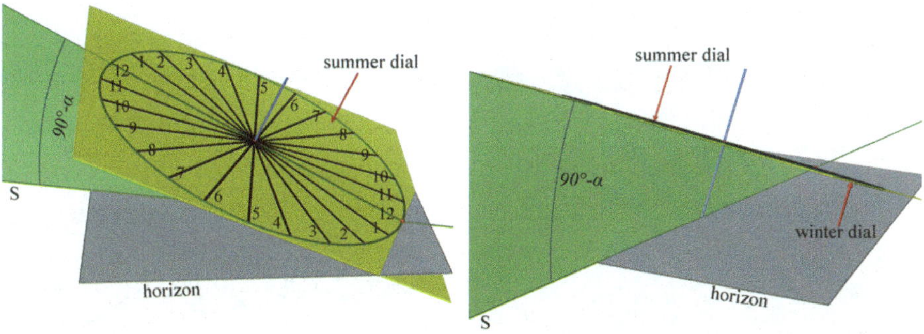

Fig. 8 Setting the equatorial sundial in the field

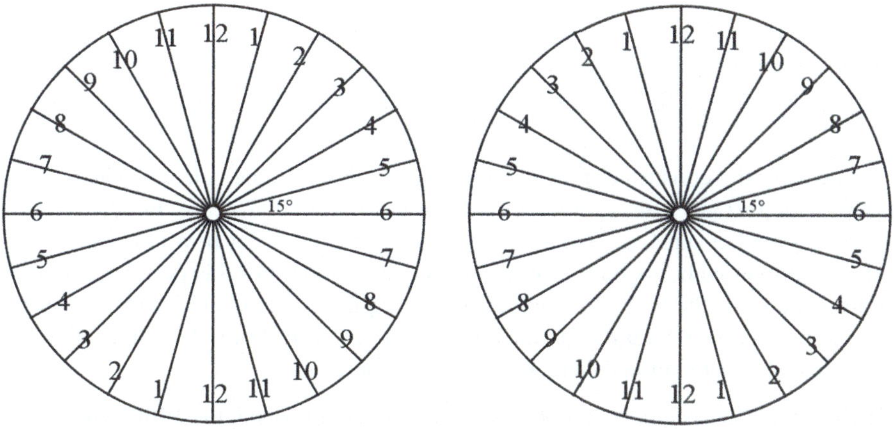

Fig. 9 Summer and winter dials of the equatorial sundial

2.2 *Horizontal Sundial*

The horizontal sundial is defined by a face that is the celestial horizon plane.

As in the case of the equatorial sundial, the planes dividing the celestial sphere into 24 equal parts indicate the lines of individual hours on the face of the horizontal sundial (Fig. 10). The Earth's axis is the gnomon.

Considerations related to the horizontal sundial in the celestial sphere allow obtaing the certain information to set it in the field: (1) the face of a horizontal sundial is placed on the plane of the local horizon or on a plane parallel to the local horizon, (2) the 12th hour line (at noon and at midnight) is located on the local meridian, (3) the gnomon is attached to the sundial face at an angle equal to the latitude of the place where the sundial is set and is parallel to the Earth's axis.

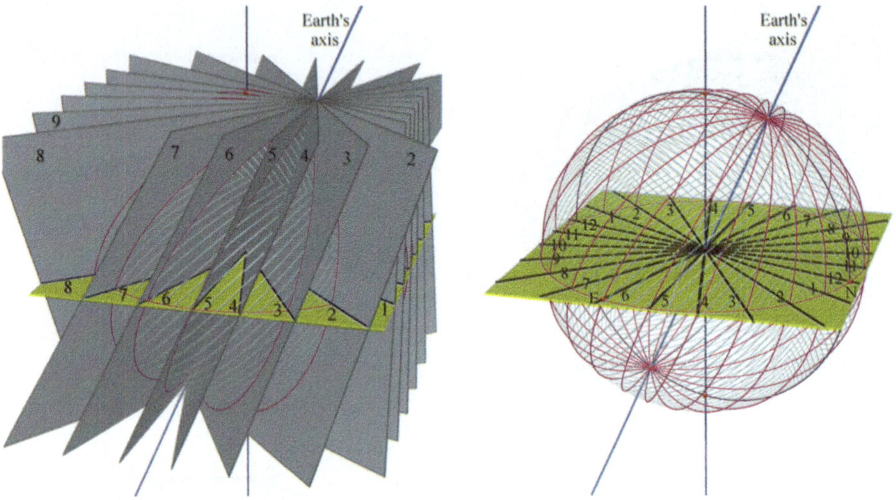

Fig. 10 Horizontal sundial in the celestial sphere

However, the construction of the horizontal sundial in the celestial sphere does not provide full guidelines for the preparation of the sundial's face. In other words, when using a flat surface, such as a marble slab, it is not clear how to construct the lines of individual hours on it. This requires clarification. For this purpose, the relationship between the horizontal sundial and the equatorial sundial is used (Tarnawski 1930, pp. 47–49), which results from the construction of both of these sundials in the celestial sphere, and is characterized by Theorem 1.

Theorem 1 If we set the equatorial sundial in the field so that the 12th hour line (at noon and at midnight) is located on the plane of the local meridian and the gnomon is parallel to the Earth's axis, and we set the horizontal sundial so that its gnomon coincides with the gnomon of the equatorial sundial, then the lines of hours on the faces of both these sundials will intersect on the line along which the faces meet (Fig. 11).

In school education in the Polish territories, the face of the horizontal sundial was prepared in two ways: using basic geometrical constructions or using trigonometry. The first method was simpler, which is why it was most often used in secondary schools. The geometric skills needed for the construction of sundial faces were as follows: moving the segment, moving the angle, dividing the right angle into six equal parts, constructing a line perpendicular to a given line and passing through the given point, and drawing a line parallel to the given line.

In order to prepare the face of a horizontal sundial by a geometric construction, the following task had to be solved:

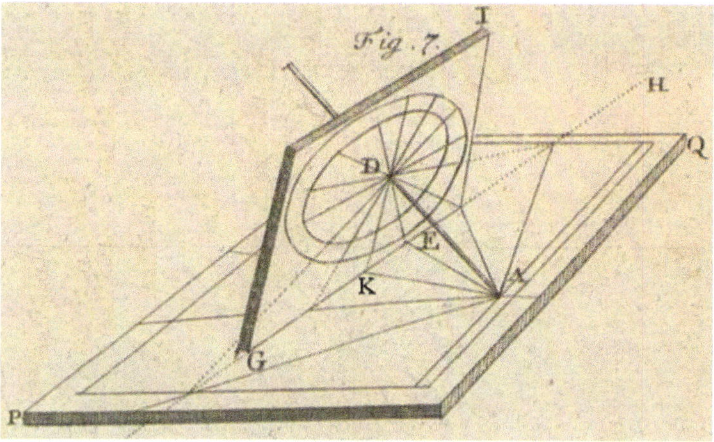

Fig. 11 The relationship between the horizontal and the equatorial sundials' faces (Wolff 1775)

Task 1 The length of the gnomon and the latitude of the place where the horizontal sundial will be set are given. Using Theorem 1, construct the lines of hours on the face of the horizontal sundial.

Solution:

(a) Using the length of the gnomon and the latitude of the place where the horizontal sundial will be set, construct the center of the horizontal sundial face and the line along which the faces of the horizontal and the equatorial sundials meet.
In order to do so, a triangle DEA should be constructed (in an arbitrary place on a piece of paper) such that $|\sphericalangle \text{EDA}| = 90°$, $|\sphericalangle \text{DAE}|$ = latitude of the place where the sundial will be set, and |DA| is equal to the length of the gnomon. Point A is then the center of the face of the horizontal sundial, and the line perpendicular to EA passing through the point E is the line in which the faces of the horizontal and the equatorial sundials meet; let us denote this line by GH.

(b) The line GH divides a sheet of paper into two half-planes. On the empty half-plane, construct the winter face of the equatorial sundial.
First the point F on the extention of the segment AE should be determined, such that |EF| = |DE|. The point F is the center of the equatorial sundial face. Then the hour lines should be constructed. To do this, we have to draw any circle with the center at point F and a radius less than or equal to |EF|, divide this circle into 24 equal parts, starting from the segment EF, and then draw straight lines through the point F and each of the points of division. These lines are the hour lines we are looking for. The line on which the segment EF is located is the 12th hour line, on the left, looking from the center of the dial, then there are successively the 1st, 2nd, 3rd hour line, etc. The points of intersection of these hour lines with GH we mark successively as e_1, e_2, e_3, etc.

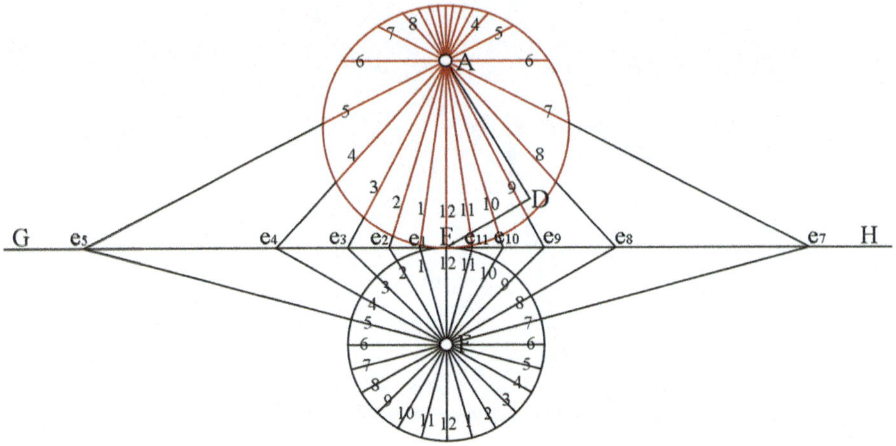

Fig. 12 Geometric construction of a horizontal sundial face

(c) Construct the horizontal sundial's hour lines.
On the half-plane where the triangle DEA is placed, the straight lines passing through the point A and each of points E, e_1, e_2, etc., should be constructed. These lines are the hour lines on the face of the horizontal sundial so that the solution is complete (Fig. 12).

In school practice, the face of the horizontal sundial was also prepared using the trigonometric method—definitions of basic trigonometric functions and trigonometric tables. Until the end of the 19th century, trigonometric functions in school education in the Polish territories were defined in a circle with a radius of 10^4, sometimes also in a circle with a radius of 10^5 or 10^6, for each of these radii there were prepared special trigonometric tables, and these were four-, five- and six-digit tables, respectively. The reason for using a circle with a radius equal to a power of 10 was to avoid calculations with fractions.

To prepare the face of a horizontal sundial using the trigonometric method, the following task had to be solved:

Task 2 Determine the angles between the meridian line (the 12th hour line) and successive hour lines on the face of the horizontal sundial, knowing that the latitude of the sundial location is equal to α. Perform the solution based on Theorem 1.

Solution (Toński 1640, pp. 258–263):

Let us assume the markings as in Fig. 11. Based on Theorem 1 and the data given in the task, we can conclude that ΔDEA is a right triangle, in which $|\sphericalangle EDA| = 90°$, $|\sphericalangle DAE| = \alpha$, |-----------------$\sphericalangle DEA| = 90° - \alpha$ (that is, the inclination of the equator to the horizon). Let us assume that $|EA| = 10^4$, then from the definition of sine in the circle with a radius equal to 10^4 (Fig. 13.) we get that $|DE| = \sin \alpha$.

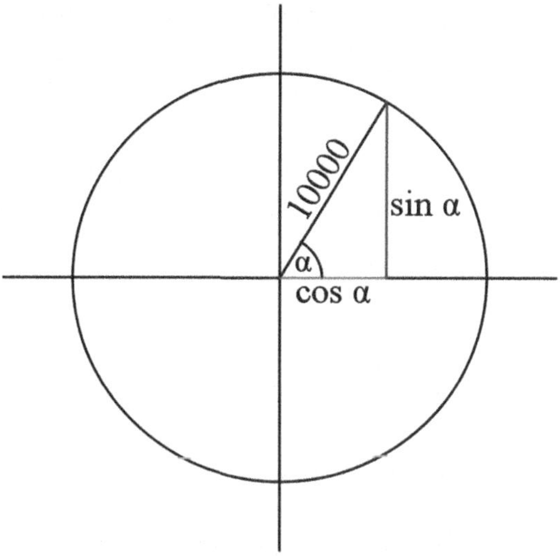

Fig. 13 Definitions of the sine and cosine in a circle with a radius of 10^4

We also know from Theorem 1 that: (1) ΔDKE is a right triangle, in which $|\sphericalangle DEK| = 90°$ and $|\sphericalangle KDE| = 15°$; hence $\tan 15° = \frac{|KE|}{|DE|}$, so $|KE| = \tan 15° \cdot |DE|$; (2) ΔEAK is a right triangle, in which $|\sphericalangle KEA| = 90°$; hence $\tan|\sphericalangle EAK| = \frac{|KE|}{|EA|}$, so $|KE| = \tan|\sphericalangle EAK| \cdot |EA|$. Therefore $\tan 15° \cdot |DE| = \tan|\sphericalangle EAK| \cdot |EA|$, so $\tan|\sphericalangle EAK| = \frac{\tan 15° \cdot |DE|}{|EA|} = \frac{\tan 15° \cdot \sin\alpha}{10000}$. If we read the values of $\tan 15°$ and $\sin\alpha$ from the trigonometric tables, then we know the value of $\tan|\sphericalangle EAK|$. Knowing $\tan|\sphericalangle EAK|$, we can use the trigonometric tables again and read the value $|\sphericalangle EAK|$. Thus, we know the angle that the meridian line (the 12th hour line) makes with the 11th hour line. The same principle is used to determine the angles between the meridian line and the remaining hour lines. The 6th hour line is perpendicular to the 12th hour line.

We can suppose that the sundial is to be set in Krakow and then use the latitude of Krakow and the trigonometric tables given (Toński 1640). The value of sin50°12' in the trigonometric tables is equal to 7683 and the value of tan15° is equal to 2679. So $\tan|\sphericalangle EAK| = \frac{2679 \cdot 7683}{10000} = 2058$. Hence the tangent of the angle between themeridian line and the 11st hour line on the horizontal sundial is equal to 2058, which, from the trigonometric tables, gives an angle measure of 11°38'. The same angle is between the meridian line and the 1st hour line. In the case of the 2nd and 10th hour lines the angles are equal to 23°55', for 3rd and 9th hour lines – 37°32', for 4th and 8th – 53°4', for 5th and 7th – 70°46'. For the 6th hour line the angle is equal to 90°.

2.3 Other Types of Flat Sundials

For all other types of flat sundials, as in the case of the horizontal sundial, the construction in the celestial sphere determines the way the sundial should face and the way the gnomon should be set in the field, while the detailed construction of the hour lines is derived from the relationship with the equatorial sundial face. In addition to the equatorial and horizontal sundials, the following types of astronomical flat sundials were discussed in school in the Polish territories:

(1) Vertical sundials—these are sundials whose faces are perpendicular to the horizon.
Vertical sundials are divided into:

   (a) north and south vertical sundials—these are sundials with faces (dials) situated on the east-west plane and facing north and south, respectively
   (b) east and west vertical sundials—these are sundials with dials situated on the plane of the meridian and facing east and west, respectively
   (c) tilted sundial—this is a sundial whose face is located on the plane perpendicular to the horizon, which is neither the plane of the meridian nor the east-west plane

(2) Inclination sundials—these are sundials whose dials are inclined to the horizon at an angle other than 0° or 90°.
These sundials are divided into:

   (a) north and south inclination sundials—these are sundials, the faces of which are located on the plane crossing the horizon along the east-west line and facing north and south respectively
   A special example of the inclination sundial, which can be faced either north or south, is the equatorial sundial
   (b) east and west inclination sundials—these are sundials, the dials of which are located on planes crossing the horizon along the line of the meridian and facing east and west, respectively
   (c) deviating sundial—this is sundial, the dial of which is located on a plane whose intersection with the horizon coincides neither with the east-west line nor with the meridian

(3) Polar sundial—this is a special type of inclination sundial; the dial of this sundial is inclined to the horizon at an angle equal to the latitude of the place where the sundial is set. The polar sundial can be faced north, south, east, west, or it can be deviating.

2.4 Practical Skills Needed to Construct Sundials

In order to prepare the sundial, in the addition to the above theory, the following practical skills are needed:

(1) determination of the local meridian line, i.e. the north-south line

The meridian line in school teaching was determined in several different ways, all of them were based on the observation of the length of the shadow cast by a rod stuck vertically into the ground. One of the easiest ways to determine a local meridian line was to capture the moment when the shadow was the shortest—at that time, the Sun was at the moment of its culmination, and the shadow marked the north-south line.

(2) determination of the latitude of the place where the sundial will be located

Let us assume that there is an observer in the place where the sundial will be located. To determine the latitude of this place, the observer should determine the height of the pole (Kowalski 1868, pp. 19–20), i.e. the measure of the angle between the local horizon and the pole. In school teaching, the height of the pole has often been approximated and identified with the height of the pole star. A quadrant was used to determine it.

The quadrant (Fig. 14) consisted of a scaled protractor measuring from 0° to 90° and two diopters (rulers with perpendicular plates with small holes) attached to it. One diopter, in Fig. 14: AB, was fixed motionless to the quadrant, and the second, CF, was mobile. In order to measure the height of the pole star, the diopter AB was situated parallel to the local horizon, and the

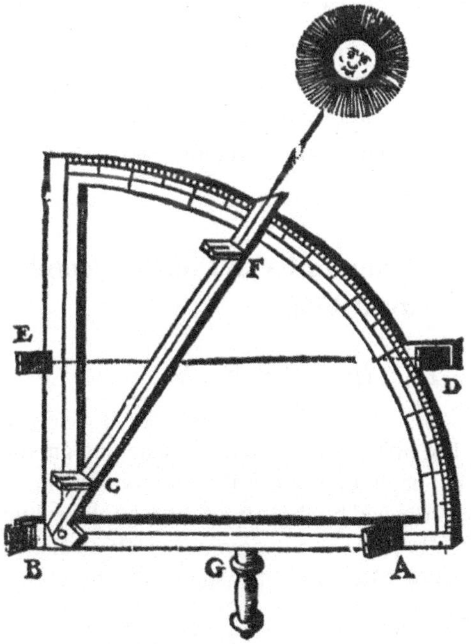

Fig. 14 Quadrant (Metium 1632, p. 128)

Fig. 15 The method of using a quadrant (Tolomeo 1574)

mobile diopter CF, looking through the holes in the plates, was aimed directly at the polar star. Then the diopter CF indicated the height of the polar star on the protractor (Fig. 15).

3 Gnomonics in Polish Secondary School Education until 1794

3.1 *Types of Schools*

The first information about teaching gnomonics in the Polish territories comes from the beginning of the 17th century. At that time, secondary education in Poland was dominated by three types of schools: colony-schools run by universities (these were Catholic schools), gymnasia (Protestant), and Jesuit schools. The first colony-schools were established in the 15th century and operated under the patronage of the oldest Polish university, i.e. the University of Kraków[4]—the teachers in the colony-schools, e.g. at the Lubrański Academy in Poznań, were often university lecturers. They used university textbooks, and students had also the opportunity to attend university lectures. The first schools of this type opened in the 15th century in Pułtusk and Warsaw. By the end of the 18th century, over 40 were established (Łukaszewicz 1851; Duda 2019, p. 62). Colony-schools were also opened by the first Polish private university—the Zamojski Academy (est. 1593); however, more of those were run by the university in Kraków, and they enjoyed a greater reputation. Catholic colony-schools were reformed, and a new type of school was created at that time—Protestant gymnasia, run according to the Sturm rules. The first were gymnasia in Elbląg (1535), Gdańsk (1558) and Toruń (1568), the Calvinist Gymnasium in Pińczów (1551), the Gymnasium of the Czech Brethren in Koźminek (1551), and a few more were established later. Jan Zamojski, the founder of the Zamojski Academy, was one of the first Poles to graduate from Johann Sturm's (1507–1589) Protestant gymnasium, established in Strasburg in 1536. Zamojski and other graduates of this gymnasium quickly began to export the knowledge gained there to Polish schools. As a result, a class-lesson system was introduced, as well as a slightly modified *trivium* and *quadrivium*, from which the beginnings of arithmetic, geometry, astronomy with astrology, and music were taught.

In 1564, Jesuits entered the Polish territories and immediately began establishing schools called Jesuit colleges, which valued mental education of a philological and historical nature, and it was the third type of school at that time. The first Jesuit colleges were established in Braniewo (1565), Pułtusk (1566), Wilno (1570) and Poznań (1570). In 1599, there were eleven of them, each attended by from one hundred to one thousand students (Bartnicka 2000, p. III). Among them were not only candidates for priests, but also lay students (sons of nobility). It is difficult to determine the percentage of lay students due to the lack of preserved data, but it was certainly significant since the nobility had a real influence on the curricula (Łukaszewicz 1851). In 1599, a unified curriculum, *Ratio Studiorum Scholarum Societatis Jesu* was prepared for this schools. In the mid-17th century, the Piarists came to Poland and they also started to establish their schools—Piarist schools. In 1755 regulations for them were prepared by Stanisław Konarski (1700–1773). In 1773, there were over 20 Piarist schools and 66 Jesuit schools (Bednarski 1933). In addition, there were several schools run by Theatines, Basilian monks, and others.

In the year 1764, the first military school was created in Poland. At that time, the School of Chivalry was established in Warsaw, which "was to provide the country with enlightened citizens for the political and social reconstruction of the state" (Lisowski 1982, p. 24). It was the first secular school institution in Poland—completely independent of the Church, where almost exclusively lay professors taught. The first Corps of Cadets was already established by the Commission of National Education—in Grodno (1774), Rydzyna (1783), and Niemirów in Podolia (1787).

[4] In the first years of the 15th century, a chair of exact sciences was established here (Duda 2019, p. 26). It was the second, after Bologna.

By the end of the 18th century, two corps of cadets in the Prussian Partition were established—one in Kalisz (1794) for students of Polish nationality and the other in Chełmno (1776) for students of Prussian nationality. Most colony-schools, Jesuit colleges, and Protestant gymnasia in 1773 were taken over by the Commission of National Education.

3.2 Intended Curricula in Gnomonics

Among all these schools, the Polish Jesuit colleges enjoyed the greatest tradition of teaching gnomonics. Although the curricula implemented in them were humanistically oriented, and the *Ratio Studiorum* did not mention gnomonics (Bartnicka 2000, p. 70), the constructions of sundials constituted a significant part of the curricula carried out there. The first extant precise information about discussing the construction of sundials in secondary schools within Poland is from 1602, and it relates to the establishment of chair of mathematics with astronomy and gnomonics as well as three chairs of Aristotelian philosophy in the Jesuit college in Kalisz. In the curriculum, which was implemented in 1636 by local teacher Tomasz Brodeski, it was noted that "After the basic lecture on arithmetic and mathematical philosophy, with certain pieces of information from its history, there was a thorough lecture in geometry with preliminary remarks about gnomonics" (Dobrzycki et al. 1975, p. 239). Gnomonics was taught there until 1774.

From the beginning of the 17th century, Polish Jesuits began to issue ordinances tailored to the needs of Polish society, which often differed significantly from the guidelines included in the *Ratio Studiorum*. This was the case, for example, with *Consuetudines Scholastiae Provinciae Polonae et Litunae S. J.* (Bednarski 1933, pp. 494–506), which (contrary to the ordinances of 1599) recommended extending theatrical events. Consequently, in Polish Jesuit colleges, theatre was a permanent part of education and a total of about 30,000 performances were staged there (Bartnicka 2000, p. V). One of the later ordinances, namely the *Catalogus Praelectionum in Scholis Societatis Jesu Provinciae Polonae* of 1711, introduced gnomonics as an obligatory element of mathematics teaching in Polish Jesuit colleges: "In Mathematica: Elementa Euclidis, Arithmetica, Geometria, Geographia, Astronomia, Gnomonica, Calendarium etc. pro arbitrio" (Bednarski 1933, p. 483). The construction of sundials was taught during the three years of study.

There are no specific curricula left with which to define the implemented and attained curricula of gnomonics. However, it is possible to sketch the intended curricula. In 1777, gnomonics appeared at the "Popis Publiczny" of students of the Jesuit college in Połock (Anon. 1777; Więsław 2007, p.147). "Popis Publiczny" was an exam that students took in front of teachers, representatives of the local elite, and parents of the students. The aim of these exams, apart from checking students' knowledge, was to present the school in the best light, to show its high level of education, and to encourage parents to send their children there. The examination questions were focused on issues that were discussed in the classroom.

During the examination in Połock, students received eight questions on gnomonics which concerned the construction of the sundial and setting the sundial in the field. Among them were the following tasks: (1) Finding the local meridian line; (2) Constructing a horizontal and vertical sundial; (3) Constructing an equatorial, polar, vertical east and west sundial; (4) Constructing a vertical tilted sundial and an inclination sundial; and (5) Constructing a sundial on a vertical plane that is tilted from south to west. Thus, the examination tasks included only the constructions of sundials on flat surfaces, i.e. the constructions of the so-called flat sundials.

To illustrate the methods of constructing sundials that were taught at Jesuit colleges, we will refer to the anonymous manuscript *Acervus selelecius* (n.d.)[5], which was most likely used at the Jesuit college in Kalisz. This is evidenced by several facts: (1) this manuscript covers the basics of mathematics (arithmetic and geometry) and astronomy, i.e. material recommended for discussion in Polish Jesuit schools in accordance with the ordinance of 1711; (2) on the first page of *Acervus selelecius* devoted to gnomonics, the author wrote: "We are not going to teach here the drawing of all sundials or [...] one in many ways, because this cannot be done even in one whole year, let alone in one month" (*Acervus selelecius*, p. 203)—this means that the material placed there was to be of help during the monthly lecture on gnomonics; (3) the author made some of the construction generally for any latitude, and some only for the latitude of Kalisz (he did not consider any other places)—this means that the textbook was most likely used in Kalisz; (4) the material placed there and the methods of construction was taken from the gnomonics textbook by Jesuit Christoph Clavius (1612), which was in common use by students and pupils of Jesuit schools; (5) the gnomonics of Clavius was often considered too extensive (Popis publiczny uczniów szkoły wojewódzkiey Lubelskiey 1823), especially for students, and therefore the author of *Acervus selelecius* simplified certain constructions, precisely by limiting the considerations only to the latitude of Kalisz; and (6) at that time, the Jesuit college was the school in Kalisz with the most extensive curriculum. In *Acervus selelecius,* the author considered four types of sundials due to the type of hours taken into account: astronomical, ancient, Babylonian, and Italian sundials. At the same time, he discussed in detail the construction and methods of setting astronomical, Babylonian, Italian and ancient horizontal, polar, equatorial and vertical south sundials, all with the lines of the zodiac on their faces. He carried out the construction of astronomical and ancient sundials in general, for any latitude, and Babylonian and Italian sundials only for the latitude of Kalisz. Although the author mentioned that the skills needed to construct the sundial are: finding the latitude of the place where the sundial will be set, that is, finding the polar star height using a quadrant, and determining the meridian line, he did not explain how to do this. In the Jesuit college in Połock examination tasks from the 1777, did not include astronomical sundials, ancient sundials, etc., nor zodiac lines. This means that the tasks: "2) Constructing a horizontal and vertical sundial", "3) Constructing an equatorial, polar, vertical east and west sundial", etc., should be understood as requiring students to construct astronomical sundials without the zodiac lines. The

[5] Kórnik Library, sign. BK 01025.

faces of astronomical sundials in *Acervus selelecius* were constructed in a purely geometrical manner, without providing explanations or proofs of correctness. The instructions for construction of the horizontal sundial face in the manuscript was as follows[6]: (1) construct a triangle DEA, such that

- $|\sphericalangle EDA| = 90°$,
- $|\sphericalangle DAE|$ = latitude of the place where the sundial will be set,
- |DA| is equal to the length of the gnomon,

(2) draw a line perpendicular to EA passing through the point E and denote it by GH, (3) extend the segment AE and designate the point F on this extension, such that |EF| = |DE|, (4) draw a circle with the center F and a radius equal to |EF|, then divide this circle into 24 equal parts, starting from the segment EF, and then draw straight lines passing through the point F and each point of division, (5) denote the points of intersection of these straight lines with GH as e_1, e_2, e_3, etc., (6) construct straight lines that pass through the point A and each of the points E, e_1, e_2, etc.—these lines are the lines of hours on the face of the horizontal sundial.

Another textbook that allows us to characterize the intended curricula in gnomonics in Polish Jesuit colleges was *Geometra Polski, tj. nauka rysowania, podziału i rozmierzania linii, angułów, figur i brył* (Solski 1686). It was the first school geometry textbook written in the Polish language[7], and it includes considerations on the construction of sundials (almost 60 pages). This textbook was used at the Jesuit college in Vilnius in 1768–1769 (Więsław 2007, p. 144–146). Solski began his considerations by explaining how to find the local meridian (he used a compass with a magnetic needle for this purpose).

The considerations carried out by Solski were more elaborate than those in *Acervus selelecius* and to some extent were modeled on the textbook by Clavius, as Solski himself stated—following Clavius's footsteps he explained how, using the face of a horizontal sundial, one can construct a sundial face on any surface (Clavius 1612, p. 532–533; Solski 1686, p. 53). Solski discussed the differences between the astronomical, Babylonian, Italian and ancient sundials; he analyzed the constructions of the horizontal, the east, west, north and south vertical sundials, the equatorial, and the polar sundial; he presented the principles of the construction of inclination and tilted sundials; he discussed the construction of the sundials on the sphere, the cylinder, and the ring; he explained the methods of drawing zodiac lines; and finally he explained how to determine the hours of sunrise and sunset and the lengths of the day and night. He carried out these considerations in a purely theoretical way—based on the geometric knowledge of the readers (in this aspect similarities with Clavius's gnomonics are visible), and with the use of sundial drawing instruments (this was not the case with Clavius). For example, Solski discussed the construction of an instrument for drawing any vertical sundial by using the lines of the zodiac (Fig. 16).

[6] Compare this construction with the solution to Task 1 in Sect. 2 of this article.

[7] The first book on geometry written in Polish was (Grzepski 1566), but it was not used in Polish secondary schools.

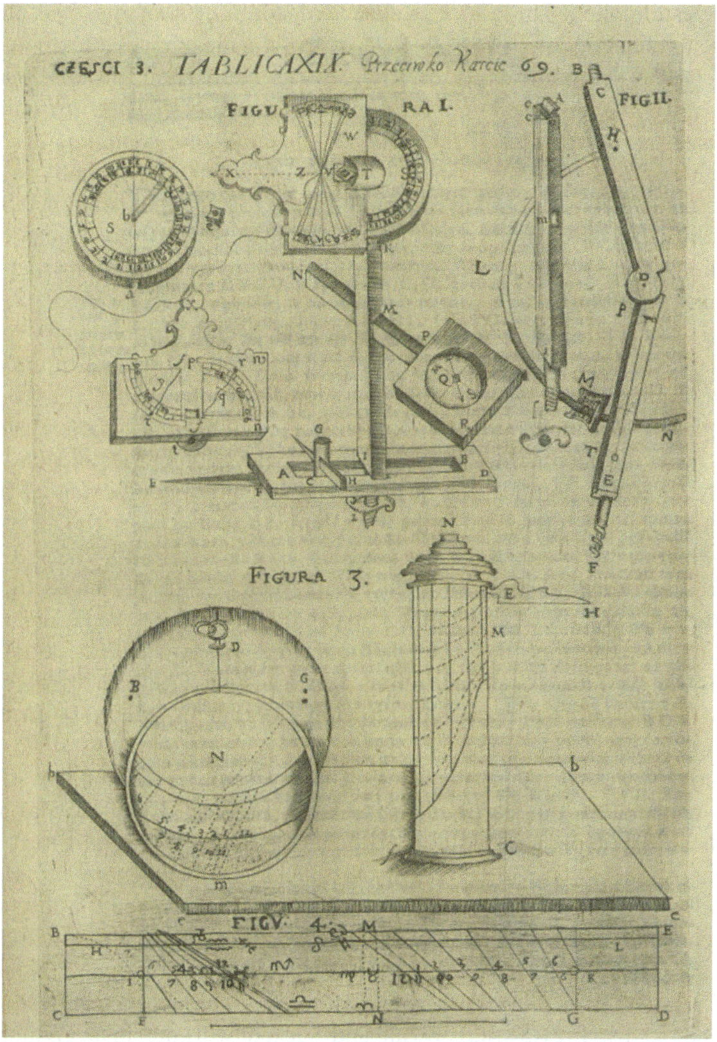

Fig. 16 Instrument for drawing any vertical sundial using the lines of the zodiac (Figura I, Figura II) (Solski 1686)

The considerations of Solski were much more practical than those in *Acervus selelecius* and the gnomonics textbook by Clavius, mainly due to the use of special instruments in the construction of sundials. Solski attached much less importance than his predecessors to Babylonian, Italian, and ancient sundials—he limited himself to constructing an instrument for drawing Italian and Babylonian vertical sundials and discussing its use, and also to the construction of ancient ones. Solski, unlike the author of *Acervus selelecius*, discussed the construction of all sundials for any latitude.

In the Jesuit colleges in Połock, Kalisz, and Vilnius, gnomonics was taught in a geometrical (constructional) way. In 1640, one of the most popular textbooks for the colony-schools of the university of Kraków was published: Toński's *Arithmetica vulgaris et trigonometria rectilineorum pro ut universae geometria Practica, alijusqve Matheseos partibus, Geographia, Architectonica, Gnomonica etc.* (1640)[8]. This textbook was used for example in the *Nowodworski* School in Kraków (Komarzyniec 2004, p. 56; Łukaszewicz 1849, pp. 231–232). It contained the basics of gnomonics, but it was the first textbook in which the plane trigonometry approach was applied. It should be emphasized that there were no regulations that required the teaching of gnomonics in colony-schools.

Toński divided his textbook into three parts: arithmetic, trigonometry (plane and spherical), and "Miscellanea", which were practical tasks related to architecture, geography and gnomonics. The main gnomonics problems that Toński included were: (1) Determine the height of the center of the Sun from the shadow of the gnomon; (2) From the declination of the Sun and the elevation of the pole, find the time of the day; (3) Construct the face of the astronomical horizontal sundial (the solution of this problem is presented in the Task 2 in Sect. 2 of this article) (Toński 1640, pp. 258–263).

Toński's solutions to these problems were sketchy—he presented the way of solving the task rather than explaining it in detail. For example, he lacked a description of how to determine the local meridian line. His procedure for solving the problems was as follows: (1) providing a theory which helped to solve a given problem (he referred to the works of Ptolemy, Copernicus and Tycho Brahe here), (2) practical tasks.

Gnomonics was also taught in selected Protestant gymnasia. What was the intended curricula in this case? In 1634, Adam Freytag (1608–1650), one of the most outstanding experts in bastion fortifications, author of *Architectura Militaris Nova Et Aucta, Oder Newe Vermehrte Fortification, Von Regular Vestungen, Von Irregular Vestungen und Aussen Wercken* (Freytag 1631) became a teacher at the Toruń Gymnasium, and immediately expanded the local curricula by including gnomonics, geography, military architecture, chronology, and physics (Kubik 1974, p. 157). However, it is not known what textbooks he based his lectures on. At the end of the 1640s, Freytag left Toruń and started working in a Gymnasium in Kiejdany, and gnomonics was removed from the curricula in Toruń for several decades. The construction of sundials only returned there thanks to Friedrich

[8]At that time, only three textbooks were used to study mathematics in the colony-schools of the university in Krakow:

(1) Brożek, Jan. 1620. *Arithmetica integrorum edita a M. Joanne Broscio etc.* Cracoviae: Ropelo Nowodworsciano.

(2) Schedel, Krzysztof. Ca.1665. *Arithmetyka to jest: nauka rachunku na trzy podzielona księgi itd.* Krakow: Schedlowie.

(3) Toński, Jan. 1640. *Arithmetica vulgaris et trigonometria rectilineorum pro ut universae geometria Practica, alijusqve Matheseos partibus, Geographia, Architectonica, Gnomonica etc.* Ingolstadt: Gregorii Hanlin (Łukaszewicz 1849, pp. 231–232).

Bormann (from 1713 to 1724) and was discussed during the subject called "foundations of mathematics" (Kubik 1972, p. 125). Bornmann based his lessons on two textbooks: Wolff (1716) and Wolff (1775). *Mathematisches Lexicon* (1716) was encyclopedic in nature. It explained gnomons, gnomonics, and people interested in details were sent by Wolff to, among others, *Elementis Gnomonica*, which was written by him (Wolff 1716, pp. 679–682). The lexicon discussed the types of sundials, but without giving details of their construction. Therefore, Friedrich Reinhold Bornmann based his lectures on gnomonics in the Toruń Gymnasium mainly on *Anfangsgründe Aller Mathematischen Wissenschaften* (1775). From the 1750s to 1762 gnomonics was also taught in Toruń by Samuel Theodor Schönwald (Kubik 1974, p. 158; Kubik 1972, pp. 132–133; in 1762 Schönwald died), on the basis of the same textbook.

In 1645–1660, gnomonics was included in the curricula of the Protestant Gymnasium in Gdańsk, too. It was taught by Wawrzyniec Eichstadt (Kubik 1959). Later, in the 18th century, it was discussed by two another teachers—Henryk Kühn (in Gdańsk: 1733–1769) and Jan Bogumił Bartoldi (1774–1788)—based, as in Toruń, on the works by Christian Wolff (Kubik 1974, p.158).

The *Anfangsgründe Aller Mathematischen Wissenschaften* was very popular in the 18th century—it was used not only in Protestant secondary schools, but also in Piarist schools, e.g. in *Collegium Nobilium* in Warsaw. It also served as an academic textbook in Jesuit universities, and Jakub Nakcyanowicz (1725–1777) based his mathematical lectures on it, which he conducted at the Vilnius Academy. Nakcyanowicz also decided to translate this textbook into Latin (Pawlikowska-Brożek 2004, pp. 261–262).

Wolff discussed the constructions of the equatorial sundial as well as the horizontal, vertical (south, north, east, and west), polar, and tilted varieties, he also explained the method for constructing sundials on any flat surface. All of the sundials he discussed were astronomical sundials, without the zodiac lines. Dealing with the equatorial sundial, he first constructed its summer and winter dials, discussed how to attach the gnomon and set the sundial in the field, and then conducted a proof of the correctness of the construction based on the basic theorems of astronomy (related to the movements of the Earth and the Sun) and optics (Wolff, pp. 1528–1532), in other words, he focused on considerations related to building the sundial in the celestial sphere. When constructing the dials of all other sundials, Wolff proceeded as follows: (1) he carried out a purely geometric construction of the sundial face (in the case of a horizontal sundial it was analogous to that presented in *Acervus Selelectius*), (2) he provided a proof of the correctness of the construction, based on the relation between the face of the constructed sundial and the face of an equatorial sundial (in the case of a horizontal sundial, this relation is presented in Theorem 1, discussed in Sect. 2 of this paper).

Wolff, in the preface to the chapter on gnomonics, announced that he would discuss only those constructions that are easily understood and implemented. In fact, the implementation of constructions presented in the *Anfangsgründe Aller Mathematischen Wissenschaften* textbook required only basic geometric skills, such as: moving the segment, moving the angle, dividing the right angle into six equal parts, constructing a line perpendicular to a given line and passing through the given point, and drawing a line parallel to the given line.

The traditions related to teaching gnomonics in Polish secondary schools were interrupted in the period between the First and the Third Partition of Poland. In 1773, the Polish Commission of National Education started its activity, taking over most of the existing schools, including Jesuit schools and colony-schools. The *Ustawy Komisyj Edukacyj Narodowej dla stanu akademickiego i na szkoły w krajach Rzeczypospolitej przepisane w Warszawie roku 1783* contained curricula recommended for implementation in Polish secondary schools run by the Commission. As part of mathematics, it included the basics of arithmetic, geometry, algebra, and logic, with particular emphasis on the use of geometry to solve tasks related to geodesy (Ustawy Komisyj Edukacyj Narodowej... 1902, p. 55). These ordinances did not mention the construction of sundials either in mathematics or physics, and these were the only subjects on which gnomonics could be taught at that time—the Commission of National Education did not provide for separate lectures in geography or astronomy in secondary schools. At that time, Polish school curricula were dictated by the invaders, and as a consequence of the partitions, gnomonics was not a compulsory subject. But there were individual schools in which this subject was taught. Paul Schwartz, in analyzing the matriculation examinations and curricula implemented in secondary schools in Prussia until 1806, mentioned only one school located in the Polish lands under Prussian rule in which gnomonics was taught—in 1788 it was included in the curricula of *Die Provinzialschule in Lyck* (Pol. Ełk) (Schwartz 1910, p. 251).

4 Intended Curricula in Gnomonics in the Years 1795–1918

In the period 1795–1918, Poland did not exist formally; its territory was occupied by Prussia, Russia, and Austria. Despite this, there were still Polish schools with Polish curricula. They functioned mainly in the Duchy of Warsaw (1807–1815) (subordinate to France), the Kingdom of Poland (1815–1832) (subordinate to Russia), and Galicia (1868–1918) (subordinate to Austria).

The first half of the 19th century saw a renaissance in teaching gnomonics in Polish schools. In 1812, an ordinance was issued, entitled *Plan nauk szkół departamentowych*, for Polish secondary schools preparing for matriculation exams functioning in the Duchy of Warsaw; it made gnomonics a compulsory subject of study. These were the first regulations concerning the teaching of gnomonics in Polish secular schools. Gnomonics was taught in the 5th grade (the whole cycle of education in these schools included six grades—from the lowest 1st to the highest 6th grade) as part of the subject "physics and chemistry," taught for 3 hours a week. The following note was included in the curriculum at that time: "*Gnomonics*. Information from the work of Wolfius' *Compendium Matheseos*" (Lipiński 1812, pp. 61, 71; Wolff 1758). The material on gnomonics in this book was identical to that in Wolff's (1775). *Plan nauk szkół Departamentowych* of 1812, a statement concerning compulsory mathematics topics in schools preparing for matriculation examinations in the Duchy of Warsaw, also introduced "the beginnings of descriptive geometry" (Lipiński 1812, p. 54).

Ministerial orders were not always implemented. For example, in 1813 in the Warsaw departmental schools of the Piarist Fathers, descriptive geometry was taught in mathematics lessons according to works by Gaspard Monge and Jean Nicolas Hachette, but there was no gnomonics in the curricula (Popis publiczny uczniów szkół departamentowych warszawskich pod nadzorem XX. Pijarów 1813). Warsaw. It appeared later—in 1814 and 1815 it was discussed in the 5th grade during the lessons on astronomical geography: "Drawing the solar compass on the equatorial plane, horizontal plane, etc." (Popis publiczny uczniów szkół departamentowych warszawskich pod nadzorem XX. Pijarów 1814). Warsaw. At the same time, gnomonics was taught in the departmental schools in Lublin during the "physics and chemistry" lessons: "At last, [...] there was a Gnomonics lecture presenting the methods of drawing sundials on variously inclined and declined planes, according to the work of Wolfius" (Na publiczny popis uczniów Szkół Departamentowych Lubelskich 1814, pp. 28–29, 1815, p.32). However, it was not included in the curricula of the departmental school in Kalisz (Popis publiczny Uczniów Szkoły Departamentowey Kaliskiey 1814) and of the departmental school in Kraków (Popis publiczny uczniów Szkoły Departamentowey Krakowskiey 1815, p. 36).

In 1820, when the Duchy of Warsaw no longer existed and part of its territory was under the Russian occupation, the ordinances *Wewnętrzne urządzenie szkół wojewódzkich* were prepared for schools ending with matriculation examinations in the Kingdom of Poland. These ordinances removed descriptive geometry from the curricula, but gnomonics remained. Gnomonics was now taught in the 6th grade of provincial schools and it was part of physics instruction (Lipiński 1820, p. 123). These ordinances did not provide the textbook on the basis of which the lecture was to be conducted.

It is significant that after 1820, the reports of most provincial schools operating in the Kingdom of Poland did not include any curricula. Curricula were not included in the reports of provincial schools in, for example, Kielce, Pułtusk, Lublin, Warsaw, or Kalisz. These schools declared that they would implement curricula recommended by the Ministry. However, there were exceptions, thanks to which we know that in 1820 in the provincial school in Płock, gnomonics was taught during physics lessons (Sprawa przed Publicznością na Popisie Rocznym Szkoły Wojewódzkiey Płockiey... 1820, p. 8); in the provincial school of Benedictine Fathers in Pułtusk during "geography and gnomonics" lessons (according to the Wolff textbook) (Popis publiczny Uczniów Szkoły Wojewódzkiey XX. Benedyktynów w Pułtusku 1821, p. 25); in the provincial school in Traszkuny during "horticulture, physics and natural history" lessons (Treść nauk dawanych uczniom w szkole na stopniu Gymnazyinym... 1823, p. 12); and in the provincial school in Szczebrzeszyn in 1825, the construction of sundials was not taught at all (Popis publiczny uczniów Szkoły Wojewódzkiey Zamoyskiey w Szczebrzeszynie 1825).

The third decade of the 19th century brought the introduction of technical education in the Kingdom of Poland. Before the first real schools were opened here, in 1833, ordinances were introduced that established the division of gymnasia into two departments: philological and technical. The technical department's curricula include a subject called "descriptive geometry"—this subject was not taught in the philological department. Thus, descriptive geometry for the first time became an autonomous subject of teaching in the Polish territories (Schiller 1998, p. 65; Sbornik administrativnykh postanovlenij... 1868, pp. 153–155). Detailed guidelines for teaching descriptive geometry in the technical department were specified in the ordinances of 1835[9]. Descriptive geometry was then taught in the three highest grades—6th to 8th. The curriculum for 7th grade included the following topics, among others: "General theory of radius of curvature of lines or known surfaces—learning shadows and their uses for drawing compasses—the principles of linear and aerial perspective, explained with numerous examples" (State archives in Lodz, unit: Szkoła Powiatowa w Rawie, sign. 12). The construction of sundials stood behind the phrase "drawing compasses". Thus, in 1835, gnomonics became a compulsory element of the descriptive geometry teaching curricula in gymnasia operating in the Kingdom of Poland.

The Ministry recommended that the study of descriptive geometry should be based on *Geometrya wykreślna* by Franciszek Sapalski[10]. It is a textbook, the full title of which read as follows: *Geometrya wykreślna z zastosowaniem do perspektywy, cieniów, gnomoniki, kamieniarstwa, ciesiołki i innych konstrukcyy, wypracowana dla użytku Szkoły woyskowey applikacyyney* [*Descriptive Geometry with Application to Perspective, Shadows, Gnomonics, Masonry, Carpentry and Other Constructions, Developed for Use in a Military School*]. Sapalski prepared this textbook based on lectures on descriptive geometry which he conducted from 1816 at the university in Kraków. He prepared to conduct these lectures during his two-year stay in Paris, where, he studied at l'Ecole Politechnique under the supervision of Jean Nicolas Hachette (Zastosowań geometryi wykreślnej wedle rozkładu pomysłów... 1839, p. 8). Sapalski's textbook was to consist of two volumes—the first was to cover theory and the second was to apply descriptive geometry. In 1822, a volume concerning theory was published. Unfortunately, the poor health of Sapalski did not allow him to prepare the second volume of this work, and, therefore, shadow theory with its application for drawing compasses was recommended by the Ministry to be taught in gymnasia based on Drzewiński (1825), which was a Polish translation of the works of Joseph Mollet *Gnomonique Graphique, ou méthode simple et facile pour tracer les cadrans solaires sur toute sorte de plans, et sur les surfaces de la sphère et du cylindre droit sans aucun calcul, et en ne faisant usage que de la régle et du kompas* (1815) (State Archives in Lodz, unit: Szkoła Powiatowa w Rawie, sign. 12).

Sapalski also discussed gnomonics during lectures on descriptive geometry which he gave in Kraków. One of the students of these lectures in 1823 was Jan Kanty Steczkowski, who later became a professor at the university of Kraków. He made notes of these lectures which have survived to this day (Jagiellonian Library

[9] State archives in Lodz, unit: Szkoła Powiatowa w Rawie, sign. 12.

[10] State archives in Lodz, unit: Szkoła Powiatowa w Rawie, sign. 12.

Rkp. 4189 IV vol. 1; *Rkp. 4189 IV vol. 2*). The notes of Steczkowski are quite sparse. They consist almost exclusively of drawings (about 200), which very rarely have any captions. Four of these drawings illustrate the construction of sundial faces (Jagiellonian Library, *Rkp. 4189 IV vol. 1*, pp. 53–55): the first is the winter dial of the equatorial sundial, the second is the face of a horizontal sundial, the third is the face of an inclined and tilted sundial illuminated at noon, and the fourth is the general construction of the inclined-tilted sundial (Fig. 17).

An attempt to recreate the constructions made by Sapalski allows us to conclude that they are analogous to those found in Drzewiński (1825). Drzewiński, however, constructed a larger number of sundials; additionally, he built: south, north, east, and west vertical sundials; an inclination sundial located on a plane perpendicular to the meridian; and a sundial on a plane tilted to the east or west, the intersection of which with the horizontal lies on the meridian plane. He also solved the problem: "From the known deviation of the Sun, and the known latitude of the place, what are the hours when the Sun begins and ceases to illuminate the vertical plane perpendicular to the meridian?" (Drzewiński 1825, p. 142).

The material dedicated to gnomonics found in the Drzewiński and Wolff's textbooks was very similar. Both textbooks focused primarily on presenting the

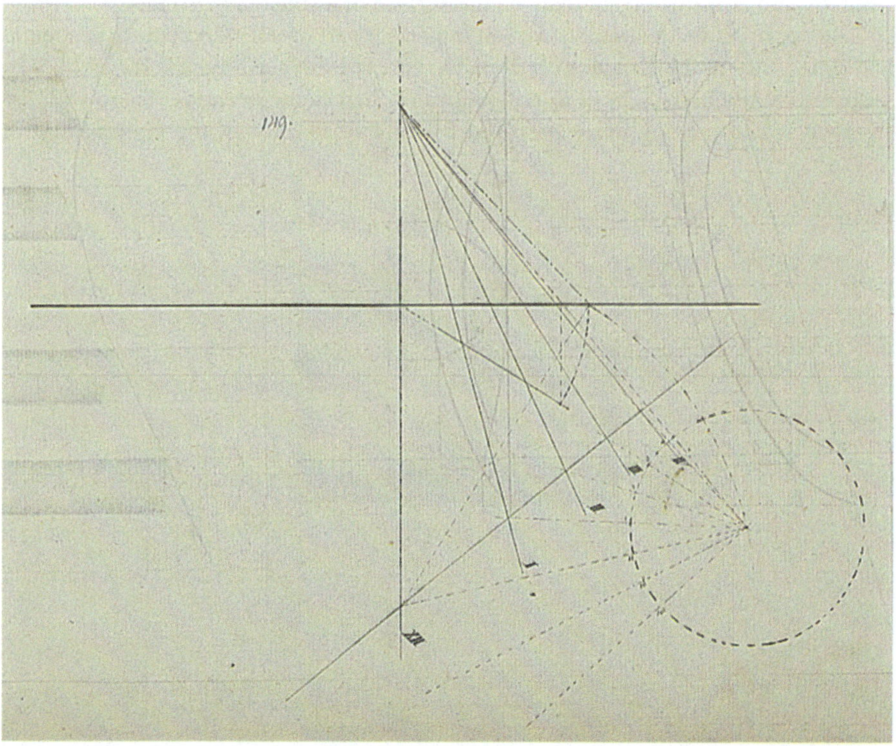

Fig. 17 The construction of the inclined-tilted sundial face (Jagiellonian Library, *Rkp. 4189 IV vol. 1*, p. 55)

constructions of selected types of flat sundials. The most important difference between the lectures of Drzewiński and Wolff are as follows: Wolff's considerations were largely carried out according to the scheme—constructing the sundial face on a piece of paper or other portable flat surface, then placing the dial on the appropriate wall and fixing the gnomon; Drzewiński constructed sundials already on the wall, on which they were supposed to be fixed (Drzewiński 1825, pp. 129–130). The differences in the construction of sundial faces by both authors were insignificant (both of them also carried out the proofs of the correctness in an analogous way) (Fig. 18).

Information on the teaching of gnomonics in the ordinances of 1835 was written quite succinctly. Due to the fact that only flat sundials were discussed in Drzewiński's textbook, it can be assumed that the Ministry limited learning gnomonics in gymnasia to the construction of sundials on flat surfaces. It turns out, however, that there were schools where teachers extended this material. For example, Bernhardt, a teacher at the Gymnasium in Piotrkowo, wrote in a report from the school year 1835/36, which he prepared at the request of the management, that in the highest 8th grade he discussed the application of descriptive geometry to "Gnomonics, learning how to make sundials on planes or curves" (State Archives in Łódź, Gimnazjum Męskie w Piotrkowie, sign. 28). Thus, the teacher went beyond the limits set by the Ministry and constructed sundials on the surface of a sphere or cylinder.

The construction of sundials was an important part of the 19th century academic and school life in the Kingdom of Poland. This is evidenced by articles and books prepared by teachers, as well as collections of their home libraries. There was, for instance, a competition entry prepared by another teacher of the Gymnasium in

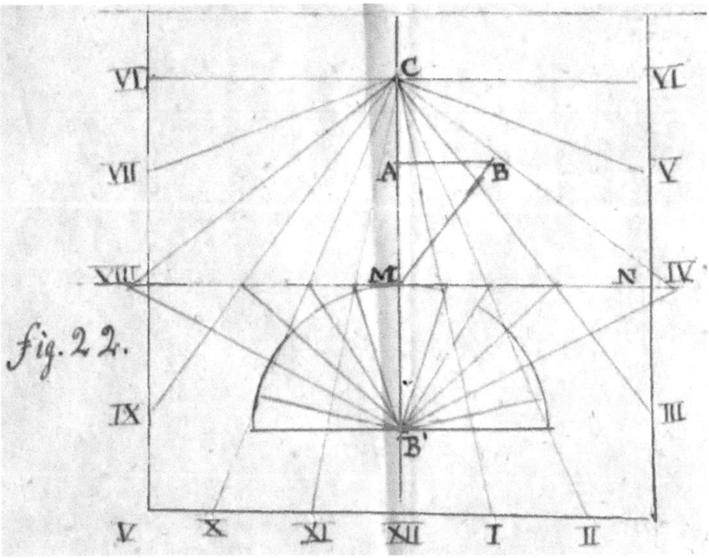

Fig. 18 The construction of the vertical south sundial face from the textbook by Drzewiński (Drzewiński 1825, Tab. II, Fig. 22)

Piotrkowo – Jędrzej Famulski. In 1827, when he was already a teacher, he wrote *Traktat Gnomoniki wraz z krótką historią jej stanu u Starożytnych i Nowoczesnych i krytyczną wiadomością o dziełach o niej traktujących* [*A Treatise on Gnomonics, along with a Brief History of Its Condition in Ancient and Modern Times, and a Critical Message about Works about It*]. He submitted it to the competition announced at the time by the university of Warsaw and won it[11] (State Archives in Łódź, Gimnazjum Męskie w Piotrkowie, sign. 28). Augustyn Frączkiewicz, who in 1820 was a teacher at the provincial school in Kielce, wrote the book *O gnomonice analitycznej* [*On Analytical Gnomonics*] (1816), in which he approached the construction of sundials in a trigonometric way and presented innovative solutions (Krzyżanowski 1820, pp. 11–12). In secondary schools, gnomonics was taught based on constructions, but teachers also had advanced-level works in their library collections. For example, on the copy of *Gnomonika uważana jako wstęp do astronomii z zastosowaniem Rozbioru i obiaśnieniem na przykładach ojczystych z przedmową Józefa Łęskiego Professora Astronomii w Uniwersytecie Jagiellońskim* (Krzyżanowski 1820), the following dedication was handwritten: "To Mr. Andrzej Trzcinski, Professor Sz: Woje: Kieleckiey, as proof of friendship and respect, this book is donated by the author." Krzyżanowski, in his textbook, approached the basic issues of gnomonics in three different ways, i.e. using trigonometry, analytical geometry and descriptive geometry, referring to the works of authors such as Dominique-François Rivard, Louis Puissant, Gaspard Monge, Nicolas Carnot, and Leonhard Euler. In addition to flat sundials, he constructed sundials on spherical and cylindrical surfaces, as well as lunar and stellar sundials, but he emphasized that his lecture was too difficult for secondary school students and, instead of encouraging them to learn, could discourage them (Krzyżanowski 1820, p. 269). The geometrical construction of the faces of flat sundials (i.e. the approach to gnomonics through descriptive geometry) was carried out by Krzyżanowski in the same way as Wolff and Drzewiński.

In 1867, schools in the Kingdom of Poland were completely Russified, and gnomonics was removed from the curricula (Programmy prepodavanija uchebnykh predmetov... 1868). The analysis of the curricula included in the reports of a dozen or so secondary schools operated in Galicia from 1868 to 1918 shows that gnomonics was not taught there either (although descriptive geometry was an important subject of study—in real schools it was taught even 6 hours a week throughout the school year). Scientific papers by local teachers on the construction of sundials were very rare; one example is the article entitled *Jaką linię zakreśli cień punktu stałego np. wierzchołek pionu oświetlonego światłem słońca w ciągu dnia na płaszczyźnie poziomej?* [*What Line Will be Made by the Shadow of a Fixed Point, e.g. the Vertex of a Vertical Rod Illuminated by the Sun during the Day, on a Horizontal Plane?*], published by Wincenty Majewski in the school report of the gymnasium in Wadowice in 1887.

---

[11] The work of Framulski was most likely lost – it could not be found while this article was being prepared.

Was gnomonics taught at that time in schools operated in the Polish territories according to Prussian, Austrian and Russian teaching principles, and that used German and Russian as the language of instruction? Analyzing the relevant ministerial regulations such as Prussian ordinances of 1859 *Unterrichts- und Prüfungs-Ordnung der Realschulen und der höheren Bürgerschulen*, Austrian ordinances of 1849 *Entwurf der Organisation der Gymnasien und Realschulen in Oesterreich*, Russian ordinances of 1868: *Programmy prepodavanija uchebnykh predmetov v muzhskikh gimnazijakh i progimnazijakh Varshavskago uchebnago okruga*, and curricula included in the reports of several dozen schools, it can be concluded that gnomonics in schools which operated acording to the Prussian, Austrian, and Russian teaching principles was almost absent. Only one such school where gnomonics was taught has been identified. It was a gymnasium in Elbląg, functioning according to Prussian rules, with German as the language of instruction. Gnomonics was taught there in 1830, 1842, 1844, and 1846 (Zu der öffentlichen Prüfung der Schüler des Gymnasiums zu Elbing 1830, 1842, 1844, Zu der öffentlichen Prüfung der Schüler des Gymnasiums zu Elbing 1846). However, neither the exact curriculum nor the textbook on which the teaching was based is known.

5 Gnomonics in Polish Schools after 1919

In the second half of the 19th century, gnomonics was removed from the curricula of schools operated in the Polish territories. It was only after Poland regained independence (1918), that the construction of sundials returned to Polish schools, and it was taught as a part of geography lessons. The construction of sundials was introduced into the curricula very slowly. Initially, it was taught by selected teachers e.g. Aleksander Tarnawski in the gymnasium in Poznań, who argued that: "[sundials] in remote, distant regions may still be useful when the mechanical measures of time measurement and regulation fail" (Tarnawski 1930, p. 46). Based on the lessons he conducted in Poznań, Tarnawski later prepared a textbook for teachers entitled *Najważniejsze pomoce przy nauce geografji matematycznej, vol. 1: Gnomon i jego użycie* (1930), with detailed information on the curricula and methodological remarks. Piotr Hrabyk and Ludomir Sawicki wrote in their book: "In relation to the exercises on the movement of the Sun, the teacher can construct a sundial on the southern wall of the school building or in the open courtyard. A pocket clock or a wall clock, and a calendar board, are needed for exercises and learning about the counting of time" (1921, p. 27). Only the ordinances entitled *Program gimnazjum państwowego. Gimnazjum niższe* from 1931 brought a return of the construction of sundials to the curricula of Polish gymnasia.

The gnomonics teaching in Polish schools in the interwar period will be discussed here based on the example of lessons conducted at the gymnasium in Poznań by Aleksander Tarnawski and discussed in Tarnawski (1930). The objects analyzed in this handbook, including sundials, were presented to the Polish Ministry of Education and were approved by the ordinance of 1926 for implementation in school education (Tarnawski 1930, p. 4). Thus, the intended curricula will be discussed once again.

Tarnawski started his textbook with the easiest issues, those which in his opinion should be discussed in primary and lower gymnasium grades (e.g. discussing the relationship between the position of the Sun and the direction and length of the shadow, determining the height of the Sun, and determining the local meridian line), then he moved to more and more difficult tasks, designed for higher level gymnasium students (among others, determining the latitude of a given place and construction of sundials). In the textbook, the author also placed the rules for converting real solar time into Central European time—this could not have been found in either Drzewiński or Wolff's textbooks, because the time zone was introduced at the end of the 19th century.

Aleksander Tarnawski, in his textbook, presented the constructions of only three sundials: the equatorial sundial, the horizontal sundial, and the vertical south sundial. During the constructions of the horizontal and vertical sundials' faces, he repeatedly referred to the construction of the equatorial sundial (Fig. 19). He was the first author to explain in his textbook that the constructions of the horizontal and

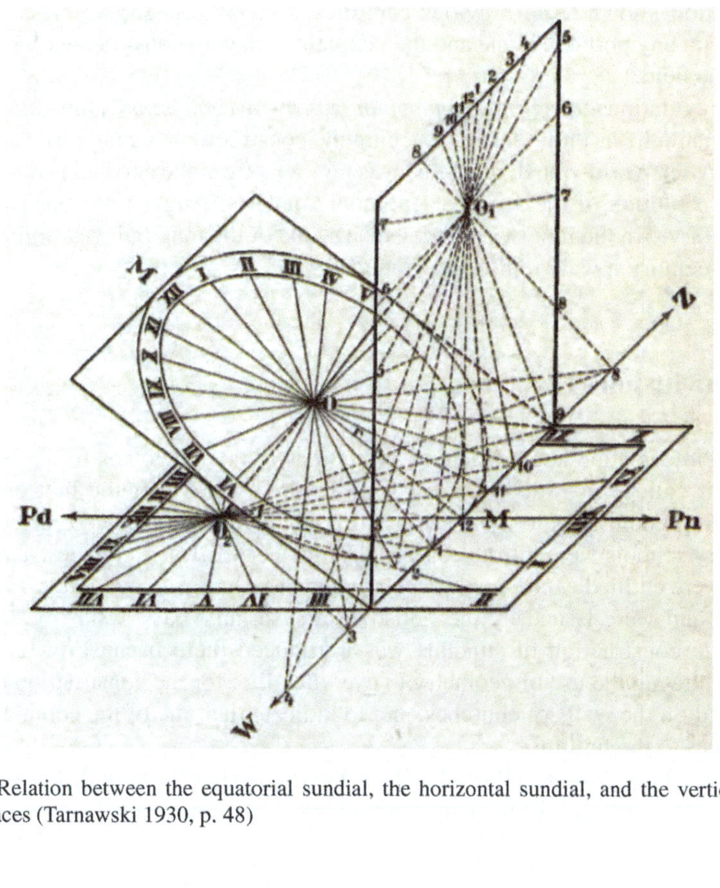

Fig. 19 Relation between the equatorial sundial, the horizontal sundial, and the vertical south sundial faces (Tarnawski 1930, p. 48)

Table 1 *Types of astronomical sundials discussed in the textbooks by Wolff, Drzewiński, and Tarnawski*

	Wolff	Drzewiński	Tarnawski
Equatorial sundial	+	+	+
Horizontal sundial	+	+	+
Vertical south sundial	+	+	+
Vertical north sundial	+	+	–
Vertical east sundial	+	+	–
Vertical west sundial	+	+	–
Polar sundial	+	–	–
Tilted sundial—deviating from the south towards the east or west direction	+	+	–
Tilted sundial—deviating from the north towards the east or west direction	+	–	–
Inclination sundial	+	+	–

vertical sundials' faces are based on descriptive geometry (on the theory of projections), but he did not use a new approach to this problem—he used methods of construction known from previous centuries. Like Wolff, Tarnawski constructed sundials on any portable plane and then explained how to construct and set the sundial in the field.

The regulations *Program gimnazjum państwowego. Gimnazjum niższe* from 1931 contain the last reference to teaching the construction of sundials in schools in Poland. After World War II, this issue was removed from the curricula. Table 1 presents the evolution of the scope of gnomonic material taught from the 18th to the 20th century—in the first two centuries the scope of the material was similar, and in the 20th century it was significantly reduced.

6 Conclusions

The first information about teaching gnomonics in Poland comes from 1602, from the Jesuit college in Kalisz. Only over a hundred years later the first ordinances introducing gnomonics to the teaching of mathematics in selected types of Polish secondary schools were introduced and they concerned Jesuit colleges. These ordinances were entitled *Catalogus Praelectionum in Scholis Societatis Jesu Provinciae Polonae* and were issued by the Polish Jesuit Adrian Miaskowski in 1711. Most likely, the construction of sundials was introduced there because of its applications—showing the use of geometry in everyday life, for the construction of sundials placed on the walls of churches, and to indicate the time of the commencement of services to the faithful.

Tadeusz Przypkowski (1905–1977) in Przypkowski (1968, p. 22) wrote that at the beginning of the 19th century there was an "unexpected" renaissance of gnomonics in Poland[12]. And in fact it was so: in 1812, gnomonics was introduced as a compulsory subject of education in Polish secondary schools preparing students for matriculation exams, and it remained so for the next several decades. Where did this renaissance come from? Gnomonics was introduced to Polish schools probably thanks to Jan Bystrzycki, a physics and astronomy teacher in Warsaw. The Ministry of Education of the Duchy of Warsaw asked him to recommend the best textbook on physics and astronomy, and he recommended *Traité élémentaire d'astronomie physique* by Jean Baptiste Biot—in the 1811 edition of this book there was a chapter on gnomonics. The Polish tradition of teaching gnomonics, initiated by the Jesuits, was also important here. Moreover, after the relative stabilization of the political situation, Poles wanted to make up for the technical backlog they had acquired over the last decades at the beginning of the 19th century. In order to regain independence, they had to develop in every respect, including science, technology, and education. Development became the driving force behind national activity, although it was undoubtedly difficult to implement—there was a shortage of Polish specialists who could construct precise watches, and their work was valued highly. For example, in 1850 the price of a chronometer produced in Poland was 2,500-3,000 złp., and if one wanted to bring one from England, the cost was 5-6,000 złp. (Czapek 1850, p. 35). It is worth noting here that the doctors' salaries at that time oscillated between 150 and 300 rubles, that is 930-1,860 złp. (Schiller 1998, p. 179), the dean of the diocese of Brzeziny earned 480 złp. (Szkutnik 2015, pp. 301-302), and a day's work in agriculture (harvesting, hay-making, etc.) paid 2 złp. (Śmiałowski 1973, pp. 306–307). This means that the purchase of a clock was beyond the reach of many Poles. These clocks, due to the fact that they were often made by incompetent persons, were also inaccurate. Sundials were an attractive alternative. Tables were added to the sundials and made it possible to convert the solar time to mean solar time (Krzyżanowski 1820, p. 3), which enabled people to read the universal time, and to check the accuracy of mechanical clocks, pendulum clocks, etc. One of the Polish constructors, Wojciech Jastrzębowski, built a universal device for the construction of sundials in 1929 (which functioned regardless of the location of the sundial and the shape of the surface on which it was to be drawn) which he described as follows:

> It is intended mainly to measure time with an accuracy which ordinary clocks often fail to do and to build in any location in an easy way, equally accurate compasses, which, especially in our country, is essential; they could do their best to establish order in facilitating matters, both public and private, which require a strict amount of time. Were it not for that, then at least the compasses arranged precisely with it could become so widespread in the country, in crowded public and private places, in cities and rural settlements, that they could replace clocks where they were missing, or serve to regulate those that are already there, and which, due to their instability, cause more disappointment and damage to various human activities than if they were not there at all (Jastrzębowski 1843, p. IV).

Another author added: "Sundials do not break easily, do not need to be repaired as often, and therefore do not cost as much as various kinds of clocks, which

[12] However, he did not explain this thesis.

sometimes hang outside watchmakers' windows for longer than they serve their owners" (Kowalski 1868, p. V).

In the first half of the 19th century, new Polish textbooks on gnomonics also began to be published, e.g. (Krzyżanowski 1820), where the author drew sundials in three ways: using plane trigonometry, planimetry, and the latest scientific achievement—descriptive geometry. In addition to the utilitarian benefits of gnomonics—reading the mean time in one of the cheapest ways, adjusting mechanical and pendulum clocks, etc.—the author draws attention to scientific benefits—discussing the principles of the theory of the Sun as well as the basics of trigonometry, planimetry, and descriptive geometry. All this meant that in the 19th century, gnomonics was introduced to Polish secondary schools as a compulsory element of the curricula. The construction of sundial faces was also important from a didactic point of view. Geometry was a field of mathematics which had been particularly valued by educators for centuries. In geometry they saw the science that develops the mind of young students, forcing them to look for connections between different geometric forms, and through the study of geometry "abstract, seemingly dry and non-absorbing claims, with a skilful lecture, become alive, arouse curiosity, and prepare the mind for self-use of one's own strength" (Dickstein 1889, p. 67). The construction of sundials taught young people independent and logical thinking. Learning of mathematics by action—the self-made constructions and the setting of the sundials—also aroused interest in the subject and helped to increase motivation for learning.

References

Acervus selelecius (n.d.), Kórnik Library, sign. BK 01025.

Anon. 1777. *EXERCITATIONES LITERARIAE Ex Jure naturali, Historiâ, Geografiâ, politica, ac naturali, Algebrâ, Geometriâ, & Trigonometriâ tam theoreticâ, quam practicâ, Gnomonicâ, Sectionibus conicîs; Mechanicâ, Staticâ, Hydrostaticâ, Hydraulicâ; Architectura militari, & civili; Astronomiâ, Physicâ, Opticâ, Catoptricâ, Rhetoricâ, Poesî, Gramaticâ, &c. &c. &c. HABITAE IN COLLEGIO POLOCENSI SOCIETATIS JESU* and *Elementa Sphericorum et Trigonometriae Sphaericae* [LITERARY EXERCISES in the Natural Law, History, Geography, politics, and nature, Algebra, Geometry, and Trigonometry, both theoretical and practical, Gnomonics, and Conics; Mechanics, Statics, Hydrostatics, Hydraulics; military and civil Architecture; Astronomy, Physics, Optics, Catoptrics, Rhetoric, Poetry, Grammar, etc. etc. etc. CONDUCTED IN THE POLISH COLLEGE OF THE SOCIETY OF JESUS and the Elements of the Sphere and Spherical Trigonometry]. Ossolineum, sign. XVIII – 8432.

Bartnicka, Kalina, ed. 2000. *Ratio atque institutio studiorum SJ czyli Ustawa Szkolna Towarzystwa Jezusowego (1599) [The Official Plan for Jesuit Education (1599)]*. Warsaw: ATENEUM.

Bednarski, Stanisław. 1933. *Upadek i odrodzenie szkół jezuickich w Polsce [Fall and revival of Jesuit schools in Poland]*. Krakow: Wydawnictwo Księży Jezuitów.

Clavius, Christoph. 1612. *Gnomonices Libri Octo, in qvibvs Non solum horologioruu solarium... [Eight books of gnomonics which include not only sundials...]*. Moguntiae: Antoni Hierat.

Czapek, Franciszek. 1850. *Słów kilka o zegarmistrzostwie ku użytku zegarmistrzów i publiczności [A few words about watchmaking for the use of watchmakers and the public]*. Leipzig: Drukiem Aleksandra Wiedego.

Dickstein, Samuel. 1889. *Gieometrya elementarna [Elementary geometry]*. Warsaw: W drukarni Józefa Sikorskiego.

Dobrzycki, Jerzy, Mieczysław Markowski, and Tadeusz Przypkowski. 1975. *Historia astronomii w Polsce [History of astronomy in Poland]*. Vol. 1. Wroclaw-Waraw-Krakow-Gdansk: Wydawnictwo PAN, Ossolineum.

Drzewiński, Feliks. 1825. *Jeografija fizyczna, z dzieła astronoma obserwatora w Imperatorskim Uniwersytecie Wileńskim, Jana Sniadeckiego, oraz Gnomonika, z drugiego wydania dzieła J. Mollet, profesora fizyki w Lyonie: wyjęte dla gymnazyów na klassę pierwszą i dla szkół powiatowych na klassę czwartą, przez Felixa Drzewińskiego* [Physical geography, from the work of the astronomer and observer at the Imperial University of Vilnius, Jan Sniadecki, and Gnomonics, from the second edition of the work of J. Mollet, professor of physics in Lyon: prepared for the first grade of gymnasia and the fourth grade of distinct schools by Felix Drzewiński]. Vilnius: Nakładem i drukiem A. Marcinkowskiego.

Duda, Roman. 2019. *Historia matematyki w Polsce na tle dziejów nauki i kultury [The history of mathematics in Poland in relation to the history of science and culture]*. Warsaw: IHN PAN, ASPRA-JR.

Freytag, Adam. 1631. Architectura Militaris Nova Et Aucta, Oder Newe Vermehrte Fortification, Von Regular Vestungen, Von Irregular Vestungen und Aussen Wercken. Leyden: Bonaventura und Abraham Elzeviers.

Gąsiorowski, J. 1871. *Gnomonika czyli nauka o kompasach* [Gnomonics or the science of sundials], Biblioteka Jagiellońska, BJ Rkp. 6988 I.

Grzepski, Stanisław. 1566. *Geometria, to jest miernicka nauka, po polsku krótko napisana z Graeckich y z Łacińskich ksiąg [Geometry, the science of measuring, briefly translated into Polish from Greek and Latin books]*. Krakow: Łazarz Andrysowic.

Hrabyk, Piotr, and Ludomir Sawicki. 1921. *Metodyka geografii dla szkoły powszechnej oparta na podręcznikach L. Sawickiego. Stopień I-III [Geography methodology for primary schools based on L. Sawicki's textbooks. Grades 1-3]*. Krakow: "Orbis".

Jastrzębowski, Wojciech. 1843. *Kompas polski czyli narzędzia służące za kompas powszechny, gnomograf, obserwatoryum przenośne i narzędzie do kreślenia sekcyi konicznych [The Polish compass or the tool used as a sundial, gnomograph, portable observatory, and a tool for drawing conics]*. Warsaw: J. Dietrich.

Karp, Alexander, and Fulvia Furinghetti. 2016. *History of Mathematics Teaching and Learning. Achievements, Problems, Prospects*. Springer.

Komarzyniec, Grażyna. 2004. *Nauczanie matematyki w krakowskiej Szkole Nowodworskiej w latach 1588–1914 [Mathematics teaching at the Nowodworski School in Kraków in 1588–1914]*. Krakow: Wydawnictwo Naukowe Akademii Pedagogicznej.

Kowalski, Tomasz. 1868. *Gnomonika czyli nauka robienia zegarów słonecznych na rozmaitych płaszczyznach za pomocą liniału i cyrkla [Gnomonics, or learning to make sundials on various planes with a ruler and a compass]*. Przemysl: Nakładem Autora.

Krzyżanowski, Paweł. 1820. *Gnomonika uważana jako wstęp do astronomii z zastosowaniem Rozbioru i obiaśnieniem na przykładach ojczystych z przedmową Józefa Łęskiego Professora Astronomii w Uniwersytecie Jagiellońskim*. [Gnomonics considered as an introduction to astronomy [...] with a preface by Józef Łęski, Professor of Astronomy at the Jagiellonian University]. Krakow: W Drukarni Akademickiey.

Kubik, Kazimierz. 1959. In *Rola Gimnazjum Akademickiego w rozwoju nauk ścisłych w Gdańsku [The role of the Academic Gymnasium in the development of exact sciences in Gdańsk]*, ed. Gdańskie Gimnazjum Akademickie, 7–78. Gdynia: Wydawnictwo Morskie.

———. 1972. Nauczanie matematyki w Toruńskim Gimnazjum Akademickim w XVI–XVIII w. [Mathematics teaching at the Toruń Academic Gymnasium in the 16th-18th centuries]. In *Księga Pamiątkowa 400-lecia Toruńskiego Gimnazjum Akademickiego*, ed. Zbigniew Zdrójkowski, vol. 1, ed. Torun: PWN.

———. 1974. Nauki ścisłe w toruńskim Gimnazjum Akademickim [Exact sciences at the Toruń Academic Gymnasium]. In *Księga Pamiątkowa 400-lecia Toruńskiego Gimnazjum*

Akademickiego, ed. Zbigniew Zdrójkowski, vol. 3. Torun: Komitet Obchodu 400-lecia Gimnazjum Akademickiego im. Mikołaja Kopernika w Toruniu.

Kukulski, Zygmunt. 1931. *Źródła do dziejów wychowania i szkolnictwa w Polsce z doby Izby Edukacji Publicznej 1807-1812 [Sources concerning the history of education in Poland from the period of the Public Education Chamber 1807–1812]*. Lublin: Dom Książki Polskiej SP. AKC. Warszawa - Plac Trzech Krzyży.

Lipiński, Józef. 1812. *Plan nauk szkół departamentowych* [Study plan of departmental schools]. Warsaw.

———. 1820. *Wewnętrzne urządzenie szkół wojewódzkich* [Principles of operation of provincial schools]. Warsaw.

Lisowski, Witold. 1982. *Polskie Korpusy Kadetów [Polish Cadet Corps]*. Warsaw: Wydawnictwo Ministerstwa Obrony Narodowej.

Łukaszewicz, Józef. 1849. *Historia szkół w Koronie i w Wielkim Księstwie Litewskim od najdawniejszych czasów aż do roku 1794 [The history of schools in the Kingdom of Poland and the Grand Duchy of Lithuania from the earliest times until 1794]*. Vol. 1. Poznan: Nakładem Księgarni Jana Konstantego Żupańskiego.

———. 1851. *Historia szkół w Koronie i w Wielkim Księstwie Litewskim od najdawniejszych czasów aż do roku 1794 [The history of schools in the Kingdom of Poland and the Grand Duchy of Lithuania from the earliest times until 1794]*. Vol. 3. Poznan: Nakładem Księgarni Jana Konstantego Żupańskiego.

Metium, Adriaan. 1632. *Astronomische ende Geographische ondervvysinghe in de welcke Door t' gebruyck der Aertschen Globi, ofte gebulte Caerten / midtsgaders t' Astrolabium Catholicum... [Astronomical and geographical education...]*. Amsterdam: Hendrick Lauwerentz.

Na publiczny popis Uczniów Szkół Departamentowych Lubelskich ... [At the public examination of the students of the Departmental Schools in Lublin...] 1814, 1815. Lublin: Drukarnia Jana Karola Pruskiego.

Pardała, Antoni. 2010. The traditions and development of mathematics education. Case of Poland. In *Russian Mathematics Education. History and World Significance*, ed. Alexander Karp and Bruce R. Vogeli. Hackensack, NJ: World Scientific Publishing.

Pawlikowska-Brożek, Zofia. 2004. Podręczniki matematyczne dla polskich szkół jezuickich [Mathematics textbooks for Polish Jesuit schools]. In *Wkład jezuitów do nauki i kultury w Rzeczypospolitej Obojga Narodów i pod zaborami*, ed. Irena Stasiewicz-Jasiukowa, 245–268. Krakow-Warsaw: Wydawnictwo WAM.

Popis publiczny Uczniów Szkoły Departamentowey Kaliskiey [Public examination of students of the Departmental School in Kalisz]. 1814. Kalisz.

Popis publiczny uczniów Szkoły Departamentowey Krakowskiey [Public examination of students of the Departmental School in Kraków]. 1815. Krakow: W Drukarni Akademickiey.

Popis publiczny uczniów szkoły wojewódzkiey Lubelskiey [Public examination of students of the Provincial School in Lublin]. 1823. Lublin: W Drukarni Rządowey.

Popis publiczny Uczniów Szkoły Wojewódzkiey XX. Benedyktynów w Pułtusku [Public examination of students of the Provincial School of the Benedictine Fathers in Pułtusk]. 1821. Warsaw: W Drukarni Józefa Węckiego.

Popis publiczny uczniów szkół departamentowych warszawskich pod nadzorem XX. Pijarów [Public examination of students of the Warsaw Departmental Schools of Piarist Fathers]. 1813. Warsaw.

Popis publiczny uczniów szkół departamentowych warszawskich pod nadzorem XX. Pijarów [Public examination of students of the Warsaw Departmental Schools of Piarist Fathers]. 1814. Warsaw.

Popis publiczny uczniów Szkoły Wojewódzkiey Zamoyskiey w Szczebrzeszynie [Public examination of students of the Zamojski Provincial School in Szczebrzeszyn]. 1825. Warsaw: W Drukarni Józefa Węckiego.

Program gimnazjum państwowego. Gimnazjum niższe [State gymnasium curriculum. Lower gymnasium]. 1931. Lvov: Państwowe Wydawnictwo Książek Szkolnych.

Programmy prepodavanija uchebnykh predmetov v muzhskikh gimnazijakh i progimnazijakh Varshavskago uchebnago okruga [Curricula in the gymnasia and lower gymnasia for boys, operated in the Warsaw School District]. 1868. Warsaw: V tipografii NRI Varshavskom uchebnom okruge.

Przypkowski, Tadeusz. 1968. The Art of sundials in Poland from the 13th to the 19th century. *Vistas in Astronomy* 9: 13–23.

Sbornik administrativnykh postanovlenij Tsarstva Polskago / Zbiór przepisów administracyjnych Królestwa Polskiego [Collection of administrative regulations of the Kingdom of Poland], vol. 3. 1868. Warsaw: W drukarni S. Orgelbranda.

Schiller, Joanna. 1998. *Portret zbiorowy nauczycieli warszawskich publicznych szkół średnich 1795-1862 [Collective portrait of teachers in public secondary schools in Warsaw 1795-1862]*. Warsaw: IHN PAN.

Schwartz, Paul. 1910. *Die Gelehrtenschulen Preußens unter dem Oberschulkollegium (1787-1806) und das Abiturientenexamen [The Prussian secondary schools under the Oberschulkollegium (1787–1806) and their matriculation examination]*. Vol. 1. Berlin: Weidmannsche Buchhandlung.

Solski, Stanisław. 1686. *Geometra Polski, tj. nauka rysowania, podziału i rozmierzania linii, angułów, figur i brył [Polish surveyor, i.e. learning to draw, divide and measure lines, angles, figures, and solids]*. Vol. 3. Krakow: W Drukarni Grzegorza y Mikołaja Schedlow.

Sprawa przed Publicznością na Popisie Rocznym Szkoły Wojewódzkiey Płockiey przez Zgromadzenie Nauczycielskie w Końcu Szkolnego Roku 1819/1820 złożona [A case submitted before the public at the Annual Examination of the Provincial School in Płock by the Teachers' Assembly at the end of the School Year 1819/1820]. 1820.

Plock. Szkutnik, Piotr. 2015. Pensja dziekanów w Królestwie Polskim do początku lat 70. XIX wieku. Egzemplifikacje [Salaries of deans in the Kingdom of Poland until the beginning of the 1870s. Exemplification]. *Archiwa, Biblioteki i Muzea Kościelne* 104: 297–309.

Śmiałowski, Józef. 1973. Płace i zarobki pozarolnicze chłopów w Królestwie Polskim (1831-1864) [Wages and non-agricultural earnings of peasants in the Kingdom of Poland (1831-1864)]. *Przegląd historyczny* 64 (2): 303–323.

Tarnawski, Aleksander. 1930. Najważniejsze pomoce przy nauce geografji matematycznej, vol. 1: Gnomon i jego użycie [The most important aids in learning mathematical geography, vol. 1: The Gnomon and its use]. In *Bibljoteka Geograficzno-Dydaktyczna*, ed. Stanisław Pawłowski, vol. 2. Lvov-Warsaw: Książnica-Atlas.

Tolomeo, Claudio. 1574. *Principe De Gli Astrologi, Et De Geografi... [Principles of astrology and geography...]*. Venice: appresso Giordano Ziletti.

Toński, Jan. 1640. *Arithmetica vulgaris et trigonometria rectilineorum pro ut universae geometria Practica, alijusqve Matheseos partibus, Geographia, Architectonica, Gnomonica etc.* [Popular arithmetic and plane trigonometry for whole practical geometry, and other parts of mathematics, geography, architectonics, gnomonics, etc.]. Ingolstadt: Typis Gregorii Hanlin.

Treść nauk dawanych uczniom w szkole na stopniu Gymnazyinym będącey Traszkuńskiey pod zaszczytem Jaśnie W. Felixa Siesickiego... przez XX. Bernardynów utrzymywaney... [The content of the curricula in the Gymnasium in Traszkuny...] 1823. Vilnius: W Drukarni XX. Bazylianów.

Ustawy Komisyj Edukacyj Narodowej dla stanu akademickiego i na szkoły w krajach Rzeczypospolitej przepisane w Warszawie roku 1783 [Regulations of the Commission of National Education for academies and schools operated in the Polish-Lithuanian Commonweealth, prepared in Warsaw in 1783]. 1902. Warsaw: M. Borkowski.

Więsław, Witold. 2007. *Matematyka polska epoki oświecenia [Polish mathematics of the Enlightenment era]*. Warsaw: Fraszka Edukacyjna.

Wolff, Christian. 1716. *Mathematisches Lexicon*. Leipzig: Joh. Friedrich Gleditschens seel. Sohn.

———. 1758. *Compendium Elementorum Matheseos Universae, In Usum Studiosae Juventutis Adornatum [Compendium of elements of whole mathematics, prepared for the use of learning youth]*. Lausannae & Genevae: Sumpt. MARCI-MIC. BOUSQUET & Soc.

———. 1775. *Anfangsgründe aller mathematischen Wissenschaften, vol. 3: Welcher die Optik, Catoptrik und Dioptrik, die Perspectiv, die sphärische Trigonometrie, Astronomie, Chronologie, Geographie und Gnomonik in sich enthalt.* [Beginnings of all mathematical sciences, vol. 3: Which contains optics, catoptrics and dioptrics, perspective, spherical trigonometry, astronomy, chronology, geography, and gnomonics]. Halle: Rengerischen Buchhandlung.

Zastosowań geometryi wykreślnej wedle rozkładu pomysłów i notat Franciszka Sapalskiego zeszyt pierwszy... [Applications of descriptive geometry according to the ideas and notes of Franciszek Sapalski, first issue...] 1839. Krakow: W Drukarni Stanisława Gieszkowskiego.

Zu der öffentlichen Prüfung der Schüler des Gymnasiums zu Elbing [The public examination of the students of the gymnasium in Elblag]. 1830. Elblag: Gedruckt bei August Albrecht.

——— [The public examination of the students of the gymnasium in Elblag]. 1842, 1844. Elblag: Gedruckt bei Agathor Wernich.

——— [The public examination of the students of the gymnasium in Elblag]. 1846. Elblag: Gedruckt bei F.W. Neumann-Hartmann.

The Transition to Modern Mathematics in Spanish Primary Education: The 1965 Syllabus

Antonio M. Oller-Marcén

Abstract In this work, we address the initial moments of the introduction of the Modern Mathematics movement in Spanish primary education. Even if the ideas of Modern Mathematics were fully introduced and implemented in the *Ley General de Educación* [General Law of Education] from 1970, we argue that, in fact, some aspects of the Modern Mathematics movement were already present in a previous, less profound reform from 1965. In order to determine how (and which) elements of this movement were introduced at that point, we analyze two types of documentary sources. First, we compare the 1965 official syllabus with the previous ones, dating from 1953. Second, we also compare two versions of the same series of textbooks edited by the same editorial company before and after the 1965 reform. At a curricular level, Modern Mathematics ideas were mostly present in the introductory section of the syllabus and had little impact on the topics. On the other hand, changes seem to have been greater at the textbook level. In particular, set theory was more deeply implemented in the textbooks than was seemingly intended in the official curriculum.

Keywords Modern mathematics · Spain · Primary education · Curriculum · Textbooks

1 Introduction

The so-called Modern Mathematics (or New Math in the USA) movement was a worldwide phenomenon that aimed to bring school mathematics closer to academic mathematics (Kilpatrick 2012). In this endeavor, a lot of attention was paid to primary education. There are many works which provide insight into this phenomenon by providing an overview and comparing the situation in different regions (Moon 1986) or by giving detailed case studies. The case of the USA, for instance, has been

A. M. Oller-Marcén (✉)
Centro Universitario de la Defensa de Zaragoza, Zaragoza, Spain
e-mail: oller@unizar.es

A. Karp (ed.), *Advances In The History Of Mathematics Education*,
International Studies in the History of Mathematics and its Teaching,
https://doi.org/10.1007/978-3-030-95235-8_7

extensively addressed by Hayden (1981) and, more recently, from a more socio-political point of view by Phillips (2014). The situation of Belgium is thoroughly studied in the recent book by De Bock and Vanpaemel (2019). In order to obtain a full view of this movement, it is particularly interesting to consider the case of the countries of the OEEC that did not participate in the Royaumont Seminar: Iceland, Portugal, and Spain (Bjarnadóttir 2017; Matos 2010).

In Spain, the first steps towards the introduction of Modern Mathematics were taken during the decade following 1960 (Ausejo 2012) but it was not until the subsequent 1970 reform that it was fully implemented (González Astudillo 2008a).

The Spanish 1970 *Ley General de Educación* [General Law of Education] replaced previous laws from 1945 (for primary education) and 1953 (for middle school). It reorganized the Spanish education system, and it also replaced the previous official syllabi. In the case of mathematics, in particular, this implied a complete implementation of Modern Mathematics both in primary and secondary education. The ideas of Modern Mathematics started to be criticized by teachers and parents by the 1980s (Ausejo and Matos 2014) but they were present in Spanish mathematics curricula until the 1990s, when a new education reform was approved. However, the influence of Modern Mathematics still remained for some time, partly because many primary and secondary education mathematics teachers had been trained in that way (González Astudillo 2010).

The process that would ultimately lead to the implementation of Modern Mathematics in Spain was not short. According to Ausejo (2012), it started in 1961, 2 years after the Royaumont Seminar, when the Ministry of Education organized a series of meetings addressed to secondary education teachers. These meetings promoted some experimental works that eventually led, in 1967, to the publication of an official syllabus for the so-called *Bachillerato Elemental* (ages 10–14) that is usually considered to be the "syllabus including, for the first time, contents of Modern Mathematics[1]" (González Astudillo 2006, p. 66). The Spanish mathematician Pedro Abellanas played an important role in this process (Ausejo 2010).

Primary education received much less attention. This different approach to Modern Mathematics in primary and secondary education might be explained by factors such as the different number of involved students or the requirement of extensive teacher training that have been described (in the context of the USA) by Hayden (1981). In Spain, we are not aware of any official political initiatives like those promoted by the Ministry of Education for secondary education. However, González Astudillo (2006) reports the presence of papers in a journal addressed to primary school teachers called *Vida Escolar* [School Life] aimed to introduce elements of Modern Mathematics as early as 1965–1968.

It is usually considered that "the first reference to Modern Mathematics in primary education is in the General Law of Education from 1970" (González Astudillo 2006, p. 68). However, the truth is that some elements of Modern Mathematics can already be found in the official syllabus for primary education (ages 6–14) that was published in 1965 and that started to be effectively applied in 1966 (ages 6–10) and

[1] All translations from Spanish have been made by the author.

1967 (ages 10–14), which was even earlier than the secondary education syllabus from 1967.

We argue that the 1965 primary education syllabus constitutes the first Spanish official curricular document in which some key elements of Modern Mathematics can be found. In spite of this fact, this syllabus does not seem to have received much attention in the research literature on the history of mathematics education in Spain. In fact, we are not aware of any work addressing the "prehistory" (before 1970) of Modern Mathematics in Spanish primary education from the point of view of curricular documents or textbooks. As we already pointed out, González Astudillo (2006) shows that, during the period 1965–1968 some elements of Modern Mathematics were being introduced to primary education teachers by means of professional pedagogical journals. The following research question naturally arises: What was the actual presence of Modern Mathematics in official documents and textbooks before its complete introduction in 1970? In order to try to answer this question, we will follow two lines of inquiry.

First, we must take into account that the 1965 official syllabus replaced a previous one dating from 1953. Consequently, in addition to analyzing the 1965 text itself in order to determine the presence of Modern Mathematics ideas, we will also compare it with the 1953 document. This comparison will allow us to assess the depth of the influence of Modern Mathematics over the new syllabus better. A similar work was carried out by Ausejo (2010) in the case of secondary education.

Second, in order to determine the impact of the 1965 reform, we will analyze and compare the editions of the same series of textbooks published just before and after the approval of the 1965 syllabus. Since "teaching practice is not so much determined by ministerial decrees and official syllabuses as by the textbooks used for teaching" (Schubring 1987, p. 41) this will lead to a better assessment of the actual presence of Modern Mathematics in Spanish classrooms at that time. It will also highlight possible differences between the intended and the implemented curriculum (van den Akker 2003).

As argued above, this work aims to give novel information about the initial moments of the introduction of Modern Mathematics in Spanish primary education. This is an interesting research topic that, according to De Bock and Vanpaemel (2018, p. 62), who approach the Belgian case, "remains underexposed in the literature." In this way, our results may contribute to place the Spanish case within the international frame and provide useful information to compare how this same phenomenon took place in other countries (Menghini 2020).

2 Some Background

Franco's fascist dictatorship in Spain lasted for almost 40 years. During all that time, and from the point of view of education, two main periods can be considered. The first one (1936–1957) was characterized by totalitarianism and National Catholicism, while the second one (1957–1975) was mostly dominated by technocracy and involved the introduction of new ideas regarding the administration and management of the education system (Viñao Frago 2014).

The number of preschool and primary education students greatly increased during the second period. It went from about 2,800,000 in 1950 to about 4,740,000 in 1970. However, it must be pointed out that by 1965 still more than half a million children did not attend primary school and it was not until 1978 that 99.8% of children ages 6–14 were schooled (Mayordomo Pérez 1997).

After the Spanish Civil War, the Law of Primary Education was enacted in 1945. This law organized primary education according to three different stages: elementary (6–10 years), perfection (10–12 years) and professional development (12–14 years). Primary education was mandatory and free only for children 6 to 12 years of age. Later, in 1965, this law was partially reformed. The previous stages were unified, and primary education became free and mandatory from 6 to 14 years of age (Egido Gálvez 1994).

The 1945 law put Primary education in the hands of the Catholic Church, and the role of the State was merely subsidiary. The first quinquennial plan to build public schools was not approved until 1956 (Viñao Frago 2014). In 1955, for instance, only 54% of primary education students attended public schools (Mayordomo Pérez 1997). With the exception of religion, physical education, political education, and a subject on housekeeping, the official syllabi (regarding both contents and methodology) were centralized and controlled by the Ministry of Education.

The control of the state over textbooks was also intense. After the Civil War, many textbooks were retired for ideological reasons (Diego Pérez 1999) and the government tried to publish its own official textbooks. This idea was soon abandoned, partly due to the pressure exerted by the Catholic Church and by specialized editorial companies (Fernández Cárcar 2020). Political and ideological censorship remained a constant throughout the dictatorship period. Textbooks had to comply with the ideological principles of the state, as well as with the official syllabi. However, over time this control grew less severe, leading to a technical and didactical improvement of textbooks (De Puelles Benítez 2007).

3 Theoretical and Methodological Issues

Recently, Hamann (2019) developed an interesting and useful theoretical framework for the description of educational reforms. In her work, she considers four layers of the educational system: the layer of theoretical scientific discipline (and the outputs—or products, as she states it—at this layer are subject-specific foundations, scientific theories, etc.), the layer of curriculum (the outputs here are curricula, syllabi, decrees, etc.), the layer of classroom concept (the outputs are textbooks, workbooks, teacher's handbooks, manipulatives, suggestions for instruction, etc.) and the layer of practical schooling (and the outputs here are lessons, learning progress, knowledge as well as tests, written tasks, exercise books, portfolios, reports, etc.). In addition, the model implicitly suggests some guiding questions:

1. What central aims and/or ideas (concerning, for example, content, didactical principles, methods, and curriculum concept) can be deduced from the products of each layer?
2. How can one characterize the process(es) of recontextualization between the different layers?
3. How did ideas and concepts develop and change over time?

This framework has revealed useful in modeling and analyzing the German primary education reform of the 1970s since it allows not only the description of the processes taking place in each layer, but also focusing on the existing interactions between different layers (Hamann 2018).

It seems plausible to use this framework in other cultural contexts. In fact, the existing research devoted to the introduction of Modern Mathematics in Spain can be interpreted in terms of Haman's model. For instance, the work of González Astudillo (2008a) illustrates possible interactions between the layer of theoretical scientific discipline and the layer of classroom concept, materialized in papers published in professional journals in which the ideas of Modern Mathematics were being disseminated among practitioners.

On the other hand, the works of Ausejo (2010, 2013) focus, among other aspects, on interactions between the layer of theoretical scientific discipline and the layers of curriculum and of classroom concept. In particular, Ausejo reports the experimental work with a new syllabus and textbooks in which mathematicians like Pedro Abellanas were actively involved. This production of textbooks implementing Modern Mathematics was not a local phenomenon and can be understood in a context of international dissemination of these ideas that has been recently described by Vanpaemel and De Bock (2019).

Before that, ideas that were somehow reminiscent of certain elements of the Modern Mathematics movement could be found in the work of another prominent Spanish mathematician, Pedro Puig Adam (González Astudillo 2008b).

If we adopt this framework and its terminology, we can say that our work focuses particularly on the curriculum and classroom concept layers. We will examine some products of each layer such as official syllabi or textbooks, so we will be mostly addressing question 1 above. However, we will also be considering question 3 since we will compare syllabi and textbooks over time.

Taking into account our previous considerations and objectives, we adopt a qualitative approach with a descriptive and comparative intention. According to the classification of Prior (2016, p. 172), our work uses documents as a resource, focusing on their content. The twofold approach that we adopt implies that our source material consists of:

- The official Primary education Mathematics syllabi from 1953 (MEN 1953) and from 1965 (MEC 1965)
- Two textbook series (for years 6–10) published by the same publishing company before and after 1965 (SM 1964a, b, c, d, 1966a, b, 1968a, b)

Scott (1990) considers four criteria which can be used to assess the documents in this type of research: authenticity, credibility, representativeness, and meaning.

Authenticity and credibility criteria are achieved since we are dealing with original legal texts and textbooks. Representativeness relies, in the case of textbooks, on the fact that they were officially approved by the government and their use was rather extensive. Finally, considering the nature of our research, we will only be concerned about literal, and not interpretative, meaning of the documents.

The selected textbooks were published by a renowned company (SM) which was close to the Christian Democratic movement and that held a relatively centrist ideological position (Martos 2019). This possibly explains why their textbook series were among the most commonly used during late Francoism, not only for mathematics but also in the case of many other subjects. This ideological moderation also partly motivates our selection since these textbooks were not expected to be particularly progressive or conservative regarding the introduction of novel pedagogical ideas.

In this work, we have followed the phases of content analysis as described by Mayring (2014) and, in particular, we carry out a text-driven analysis (Krippendorff 2014), both of the syllabi and the selected textbooks. The changes related to the Modern Mathematics movement involved both topics and teaching methods. Consequently, our analysis will focus on mathematical contents as well as on methodological and didactical indications. The descriptions of the mathematical contents in the syllabi are brief enough so that we are able to transcribe them verbatim (see Tables 1, 2, and 3, for instance). In the case of textbooks, the indices have been helpful in a first approach (Table 4, for instance) that was necessarily followed by a detailed content analysis.

In our analysis, we take into account important features (Kline 1973) of the Modern Mathematics movement methodology such as the rigor, the abstraction, the emphasis on deductive approaches and on the use of specialized mathematical language, etc. In addition, we are particularly interested in the presence of some particular topics that Modern Mathematics promoted in primary education such as set

Table 1 Contents for the first grade of primary education in the 1965 syllabus (MEC 1965, p. 20)

Exercises	Learning outcomes
Counting forward and backward from 1 to 100…, 1 by 1, 2 by 2, 5 by 5… Adding and subtracting two-digit quantities. Addition of one-digit numbers leading to the same result [as 2+6=3+5]. Composing and decomposing numbers [say, introducing 7 as 1+6, 2+5, 3+4, etc.]. Easy problems of addition and subtraction. Measurements using natural conventional units (only the most usual ones). Ordering things by size. Grouping objects by their shape.	Idea of "set." Belonging to a set. Intuitive knowledge of the most usual units: kilograms…, liters…, hours…, meters and *pesetas* [currency]. Identification and writing of the first 100 numbers. Idea of ten and of dozen. Evaluation of distances in elementary cases. Recognition of elementary geometric figures. Recognition of halves and doubles. Idea of multiplication as the addition of equal addends.

Table 2 Treatment of geometry in both syllabi during grades 1 to 4 (MEN 1953, pp. 65–70; MEC 1965, pp. 20–21)

	1953	1965
Grade 1	Observation of geometric figures: the sphere and the cube (Term 2). Observation of geometric figures: the prism and the cylinder (Term 3).	*Exercises*: n/a *Learning outcomes:* Recognition of elementary geometric forms.
Grade 2	Observation of forms: the pyramid and the cone (Term 1). Idea of line and curve and their types (Term 1). Closed curved lines. The circumference (Term 1). Observation of the regular polyhedral: the cube, the octahedron, and the icosahedron. Notion of square and triangle (Term 2). Parallel lines (Term 2). General idea of a polygon. Classification of triangles according to their sides. (Term 3). Idea of angle and its elements. Measurement of angles (Term 3).	*Exercises*: Recognition of fundamental geometric forms. *Learning outcomes*: Lines and curves and their types. Perception of forms: sphere, cube, pyramid, cone; circumference and circle.
Grade 3	Introduction of the notion of quadrilaterals and their classes, starting from the cube and the prism (Term 1). Measurement of angles. Addition and subtraction of angles. The circumference and remarkable lines on it. Arcs and their measurement (Term 1). Classification of triangles according to their angles. Notion of parallelogram and its classes. Value of the angles in a triangle and in a quadrilateral. Area of the square and of the rectangle (Term 2).	*Exercises*: Comparison of angles. Area of the square and rectangle. Addition and subtraction of angles. *Learning outcomes:* Angles. Triangles. Parallelograms and their classes.
Grade 4	Types of lines and angles (Term 1). Length of the circumference. Area of the triangle and of regular polygons (Term 1). Revision of plane geometry. Volume of the cube and of quadrangular prisms. Regular polyhedra (Term 2). Area of plane figures, including irregular polygons and the circle. Round bodies and polyhedra. Exercises about areas and volumes of cubes, prisms, and pyramids (Term 3).	*Exercises*: Measurement of perimeters. Graphical problems about addition and subtraction of angles. Problems about areas of parallelograms and triangles. *Learning outcomes*: The circumference. Relative positions. Rectification. The circle. Measurement of arcs in a circumference. Quadrilaterals.

theory, the number line, number bases, etc. as well as the treatment of geometry, due to the debates that existed around its teaching (De Bock et al. 2019).

Table 3 Treatment of Multiplication in Both Syllabi during Grades 1 to 4 (MEN 1953, pp. 65–70; MEC 1965, pp. 20–21)

	1953	1965
Grade 1	Idea of double (Term 3).	*Exercises*: n/a.
		Learning outcomes: n/a.
Grade 2	Idea of double and triple (Term 1). Idea of multiplication and formation of the multiplication table (Term 1). Continuation formation and memorization of the multiplication table (Term 2). Multiplication of digits (Term 2). Multiplication of natural numbers when the multiplier has only one digit and it is not greater than 5 (Term 3).	*Exercises*: Multiplication by numbers with one digit.
		Learning outcomes: Idea of double and triple. Commutative property of multiplication.
Grade 3	Multiplication of natural numbers when the multiplier has only one digit (Term 1). Multiplications of natural numbers with any number of digits (Term 2). Strategies for fast multiplication (Term 2).	*Exercises*: Multiplication when no digit of the multiplier is zero. Multiplication by 10, 100, 1000, etc.
		Learning outcomes: Associative property of multiplication.
Grade 4	n/a	*Exercises*: Multiplication of natural numbers (in general).
		Learning outcomes: Strategies for fast multiplication.

4 The Layer of Curriculum

In this section we analyze and compare the official primary education mathematics syllabi from 1953 and 1965. The 1953 syllabus corresponded to the Law of Primary Education from 1945, which had been enacted 8 years before (Egido Gálvez 1994). This document was later republished, without any changes, in 1955. The reform from 1965 did not involve any deep reorganization of the Spanish education system, but the syllabi of all subjects (not only of mathematics) were completely revised in an attempt to "adapt primary education to the requirements of the new socio-cultural structures" (MEC 1965, p. 1).

4.1 Overview of the 1953 Syllabus

The 1953 syllabus (see Fig. 1) began with a short section of about 2 pages called *Normas Didácticas* [Didactical Rules] which was followed by a description of the contents that had to be covered in each grade of primary education. These contents

Table 4 Summary of the contents of the textbooks

1964		1966	
Numbers 1 to 10.	18 pages.	Sets. Subsets. One-to-one correspondence. Definition of natural number.	4 pages.
Concept of 'ten.' Numbers 11 to 19.	15 pages.	Numbers 0 to 10. Concept of 'ten.'	14 pages.
Review.	6 pages.	Numbers 11 to 19.	10 pages.
Units of time (hours, weeks, months).	3 pages.	Review.	8 pages.
The liter.	1 page.	Numbers 20 to 30.	2 pages.
The gram and the kilogram.	1 page.	Numbers 1 to 100. Concept of 'hundred.'	6 pages.
Units of length (dm, m, km).	1 page.	Numbers 100 to 1000.	2 pages.
Numbers 20 to 30.	1 page.	Absolute and positional value of digits.	3 pages.
Numbers 1 to 100.	2 pages.	Addition and subtraction problems.	5 pages.
Review and problems.	5 pages.	Units of length (dm, m, km).	3 pages.
Numbers from 100 on.	4 pages.	Units of time (hours, weeks, months).	3 pages.
Addition.	6 pages.	The liter.	1 page.
Subtraction.	5 pages.	The kilogram.	2 pages.
Review and problems.	4 pages.	Addition.	12 pages.
Idea of multiplication and division.	3 pages.	Subtraction.	9 pages.
Multiplication by 2, 3, 4, and 5.	20 pages.	Review and problems.	6 pages.
Multiplication by a digit followed by zeroes.	1 page.	Plane geometry (lines, circumference and circle, triangle, square).	5 pages.
Multiplication of numbers with two and three digits not bigger than 5.	2 pages.	Space geometry (cube, sphere, pyramid, cone).	4 pages.
Multiplication by 6, 7, 8, and 9.	15 pages.	Idea of multiplication and division.	2 pages.
Problems.	5 pages.	Multiplication and division by 2, 3, 4, and 5. Commutativity of multiplication.	19 pages.
Exercises.	13 pages.	Multiplication and division by 6, 7, 8, and 9. Multiplication table.	13 pages.
		Review and problems.	3 pages.
		Division by one-digit numbers.	5 pages.
		Currency.	3 pages.

were further divided into terms, three for each course, and they were presented in the form of a list of items without any further structure. All didactical and methodological indications were provided in the document's introductory section, with very little reference to particular mathematical topics.

The implicit structure of the introductory section was quite clear, being divided into three parts. The first part was devoted to explaining the presence of mathematics in the curriculum and to highlight its formative value. The second part was devoted to providing some general indications about cognitive aspects of the learning of mathematics. Finally, the last part was devoted to supplying methodological comments regarding some mathematical contents with reference to the different stages of primary education.

The presence of mathematics was primarily grounded on tradition: "traditionally Spanish school has devoted special attention to the teaching of mathematics without being influenced by innovations relegating it to a complementary part" (MEN 1953, p. 63). The importance of the subject was motivated by its formative value and by its applicability.

Regarding cognitive aspects, a lot of emphasis was made on the requirement of repetition and constant exercise of techniques, the scaffolding of learning, the stepwise reasoning, and the need for generalization. In spite of stating that "a fundamental rule in the teaching of mathematics will be the solid foundation of knowledge" (MEN 1953, p. 63), we see that the appeal to authority was the ultimate foundation

Fig. 1 Cover of the 1953 Spanish official primary education syllabus

in some situations: "imposing a property whose explanation is not within the pupil's grasp will be preferred over [...] giving a wrong definition. In such case it will be enough to confess the difficulty and trust the authority of scientists" (MEN 1953, p. 63).

Some contents deserve particular comment. For example, both mental and written arithmetic were considered "indispensable." Elementary arithmetic problems had to be introduced in increasing order of difficulty, and they had to be grouped "not according to the required operations, but to the domain of application." Graphical representations to solve such problems were encouraged, but only as an "auxiliary and complementary means." Regarding geometry, teachers were advised "to avoid difficult proofs that use too many letters or symbols" (MEN 1953, p. 64).

After this introduction, the document proceeded to present contents that had to be taught during specific grades of primary education. A list for each of the three terms in each grade was given in a merely expositive way. As an example, the contents corresponding to the first term of the first grade were the following (MEN 1953, p. 65):

> Exercises about forming groups and series. Ordering of objects and counting. Composition and decomposition of one-digit numbers. Monographic treatment of each of them. Their graphical representation. Counting forward and backward one by one up to 20. Reading and writing of numbers up to 20. Exercises to introduce measurement using natural units. The same to compare distances by sight. Representation of the resulting integer numbers. Exercises about classification and grouping of bodies with diverse shapes.

As we mentioned, the first stage of primary education covered the initial four grades (ages 6 to 10). During these grades, the students were taught the number system, elementary arithmetic with natural numbers, decimal numbers, and fractions. The metric system was also introduced, together with some basic geometry concepts.

4.2 *Overview of the 1965 Syllabus*

The 1965 syllabus also began with a brief introduction that had a clear and explicit structure. This introduction did not contain any specific methodological comments, and it mostly focused on describing the objectives of mathematics teaching. The first two paragraphs, however, were devoted to introducing and explaining the new organization of the syllabus (see below) stating that "the learning of mathematical concepts and relations must be active" and that "concepts can be reached only through a series of exercises" (MEC 1965, p. 19).

After that, the objectives were introduced, distinguishing between (five) mathematical and (three) social objectives. Regarding the five mathematical objectives, we provide an excerpt of the more relevant fragments (MEC 1965, p. 19):

(a) To develop the understanding of mathematics as a system of interrelated principles. This means, on one side, the acquisition of concepts [...] through operations such as counting, measuring, and comparing and, on the other side, the understanding of the base 10 number system…

(b) To promote the understanding and use of mathematical vocabulary…
(c) To promote exactness and ease in the use of the four operations with integers, fractions, and decimal numbers…
(d) To introduce a series of functional concepts that aim to familiarize the student with current procedures in the study of mathematics […] rudiments of topology, such as the graphical representation of sets moving to the relations of "inside" and "outside" reaching the intuition of "common" …
(e) To introduce the ideas of "translation," "rotation," "symmetry," and "projection." In this way it is easy to develop the spatial understanding in our students…

The social objectives received less attention. They were essentially to show the student the "use of numbers in real-life situations" or the "way in which the number system has contributed to the progress of humanity" and to "promote positive attitudes towards the use of mathematics in problem solving" (MEC 1965, p. 19).

After this introduction, the contents for each course were given, with the distinction between exercises and learning outcomes that was presented in the introduction. As an example, in Table 1 we present the contents corresponding to grade 1.

In general terms, for the initial four grades (ages 6 to 10), the 1965 syllabus covered the number system, elementary arithmetic with natural numbers, decimal numbers and fractions, the metric system, and some basic geometry concepts. It also included a very brief mention of set theory in grade 1.

4.3 Comparing Both Documents

The general structure of both documents was rather similar. An introduction of an expository nature was followed by an enumeration of the contents. The organization of these contents radically differed in the two documents. In 1953, the contents were sequentially organized in terms, and they were introduced just as a list of topics. This was not the case in 1965 when a novel distinction between two types of contents (exercises and learning outcomes) was introduced.

4.3.1 The Introductory Section

The introductory sections addressed rather different aspects. The usefulness and the formative value of mathematics were present in both documents, and we can also find cognitive considerations in both of them. However, the 1953 syllabus emphasized methodological aspects, while the 1965 document focused on the objectives of teaching and learning mathematics. This latter document also made explicit, even if not very clearly, a social and an affective dimension of mathematics learning. The reason to learn mathematics in primary education was somewhat shifted from a traditional foundation to a social requirement.

Different cognitive assumptions can be detected in the introduction of both documents. In 1953, we read, "the repetition, the constant exercise of each acquired mechanism are indispensable didactical means, together with the scaffolding of the successive learning steps" (MEN 1953, p. 63). In 1965, the idea of exercising is also present, but as a means to reach a concept. The emphasis is not made on repeating an exercise to acquire a mechanism, but rather on doing a series of exercises as a pathway to conceptual understanding. In some sense these different assumptions had an impact on the list of topics and on their presentation. For example, in 1953, grade 1 students were required in the first term to count forward and backward one by one up to 20, in the second term up to 50, and in the third term up to 100. In turn, the 1965 syllabus for grade 1 only presented the exercise "count forward and backward from 1 to 100" that lead to the learning outcome "identification and writing on the first 100 numbers" (MEC, 1965, p. 20).

Finally in the 1965 syllabus we can find reference to some elements that are directly related to Modern Mathematics ideas: a structural view of mathematics, abstract topological or set-theoretical concepts, geometric transformations, etc.

4.3.2 Topics

As we said, both documents covered elementary arithmetic (natural numbers, decimal numbers, and fractions), the metric system and elementary geometry. Regarding the metric system, we have not identified any real difference between both syllabi, aside from the greater level of detail in which the 1953 curriculum described the corresponding topics.

In Table 2 we summarize the appearance of geometry in both documents. We can see that the 1953 syllabus was more detailed and, in some sense, more extensive. The only topics present in 1965 that were not explicitly mentioned in 1953 were perimeters and relative positions of circumferences. Both documents seemed to propose the recognition and observation of forms as the starting point to geometry. The 1965 document was more ambiguous in this regard, only referring to "geometric forms" in grade 1, while the 1953 curriculum explicitly mentioned the sphere, the cube, etc.

In fact, the 1953 syllabus paid more attention to three-dimensional geometry. For example, the notion of quadrilateral was meant to be introduced in grade 3 using cubes and prisms as a starting point (so, for example, quadrilaterals were supposed to be introduced as the faces of prisms using models for demonstration). In addition, regular polyhedra (missing in the 1965 curriculum until grade 8) were introduced in 1953 as early as in grade 2. Finally, from the metric point of view, the computation of volumes (at least for cubes, prisms, and pyramids) was introduced in grade 4 in the 1953 document and in grade 6 in 1965. Regarding plane geometry, the topics were approximately the same in both documents. However, some concepts were introduced earlier in 1953. This was the case, for instance, for angles (grade 2 in 1953 and grade 3 in 1965) and quadrilaterals (grade 3 in 1953 and grade 4 in 1965). Regular polygons, which in 1953 were introduced in grade 4, were not covered until grade 5 in 1965.

Now, for the rest of topics, we conclude again that they were practically the same in both documents. As an example, in Table 3 we summarize and compare the contents related to multiplication of natural numbers in both documents.

As we discussed before, the 1953 document presented a higher degree of scaffolding and it emphasized "traditional" elements such as the construction and memorization of the multiplication table. In 1965 more conceptual issues were addressed. The absence of an explicit reference to the multiplication table in 1965 is, in fact, quite remarkable. We might conclude that multiplication was introduced later in the 1965 document, but its treatment was also deeper than in the 1953 curriculum.

Finally, the 1965 syllabus contained some noteworthy additional elements. In particular:

- In grade 1, the idea of set was introduced together with the idea of belonging to a set.
- In grade 2, the commutative and associative properties of addition and the commutative property of multiplication were explicitly stated.
- In grade 3, the associative property of multiplication was introduced explicitly.
- In grade 4, the idea of 10 as the base of the number system was presented.

The presence of these topics, which clearly relate to Modern Mathematics ideas, corresponds to issues covered in the introductory section such as the use of "mathematical vocabulary" or the introduction of "functional concepts" related to the "current procedures in the study of mathematics" (MEC 1965, p. 19).

5 The Layer of Classroom Concept

In order to address the layer of classroom concept, we are going to analyze two series of textbooks published before and after the 1965 reform. Both series were published by the same company (SM), and they were entitled *El Cálculo en la Vida Diaria* [Calculation in Daily Life].

Just by looking at the covers (see Fig. 2), we can detect a change in the tone between both series. In 1964 the illustrations showed "real life" situations in which mathematics may play a role: cutting something in half (bottom right), counting money (bottom left), etc. In 1966, however, the illustrations referred to purely mathematical elements: sets and applications, the decimal number system, and additive decompositions of the number 15.

To compare both series, we will first carry out a detailed comparison of the grade 2 textbooks. This will give us a general overview of the differences among them. After that, due to its importance, we will describe the treatment of set theory by the post 1965 textbooks.

Fig. 2 Covers of the grade 2 textbooks: 1964 (left) and 1966 (right)

5.1 Comparison of the Grade 2 Textbooks

In Table 4 we summarize and compare the topics and their ordering in both textbooks, providing as a reference the number of pages devoted to each topic (the 1964 book had 135 pages and the 1966 book had 152 pages).

As we see, there are changes in the list of topics as well as in their ordering and extension. From the simple comparison of the tables of contents, shown in Table 4, the most notable differences were:

- The presence of "set theory" in the 1966 textbook.
- The complete absence of geometry in the 1964 textbook.
- The bigger number of problems and exercises in 1964.
- The different treatment of multiplication in both texts.

In order to identify differences in the treatment of the topics, we must have a closer look at the contents of both texts. For example, a subtle, yet interesting, difference between both texts can be found in the definition of the concept of 'ten.' In 1964 we read: "One single thing is a unit. Ten units form a ten" (SM 1964b, p. 24), while in 1966 we read: "One single thing is a simple unit. Ten units form a ten, or a unit of second order" (SM 1966b, p. 21). The same distinction was made when introducing the idea of 'hundred,' which in 1966 was referred to as a unit of third order.

Regarding the treatment of multiplication, it is noteworthy that in the 1966 textbook, multiplication and division were explicitly presented together throughout all the text. In 1964, after giving the definition of division as "sharing in equal parts," the concept was never revisited again. In 1966, much more explicit emphasis was given to the idea of division as the inverse of multiplication.

In any case, for the most part, the treatment of the contents was very similar or even almost identical. In Fig. 3, for example, we can compare the page devoted to the liter in both editions. They have the exact same layout, although the drawings were slightly modernized in 1966 and used colors. One problem was removed and, for some unknown reason, the data of another were changed. In some other situations, the aesthetic changes were accompanied by the introduction of novel resources. In Fig. 4, for example, we can see how the additive decompositions of the number 4 were illustrated by the means of dominoes and squares in the 1964 edition and using Cuisenaire rods in the 1966 edition. The use of these rods was extensive among the four textbooks published after 1965.

Finally, it is also noteworthy that, like in the examples shown in Figs. 3 and 4, the number of verbal problems decreased in 1966. This phenomenon was present throughout the textbook. Moreover (recall Table 4) the 1964 text ended with 13 pages devoted to problems and exercises for student training. Those pages included 31 verbal arithmetic problems, 364 additions of five-digit numbers (with two, three, and four addends), 240 subtractions of five-digit numbers, and 525 multiplications

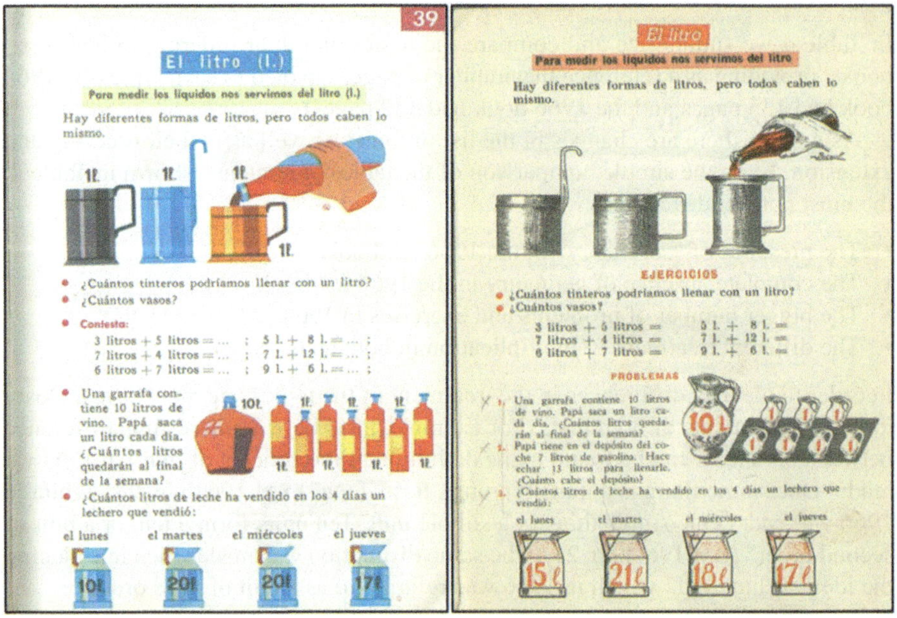

39

El litro (l.)

Para medir los líquidos nos servimos del litro (l.)

Hay diferentes formas de litros, pero todos caben lo mismo.

- ¿Cuántos tinteros podríamos llenar con un litro?
- ¿Cuántos vasos?
- Contesta:

3 litros + 5 litros = ... ; 5 l. + 8 l. = ... ;
7 litros + 4 litros = ... ; 7 l. + 12 l. = ... ;
6 litros + 7 litros = ... ; 9 l. + 6 l. = ... ;

- Una garrafa contiene 10 litros de vino. Papá saca un litro cada día. ¿Cuántos litros quedarán al final de la semana?
- ¿Cuántos litros de leche ha vendido en los 4 días un lechero que vendió:

el lunes · el martes · el miércoles · el jueves

El litro

Para medir los líquidos nos servimos del litro

Hay diferentes formas de litros, pero todos caben lo mismo.

EJERCICIOS

- ¿Cuántos tinteros podríamos llenar con un litro?
- ¿Cuántos vasos?

3 litros + 5 litros = 5 l. + 8 l. =
7 litros + 4 litros = 7 l. + 12 l. =
6 litros + 7 litros = 9 l. + 6 l. =

PROBLEMAS

1. Una garrafa contiene 10 litros de vino. Papá saca un litro cada día. ¿Cuántos litros quedarán al final de la semana?
2. Papá tiene en el depósito del coche 7 litros de gasolina. Hace echar 13 litros para llenarle. ¿Cuánto cabe el depósito?
3. ¿Cuántos litros de leche ha vendido en los 4 días un lechero que vendió:

el lunes · el martes · el miércoles · el jueves

Fig. 3 The liter: 1966 (left) and 1964 (right)

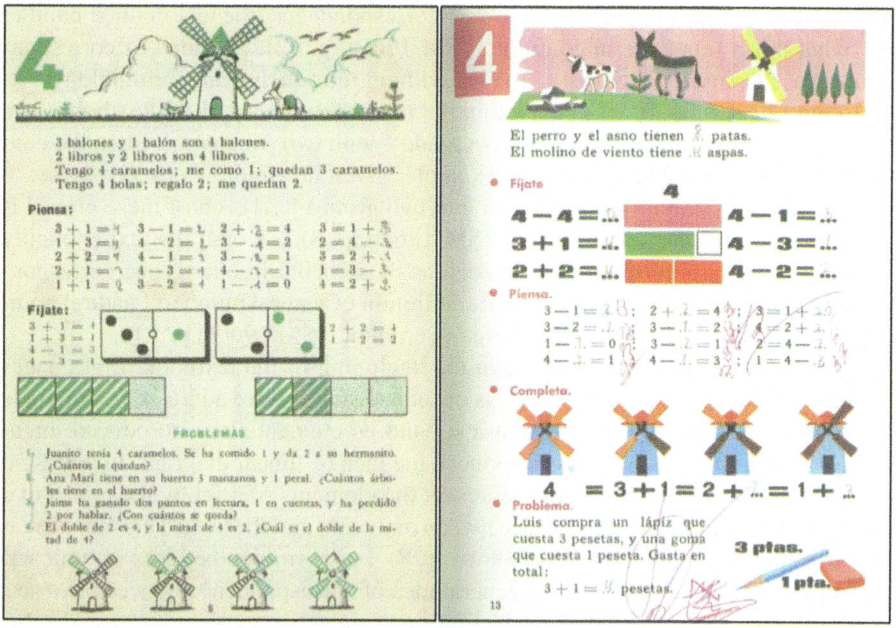

Fig. 4 The number 4: 1964 (left) and 1966 (right)

in which one factor had five digits and the other one had two or three digits. These final pages were removed from the 1966 textbook.

5.2 Set Theory *in the Textbooks Published After 1965*

As we have seen in the comparison of the grade 2 textbooks, the presence of set theory in the 1966 edition was probably the most remarkable difference. This motivates a closer look at the treatment of this topic in the textbooks for grades 1 to 4.

In Grade 1 four pages were devoted to introducing set-theoretical concepts. They were introduced after having covered the numbers 0 to 10, the ideas of addition, equality, subtraction and multiplication, and some topics of elementary geometry (from lines and segments to the sphere).

The ideas of 'set,' 'element,' and 'belonging to' were introduced rather informally using examples. After that, the idea of 'one-to-one correspondence' was introduced (see Fig. 5) joining with arrows the elements of two particular sets with the same number of elements (SM 1966a, p. 41):

> Look at the set of bottles and at the set of cakes: from each bottle there is a line joining it to a cake, and from each cake there is a line joining it to a bottle. This is a one-to-one correspondence.

Then, given two sets having a one-to-one correspondence, the text defined number as "what those sets have in common" (SM 1966a, p. 42). The idea of correspondence was also used to compare sets, but without introducing any terminology.

In grade 2 set theory was covered in the initial four pages of the textbook. The treatment was essentially the same as in grade 1 with two exceptions. First, the concept of 'subset' was addressed using examples given in a figure: "observe the set of apples. We are going to separate them into two groups [...] each of these groups is called a subset of the set of apples" (SM 1966b, p. 8). Second, the text explicitly introduced the notion of 'equivalent sets' as those being in one-to-one correspondence, leading to the following abstract definition of natural number: "natural number is what all the equivalent sets share in common" (SM 1966b, p. 10).

In grade 3, set theory was found at the beginning of the textbook. Eight pages were devoted to these topics. The ideas of set and subset were addressed as in grade 2, but the use of braces to represent a set and its elements was introduced intensively. In addition, the textbook introduced the idea of 'linear diagram' to represent the elements of set: "It is usual to represent the elements of a set by the points of a straight line, taking equal distances between one point and the next. This representation is called linear diagram" (SM 1968a, p. 8). Furthermore, the ideas of 'unit set' and 'empty set' were defined. The general idea of correspondence between two sets was introduced, together with the definition of injective correspondence. In this way, the concept of one-to-one correspondence is revisited and redefined as a correspondence which is injective "in both directions." Finally, the distinction between cardinal and ordinal numbers was introduced. The cardinal number was defined as "the number of elements of a set and of all the sets that are equivalent to it" (SM 1968a, p. 13), while the ordinal number was informally defined using a particular example of an ordered set.

In grade 4, twelve pages were devoted to set theory at the beginning of the book. In addition to all the material covered in grade 3, a few new ideas could be found. For example, two possibilities to define a set were given: "A set can be defined

Fig. 5 One-to-one correspondence in grade 1

either by providing a characteristic property of it, or by enumerating its elements" (SM 1968b, p. 7) and Venn diagrams were introduced. The notions of disjoint sets and their addition were defined. Regarding correspondences, the notions of 'initial' and 'image' sets and the definition of 'inverse correspondence' were given. Also, the notation of a correspondence as ordered pairs was introduced without any further explanation. The following abstract definition of 'counting' was provided: "Counting a set consists in placing it in one-to-one correspondence with a part of the set of natural numbers" (SM 1968b, p. 17). A naïve idea of infinity was introduced by stating that "the series of natural numbers is unlimited; every number is followed by another one" (SM 1968b, p. 18). Finally, in this grade some symbolism was introduced for the first time. In particular we can find the symbols $\in$, $\notin$, $\varnothing$ and $\subset$, and the notations f^{-1} and card(A).

With the exception of grade 1, the ideas of set theory were always presented in the first pages of the textbook. The previous paragraphs clearly show that the underlying idea was to cover the material in some kind of spiral curriculum, revisiting the already covered material and introducing a few novel ideas, notation, etc. in each grade. However, after being introduced in the initial chapters, the set-theoretical terminology was never used throughout the textbook again in any grade.

6 Discussion and Conclusions

In this work, we have had a look at the first steps of the process that led to the implementation of Modern Mathematics in the first stage of Spanish primary education (Grades 1 to 4, 6–10 year old students). The individual modifications that were introduced during these initial moments in the 1965 syllabus may not have been very important. However, they reflected a deep change in mentality and a step towards modernization with respect to the previous syllabus. This change was shown both at a methodological and a philosophical level (in the introductory section of the document) and at a mathematical level (in the selection and organization of the topics). The differences between the textbooks were bigger and more explicit. This shows their important role in the education system and the need to complement the analysis of curricular documents with that of textbooks (Schubring 1987).

At the layer of curriculum, we identified references to ideas related to Modern Mathematics such as a structural view of mathematics, abstract topological or set-theoretical concepts, geometric transformations, etc. We also noted some changes in vocabulary that reflected social aims. For example, the distinction between "exercises" (described by action words) and "learning outcomes" (more conceptually described), which can be identified using more modern terminology as "procedural" and "conceptual" knowledge (Hiebert and Lefevre 1986), was a radical change with respect to the older presentation of contents as a mere list of topics.

Regarding particular topics, the biggest change was related to the introduction in the 1965 syllabus of set theory ideas and terminology (only in grade 1). Some abstract concepts such as the commutative or associative properties of operations or the base of a number system were explicitly mentioned in 1965. In the case of geometry, very few changes were introduced during the first stage of primary education. Some contents were removed, and others were introduced later. The shift from a static geometry to a more modern approach using the more abstract concepts of translation, rotation, symmetry, and projection only was explicitly addressed starting from grade 5.

In the layer of classroom concept, which we have approached through textbook analysis, changes were more profound. Not only were new topics introduced, but there was also a change in their ordering and in the depth with which some of them were treated. Also, some novel resources were introduced. The most remarkable case was that of Cuisenaire rods. Even though they had been invented more than 10 years before (Cuisenaire 1952), they were presented as an innovation in grade 1: "Since the system of numbers in color is being spread, not only in Spain, but also among the most advanced countries, we decidedly use it from the first page of the book" (SM 1966a, p. 4). However, as De Bock (2020) points out, the use of this material did not seem to catch on since it was no longer present in the editions of the textbook that were published following the 1970 reform (SM 1982a, b).

The most notable difference between the two series of textbooks was the presence of set theory in all grades. In this regard we note that, even if set-theoretical concepts and terminology were introduced in the first chapters of the textbooks, they were absent from the remaining chapters. In fact, this incomplete integration of set theory with the remaining topics was a particular phenomenon associated with this stage of the introduction of Modern Mathematics in Spain. Once the 1970 reform was approved, the use of set-theoretical terminology and concepts was extended throughout all the text. As an example, Fig. 6 shows the evolution of the same page of a textbook from 1964 to 1966, and then to 1982.

In comparing the layer of classroom concept and the layer of curriculum, we have observed that textbooks did not faithfully reflect the official syllabus. A notable example was the treatment of set theory. The 1965 syllabus only included the idea of set and of belonging to a set in grade 1. In contrast, as we have seen, the analyzed textbooks included set-theoretical ideas in grades 1 through 4, covering far more concepts than those stated in the official curriculum. Also, some traditional topics were not abandoned. While the multiplication table was removed from the 1965 syllabus (recall Table 3), it was still present in the grade 2 textbook from 1966. This gap between the intended curriculum and the implemented curriculum (van den Akker 2003) in the process of implementation of Modern Mathematics has also been identified by other authors (Prytz 2018).

In general, we see quite a few elements of Modern Mathematics in the analyzed textbooks. Compared to other European countries, it is in fact an early presence. This is more noteworthy if we take into account that Spain was not present in the international meetings of Royaumont and Zagreb-Dubrovnik (OEEC 1961a, 1961b) and is a possible avenue for further research.

In order to improve and enrich our analysis, the other two layers (scientific discipline and practical schooling) should be taken into consideration. In fact, González

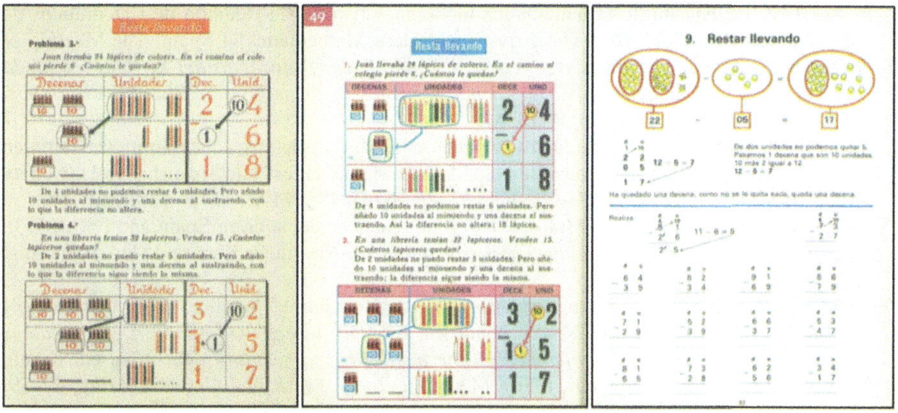

Fig. 6 Subtraction with borrowing: 1964 (left), 1966 (center) and 1982 (right)

Astudillo (2008a) has shown that professional journals played an important role in introducing Modern Mathematics ideas among primary education teachers. In a more recent work, González Astudillo (2018) has approached the impact of Modern Mathematics on in-service teacher training by analyzing the notebook of a primary education teacher. We think that integrating all these findings would provide a more complete picture of the early stages of the transition to Modern Mathematics in Spain.

Finally, we must point out that many of the features identified in our analysis of the syllabi and of the textbooks are analogous or can be related to results obtained in similar research conducted in different geographical contexts (Bjarnadóttir 2017; De Bock et al. 2019; Hamann 2018; Prytz 2018). Modern Mathematics was a worldwide phenomenon (Kilpatrick 2012), and we agree with Menghini (2020, p. 388) who points out that "collaborations between authors from different countries (which rarely happens in this field) … could include such themes as … New Math" We think that a detailed cross-national comparison of the process of its introduction, implementation and eventual disappearance is still to be made, and we hope that our work might be useful in this regard.

Acknowledgement This work has been partially funded by the Government of Aragón and the European Regional Development Fund via the research group S60_20R "Investigación en Educación Matemática." The author thanks the reviewers for their useful comments and suggestions, which improved the paper, and Alexander Karp for his help in preparing the final version of the manuscript.

References

Ausejo, Elena. 2010. The introduction of "modern mathematics" in secondary education in Spain (1954–1970). *International Journal for the History of Mathematics Education* 5 (2): 1–14.

———. 2012. La llegada de la matemática moderna a España: la Reunión de Catedráticos de Matemáticas (Madrid 1961) [The arrival of Modern Mathematics in Spain: The Meeting of Secondary Education Mathematics Teachers (Madrid 1961)]. In *Actas del XI Congreso de la Sociedad Española de Historia de las Ciencias y de las Técnicas*, ed. José María Urkia Etxabe, 519–531. San Sebastián, Spain: RSBEAP.

———. 2013. La introducción de la "matemática moderna" en la enseñanza no universitaria en España (1953–1970) [The Introduction of "Modern Mathematics" into Spanish Non-University Education (1953–1970)]. *La Gaceta de la RSME* 16 (4): 727–747.

Ausejo, Elena, and José Manuel Matos. 2014. Mathematics education in Spain and Portugal. In *Handbook on the history of mathematics education*, ed. Alexander Karp and Gert Schubring, 283–302. New York: Springer.

Bjarnadóttir, Kristín. 2017. Recommendations of the Royaumont Seminar on primary school arithmetic. Influences in the Nordic countries. In *"Dig where you stand" 4. Proceedings of the fourth international conference on the history of mathematics education*, ed. Kristín Bjarnadóttir, Fulvia Furinghetti, Marta Menghini, Johan Prytz, and Gert Schubring, 47–59. Turin, Italy: Edizioni Nuova Cultura.

Cuisenaire, Georges. 1952. *Les nombres en couleurs: nouveau procédé de calcul par la méthode active applicable à tous les degrés de l'école primaire [Numbers in color: a new calculation procedure through the active method applicable to all the degrees of Primary school].* Tamines, Belgium: Duculot-Roulin.

De Bock, Dirk. 2020. Georges Cuisenaire's numbers in colour. A teaching aid that survived the 1950s. In *"Dig where you stand" 6. Proceedings of the sixth international conference on the history of mathematics education*, ed. Évelyne Barbin, Kristín Bjarnadóttir, Fulvia Furinghetti, Alexander Karp, Guillaume Moussard, Johan Prytz, and Gert Schubring, 355–368. Münster, Germany: WTM-Verlag.

De Bock, Dirk, and Geert Vanpaemel. 2018. Early experiments with modern mathematics in Belgium. Advanced mathematics taught from childhood? In *Researching the history of mathematics education. An international overview*, ed. F. Furinghetti and A. Karp, 61–77. New York: Springer.

———. 2019. *Rods, sets and arrows. The rise and fall of modern mathematics in Belgium.* Cham, Switzerland: Springer.

De Bock, Dirk, Michel Roelens, and Geert Vanpaemel. 2019. Mathématique moderne: A pioneering Belgian textbook series shaping the New Math reform of the 1960s. In *"Dig where you stand" 5. Proceedings of the fifth international conference on the history of mathematics education*, ed. Kristín Bjarnadóttir, Jenneke Krüger, Johan Prytz, Gert Schubring, and Harm Jan Smid, 129–145. Utrecht, the Netherlands: Utrecht University.

De Puelles Benítez, Manuel. 2007. La política escolar del libro de texto en la España contemporánea [School policy on textbooks in contemporary Spain]. *Avances en supervisión educativa* 6: 1–18.

Diego Pérez, Carmen. 1999. Intervención del primer Ministerio de Educación Nacional del franquismo sobre los libros escolares [Intervention over textbooks of the first Ministery of National Education during Francoism]. *Revista complutense de educación* 10 (2): 53–72.

Egido Gálvez, Inmaculada. 1994. La evolución de la enseñanza primaria en España: organización de la etapa y programa de estudio [Evolution of primary education in Spain: organization and syllabus]. *Tendencias pedagógicas* 1: 75–86.

Fernández Cárcar, Miguel. 2020. Los manuales escolares durante el primer franquismo (1939–1959), un acercamiento al caso navarro [Textbooks during the early Francoism (1939–1959), an approach to the case of Navarra]. *Príncipe de Viana* 277: 613–641.

González Astudillo, María Teresa. 2006. La matemática moderna en España [Modern Mathematics in Spain]. *Unión. Revista Iberoamericana de educación matemática* 6: 63–71.

———. 2008a. La transición hacia la matemática moderna en España: la revista Vida Escolar [The transition towards Modern Mathematics in Spain: the journal School Life]. *Revista Diálogo Educacional* 8 (25): 615–631.

———. 2008b. Las ideas sobre la educación matemática de Pedro Puig Adam ¿precursoras de la Matemática Moderna? [The ideas of Pedro Puig Adam on Mathematics Education Forerunners of Modern Mathematics?]. *Quadrante* 17 (1): 93–108.

———. 2010. La matemática moderna en España [Modern Mathematics in Spain]. In *A reforma da Matemática Moderna em contextos ibero-americanos*, ed. José Manuel Matos and Wagner Rodrigues Valente, 41–71. Lisboa, Portugal: UIED.

———. 2018. Formación de un maestro para la enseñanza de la matemática moderna: su cuaderno de apuntes [Training of a primary education teacher for modern mathematics teaching: his notebook]. In *Actas del IV Congreso Iberoamericano de Historia de la Educación Matemática*, ed. Dolores Carrillo Gallego, Encarna Sánchez Jiménez, José Manuel Matos, Pedro Luis Moreno Martínez, and Wagner Rodrigues Valente, 292–302. Murcia, España: CEME. Universidad de Murcia.

Hamann, Tanja. 2018. *Die "Mengenlehre" im Anfangsunterricht – historische Darstellung einer gescheiterten Unterrichtsreform in der Bundesrepublik Deutschland [Set theory in Elementary Education – historical representation of a failed teaching reform in West Germany]*. Siegen, Germany: Universi-Universitätsverlag Siegen.

———. 2019. New Math at primary schools in West Germany – a theoretical framework for the description of educational reforms. In *Proceedings of the Eleventh Congress of the European Society for Research in Mathematics Education*, ed. Thomas Jankvist Uffe, Marja van den Heuvel-Panhuizen, and Michiel Veldhuis, 2117–2124. Utrecht, the Netherlands: Freudenthal Group & Freudenthal Institute, Utrecht University and ERME.

Hayden, Robert W. 1981. A history of the "New Math" movement in the United States (Doctoral dissertation, Iowa State University, Ames, Iowa, USA). Retrieved from https://lib.dr.iastate.edu/rtd/7427

Hiebert, James, and Patricia Lefevre. 1986. Conceptual and procedural knowledge in mathematics. An introductory analysis. In *Conceptual and procedural knowledge: The case of mathematics*, ed. James Hiebert, 1–28. Hillsdale, NJ: Lawrence Erlbaum Associates.

Kilpatrick, Jeremy. 2012. The new math as an international phenomenon. *ZDM – Mathematics Education* 44 (4): 563–571.

Kline, Morris. 1973. *Why Johnny can't add: The failure of the New Math*. New York, NY, USA: St. Martin's Press.

Krippendorff, Klaus. 2014. *Content analysis. An introduction to its methodology*. London: SAGE.

Martos, Juan Antonio. 2019. Paleolítico y evolución humana en los manuales de segunda enseñanza del final del franquismo (1967–1976): elaboración de un discurso legitimado [Paleolithic and human evolution in Secondary education textbooks at the end of Francoism (1967–1976): building up a legitimated discourse]. *Trabajos de Prehistoria* 76 (1): 11–28.

Matos, José Manuel. 2010. Elementos sobre o ensino e a aprendizagem da Matemática Moderna em Portugal no final dos anos 70 [Elements of teaching and learning of Modern Mathematics in Portugal at the end of the 1970s]. In *A reforma da Matemática Moderna em contextos ibero-americanos*, ed. José Manuel Matos and Wagner Rodrigues Valente, 137–174. Lisboa, Portugal: UIED.

Mayordomo Pérez, Alejandro. 1997. Nacional-catolicismo, tecnocracia y educación en la España del franquismo: (1939–1975) [National Catholicism, technocracy and education in Spain during Francoism (1939–1975)]. In *Los caminos hacia la modernidad educativa en España y Portugal (1800–1975)*, ed. Agustín Escolano and Rogério Fernandes, 147–174. Zamora, Spain: SEDHE-SPCE.

Mayring, Philipp. 2014. *Qualitative content analysis. Theoretical foundation, basic procedures and software solution*. Klagenfurt: Beltz.

MEC. 1965. Cuestionarios Nacionales para la Enseñanza Primaria [Primary education syllabi]. *Vida Escolar* 70–71: 1–96.

MEN. 1953. *Cuestionarios Nacionales Para la Enseñanza Primaria [Primary education syllabi]*. Madrid, Spain: Servicio de Publicaciones del Ministerio de Educación Nacional.

Menghini, Marta. 2020. Book review: digging into the memory of mathematics education. Fulvia Furinghetti, Alexander Karp (Eds.) (2018). Researching the history of mathematics education. An international overview. *Educational Studies in Mathematics* 103 (3): 383–389.
Moon, Bob. 1986. *The "New Maths" curriculum controversy: An international story*. London, UK: The Falmer Press.
OEEC. 1961a. *New thinking in school mathematics*. Paris: OEEC.
———. 1961b. *Synopses for modern secondary school mathematics*. Paris: OEEC.
Phillips, Christopher J. 2014. *The new math: A political history*. Chicago, IL, USA: University of Chicago Press.
Prior, Lindsay. 2016. Using documents in social research. In *Qualitative research*, ed. David Silverman, 171–186. London, UK: SAGE.
Prytz, Johan. 2018. The new math and school governance: An explanation of the decline of the new math in Sweden. In *Researching the history of mathematics education. An international overview*, ed. Fulvia Furinghetti and Alexander Karp, 189–216. Cham, Switzerland: Springer.
Schubring, Gert. 1987. On the methodology of analyzing historical textbooks: Lacroix as textbook author. *For the Learning of Mathematics* 7 (3): 41–51.
Scott, John. 1990. *A matter of record, documentary sources in social research*. Cambridge, MA: Polity Press.
SM. 1964a. *Cálculo, iniciación [Calculations, beginning]*. Madrid, Spain: Ediciones S.M.
———. 1964b. *El Cálculo en la vida diaria, 2° Grado [Calculations in daily life, 2nd Grade]*. Madrid, Spain: Ediciones S.M.
———. 1964c. *El Cálculo en la vida diaria, 3° Grado [Calculations in daily life, 3rd Grade]*. Madrid, Spain: Ediciones S.M.
———. 1964d. *El Cálculo en la vida diaria, 4° Grado [Calculations in daily life, 4th Grade]*. Madrid, Spain: Ediciones S.M.
———. 1966a. *Cálculo, 1er Curso [Calculations, 1st Grade]*. Madrid, Spain: Ediciones S.M.
———. 1966b. *El Cálculo en la vida diaria, 2° Curso [Calculations in daily life, 2nd Grade]*. Madrid, Spain: Ediciones S.M.
———. 1968a. *El Cálculo en la vida diaria, 3° Curso [Calculations in daily life, 3rd Grade]*. Madrid, Spain: Ediciones S.M.
———. 1968b. *El Cálculo en la vida diaria, 4° Curso [Calculations in daily life, 4th Grade]*. Madrid, Spain: Ediciones S.M.
———. 1982a. *El Cálculo 1 [Calculations 1]*. Madrid, Spain: Ediciones S.M.
———. 1982b. *El Cálculo 2 [Calculations 2]*. Madrid, Spain: Ediciones S.M.
van den Akker, J. 2003. Curriculum perspectives: An introduction. In *Curriculum landscape and trends*, ed. J. van den Akker, W. Kuiper, and U. Hameyer, 1–10. Dordrecth: Kluwer Academic Publishers.
Vanpaemel, Geert, and Dirk De Bock. 2019. New Math, An international movement? In *Proceedings of the Eighth European Summer University on History and Epistemology in Mathematics Education ESU 8*, ed. E. Barbin, U.T. Jankvist, T.H. Kjeldsen, B. Smestad, and C. Tzanakis, 801–812. Oslo: Oslo Metropolitan University.
Viñao Frago, Antonio. 2014. La educación en el franquismo (1936–1975) [Education during Francoism (1936–1975)]. *Educar em Revista* 51: 19–35.

Toward a History of Mathematics Education for Young Women: 1890–1920

Yana Shvartsberg

Abstract The main aim of this paper is to provide a historical review of female mathematics education in the United States during the Progressive Era, from 1890 through 1920. This time period in the United States is known for multiple social reforms. Secondary schools experienced rapid expansion in enrollment, and the purpose and direction of education underwent development and change. During this era, secondary education, which had been available to few, came to be accepted as a necessity for a majority of children in the United States. Through this period of flux, the educational system encountered several challenges. One such challenge was to tailor curricula to the needs of different students. In parallel with increased educational enrollment, labor market demands experienced changes as well, and these changes were especially prominent in urban areas. Historical evidence documents that, in connection with these changes, the purpose of girls' and boys' education was often considered to be distinct. This belief stemmed from the idea that girls and boys had different career paths upon high school graduation. Therefore, differentiation of curricula, which allowed high schools to provide an elective system of subjects, was perceived as necessary by many education leaders and policymakers. This paper provides historical analysis of the mathematics education available for girls in the Progressive Era, focusing on the purpose of mathematics education, on curricular differentiation, and on social factors that affected girls' enrollment in mathematics classes when the election of mathematics classes was allowed.

Keywords Elective courses · Female education · Mathematics education · Progressive era · Secondary schools · Social efficiency

Y. Shvartsberg (✉)
Pace University, New York, NY, USA
e-mail: ys3016@tc.columbia.edu

A. Karp (ed.), *Advances In The History Of Mathematics Education*,
International Studies in the History of Mathematics and its Teaching,
https://doi.org/10.1007/978-3-030-95235-8_8

1 Introduction

The complexity of gender issues connected to mathematics education and to the lack of women in mathematics-related careers has been well documented and has been on educators' radar since the second half of the twentieth century (Eisenmann 1998; Fennema 1974; Fennema 1979; Fennema and Peterson 1985; Eccles 1984; Leder 2001). Though many researchers and writers have focused their works on the history of female education, extensive exploration of publications on the history of female mathematics education shows a scarcity of comparative analysis and research dedicated to the period of 1890 through 1920, a period known in the United States as the Progressive Era. The Progressive Era marked a turning point in public and private education. Wide-ranging reforms affected almost every aspect of American society, and education was recognized as one of the areas that required improvement (Eisenmann 1998, p. 328). By examining various primary sources such as minutes of meetings and conferences, collections of bulletins and reports issued by government agencies, curriculum records, and teacher handbooks, this work will analyze practices and changes that took place in mathematics education during the period, especially practices in mathematics education for girls.

Furthermore, research of the history of mathematics education as a part of social history becomes meaningful only when it includes social analysis and examines mathematics education in connection with the events current in the society of the time (Karp 2014; Schubring 2014). Therefore, research regarding women's education should include inquiry into the economic and societal changes that influenced women's choices concerning mathematics education. Within this context, opposing viewpoints of teachers, educational reformers, and public figures is cited to create a multifaceted representation of social developments as well educators' positions on the topic of mathematics education for girls.

This work focuses particularly on urban areas. Though most states had established public schools by 1860, gender differentiation was found to be most problematic in the urban environment, "[partly because] adolescents were recognized as sexual beings and partly because educators and parents saw the school as the portal to adult work that differed for women and men" (Hansot and Tyack 1988, p. 751). Also, urban areas such as New York, Chicago, Boston, and Philadelphia were all struggling with the unprecedented increase in public school enrollment brought on by waves of immigration (Cremin 1961). The northeastern region of the United States primarily experienced immigration of groups coming from Europe, whereas other groups (Hispanic and Asian) were concentrated on the West Coast and were limited in number during the period under review (Rury 1991). Thus, this study will focus on the northeastern United States, especially New York City, as it had become one of the largest cities in world during this time period, largely because of its increased immigrant population. It therefore offers both demographic diversity and socioeconomic diversity for research.

2 Progressive Era

The Progressive Era is known for its social, political, and educational reforms aimed at improving the worsening living conditions and aftereffects of the Industrial Revolution. The theme of this period was "awakening of social conscience," which led to the belief that the road to alleviating growing poverty and despair was "neither charity nor revolution, but in the last analysis, education" (Cremin 1961, p. 59). The poverty and desperation that was particularly prominent in large cities was not new to the United States at that time; however, what was new was that the individuals in power began to get involved and look for ways to change conditions for the better.

Upon reading Jacob Riis's work *How the Other Half Lives in 1890* (1890), Theodore Roosevelt called it "an enlightenment and an inspiration for which he would be forever grateful" (*Theodore Roosevelt: An Autobiography* [1926], as cited in Cremin 1961, p. 58). Social awakening and a desire for change started to manifest in many institutions—foundations, educational programs, and settlements intended to help immigrants and other poor workers gain skills began to appear. The Progressive Era became known for collective movements rather than isolated efforts. Education started to become accepted as a necessity for most children in the United States, and by 1914, compulsory education laws had been adopted by almost every state (Eisenmann 1998, p. 329).

Even before the Progressive Era, reforms were bringing advances to formalized public education. After 1865 the number of public high schools in the United States increased, and by the end of the decade, there were at least 160. In 1880, there were nearly 800; in 1890, there were 2526; by 1900, there were 6005; by 1910, there were 10,213; and in 1920, there were 14,326. With the population of the United States nearly doubling from 1890 to 1920, the number of children attending public high schools multiplied by nearly 10 (Woody 1929, p. 545). Also, as secondary education opportunities for girls expanded, some young women began to seek collegiate opportunities. Oberlin Collegiate Institute was the first to admit females, in 1833. Then, higher education opportunities for women expanded, starting in the Boston area—not only with women's colleges[1] and normal schools but also with coeducational universities. By the end of the nineteenth century, although collegiate education continued to be unavailable to many young women, those who could afford it and whose determination was strong were able to attain a higher education from some single-sex and coeducational institutions (King 1981).

As secondary public schools started to become more accessible, the only cost of attending them was time spent in school, as opposed to working. During this period, there were relatively few job opportunities for young middle-class girls as compared to those for boys, which led the number of young girls who enrolled in high school to increase, not only in actual numbers, but also proportionally in relation to

[1] In this work, the term college refers to programs that offer undergraduate programs—either as an independent institution or as the undergraduate program of a university.

boys' enrollment (Rury 1991, p. 16). And, by the end of the nineteenth century, when the debate over women's access to basic education had been settled, the academic curriculum debate took center stage, focusing on a question: "Shouldn't girls be studying subjects that would fit them for their life work?" (Powers 1992, p. 12). Therefore, patterns and major developments in gender and education encompass inquiries that go further than a straightforward analysis of female headway over time in accessing education. Educational processes ranging from formal and informal practices to structure of the curricula geared for diversified groups of students should be topics of inquiry.

Some classic publications such as the contributions of Cremin (1961), Ravitch (1974), Boorstin (1973), Woody (1929), and Krug (1969) address a complex series of developments that took place in education during the reviewed time period. Among the most important topics covered in the literature are the increase in immigrant children attending schools in urban areas, the aim of education, the alignment of curricula with changing economic conditions, and the subsequent rationale for expansion of vocational[2] education within public schools. But most importantly, the crucial works mentioned above discuss calls for differentiation of the curriculum and subsequent elimination of certain subjects. Subsequently, this research will try to understand how these developments influenced female mathematics education.

2.1 Key Educational Ideas

In the 1880s, before the Progressive Era began, high schools experienced major problems: low attendance rates, less than 1% of the population attending high school, high dropout rates, and the inability of schools to meet obscure university admissions requirements and prepare students for adulthood (Krug 1969). These problems led to the formation of the Committee of Ten, also referred to as the Committee on Secondary School Studies. The Committee of Ten was created by the National Education Association in response to the ongoing debates concerning the development of an organized education system and uniform standards. It was the first national attempt to formulate recommendations for schools and colleges on the national level. The work of the Committee of Ten culminated in a report that was published in 1893, in which the committee concluded that all students should receive an academic education without differentiation based on a predetermined future. It was acknowledged in the report that even though a small proportion of

[2] In the second decade of the twentieth century, the term *vocational education* did not encompass all types of professions, but rather, it was used to distinguish commercial, industrial, homemaking, and agricultural studies from a general course of education (Committee on Vocational Education and Vocational Guidance 1916). The High Schools of New York City: A Hand-book of Procedure and Personnel (Walker and Jeffords 1921) refers to general course of study as a course of study that "offers thorough instruction and training in general subjects; it also prepares for admission to college, and to training schools for teachers, and to nearly all professional schools (p. 25).

children went to college, the schools of the United States "do not exist for the purpose of preparing boys and girls for college" (Krug 1969 p. 72).

Though this report was "influential for a decade or so after it was issued," public schools did not move away from classifying students into the ones who were preparing for college and those who were not (Ravitch 2000, p. 47). In fact, the efforts of the Committee of Ten were criticized based on the belief that endeavors to impose the same rigid preparatory requirements without consideration of the social differences and needs of the students was old-fashioned and undemocratic (Ravitch 2000).

Consequently, in the early years of the twentieth century, many educators unified with the view that schools should align themselves with social developments and insisted that the education of very diverse groups of students required efficient education coordinated with the needs of the labor market. Differentiated curricula that allowed schools "to devise programs of study that prepared individuals specifically and directly for the role they would play as adult members of the social order" became attractive to the curriculum-makers (Kliebard 2004, pp. 76–77):

> [I]nstead of a ladder that stretched from kindergarten to the university and was open to all students, there would be many paths leading to different destinations: the future professional would prepare for college; the future farmer would study agriculture; the future housewife would study household management; the future clerk would study commercial subjects; the future industrial worker would study metalworking and woodworking; and so on (Ravitch 2000, p. 95).

Hence, the discussions put forth by the Committee of Ten, that all students should have access to the same academic courses, were replaced by the idea of social efficiency (Ravitch 2000). The term *social efficiency* is invoked in the majority of sources relevant to education during the Progressive Era. This term was advocated by David S. Snedden, State Commissioner of Education in Massachusetts (1909–1916) and professor of educational sociology at Teachers College (1916–1935), in support of aimed and differentiated education (Knoll 2009).

The objective of education became to "to serve society as efficiently as possible" and, to achieve that goal, educational experts advised dividing students into appropriate programs (Ravitch 1974, p. 88). Moreover, debates on academic curricula for high schools were especially contentious because in elementary schools, students were taught "something which could be profitably consumed by all citizens" (Boorstin 1973, p. 491). But high schools were viewed differently: "If it was to be the road to a 'higher' learning, then perhaps all citizens might not be equally qualified to go up the road" (p. 491). Thus, differentiated education was put into place and high schools started to guide students into the programs that matched their intended occupations.

3 Discussions on the Role of Mathematics in Schools

As debates about the academic curricula became more acute, the premise of mathematical education started to change as well. The dominant theory of the nineteenth century, mental discipline theory, was that mathematics improved and encouraged people's reasoning and thinking (Stanic 1986, p. 39). However, attaching the theory of mental discipline to the study of mathematics came under scrutiny during the first decades of the twentieth century. Increasing support of differentiated education led educators to believe that they had to "dethrone the academic curriculum and that they would not succeed unless they were able to break the tenacious hold of the doctrine of mental discipline" (Ravitch 2000, p. 61).

William Betz, a mathematics textbook author, teacher, and school administrator from Rochester, New York, in his article "The Teaching of Geometry in Its Relation to the Present Educational Trend" (Betz 1908) states that modern industrialism caused a high demand for tangible skills. Explaining the rise of commercial, trade, and technical schools, Betz displays concern:

> In the first place, modern industrialism, with its demand for tangible success, has led to a great outcry for more practical schoolwork. There is an increasing contempt of "mere theory." This feeling finds its expression in the establishment of trade and technical schools... Geometry, as usually taught, furnishes a welcome target to the utilitarian educator. As a result, there is a growing fear that we may rift too far from the ideal liberal culture and that the direct bread-winning power of a subject may be made the sole criterion of its usefulness (p. 626).

With the primary goal of understanding demands of society so as to advise schools on how such changes should be reflected in the objectives of secondary education, the Commission on the Reorganization of Secondary Education (CRSE) was formed in 1911. The focus of the CRSE was to deal with the problems of the changing environment in relation to the demands of employment, of the diverse population of pupils, and of other social influences that existed in secondary education during the first decades of the twentieth century. The CRSE was composed of 16 committees whose task was to evaluate and report on secondary school subjects. One of the component committees, the Committee on the Problem of Mathematics in Secondary Education (CPMSE), headed by William Heard Kilpatrick, a professor of education at Teachers College, was organized to gain better insight into the mathematics education being offered and to evaluate its value in the environment of the time.

The CPMSE's findings were published in the form of a bulletin, *The Problem of Mathematics in Secondary Education* (U.S. Department of the Interior, Bureau of Education 1920). In the report, the CPMSE stated the following: "The doctrine of 'mental discipline' has been attached to the education of mathematics, in arithmetic to an appreciable extent, more in algebra, most of all in geometry" (p. 10). Part of the CPMSE's inquiry was to determine if mathematics courses still exemplified that principle or if the curricula and purpose should be changed. Although the CPSME made some general recommendations, it was not able to provide specific recommendations for course reconstructions. However, it did fulfill the intention of the publication — that of providing an analysis to teachers, directors of mathematics departments, and students of mathematics with the hope that it would "give rise to such discussion and experimenting as will enable other committees to carry forward the work [to] the point of definite reconstruction of courses of study in this subject" (p. 7).

In its analysis, the CPSME reported that secondary schools' mathematics curricula should be differentiated based on the needs of the following groups:

1. General readers: The CPSME reported that this group would need to use mathematics in a practical fashion, but would need little mathematics other than ordinary arithmetic.
2. Those preparing for certain trades: According to the CPSME, ordinary high school mathematics was "ill-adjusted" for this group, and it recommended that the focuses of the curricula for this group be: (1) arithmetic; (2) the "interpretative" mathematics, defined as simple equations, the use of formulas, the measurement of angles, the measurement of areas and volumes, square roots, the creation and interpretation of graphs, right triangles, and the geometry of circles; and (3) special applications according to the subject matter of the particular vocation.
3. Those preparing for engineering: The CPSME reported that this group, which consisted mostly of boys, would require more mathematics, but it recommended that the curriculum be limited by the specific needs of the future engineering profession (U.S. Department of the Interior, Bureau of Education 1920, p. 18).

The CPSME emphasized that high schools should not attempt to make full course offerings available to all. Their recommendation for mathematics education endorsed differentiation of studies based the idea that students' probable destinations or activities should determine "the extent and type of their future mathematical needs" (U.S. Department of the Interior, Bureau of Education 1920, p. 14). And according to this recommendation, those who needed moderate amount of mathematics education should not be forced into taking courses beyond "ordinary arithmetic".

Not all educators supported the notions that the goal of mathematics education should be tied to social efficiency and that students should be allowed to elect their course of study. David Eugene Smith, the chair of the American delegation of the International Commission on the Teaching of Mathematics (ICTM[3]) and a professor at Teachers College, was not supportive of the CPSME report and tried to persuade the CRSE not to release it, although his attempt was unsuccessful (Ravitch 2000).

Moreover, in the same year that the CPSME released its report, another publication became available to educators. Under the auspices of the Mathematical Association of America, a preliminary report was released by the National Committee on Mathematical Requirements (NCMR), *The Reorganization of the First Course in Secondary School Mathematics* (National Committee on Mathematical Requirements 1920). The chair of the NCMR was John Wesley Young; Smith, then a former chair of the American delegation of the ICTM, was one of its members. The report recommended that the first year (ninth grade) be devoted

[3] Now known as ICMI. In the first decades from its establishment, the Commission was mainly referred to by the French acronym CIEM (Commission Internationale de l'Enseignement Mathématique), and the German IMUK (Internationale Mathematische Unterrichtskommission) (Furinghetti 2003).

to algebra, the numerical topics of intuitive geometry, and the beginning of trigonometry, and that the second year (10th grade) cover further work in algebra, trigonometry, and demonstrative geometry. The report notes that the required material "is of sufficient importance for every educated person to warrant making it a requirement for all pupils" (National Committee on Mathematical Requirements 1920, p. 10).

Though support for differentiated education and for the belief that mathematics courses should become categorized as elective as early as high school was challenged, the curricula offered to students in public schools started to change. In New York City, indication of this change is seen by tracing annual reports of the City Superintendent of Schools published during the first two decades of the twentieth century. According to the Third Annual Report of the City Superintendent of Schools to the Board of Education for the Year Ending July 31, 1901 (Department of Education, City of New York 1901), the courses of study pursued in high schools, adopted in December 1901, were divided into Classical, One Foreign Language, Two Foreign Languages, and Commercial. All the courses of study required students to take mathematics during their first and second years, with algebra being taught during the first year and geometry in the second year. After the second year, mathematics was required for the students who had selected courses of study other than Commercial courses, such that the combination of algebra and geometry was taught during the third year, and algebra, geometry, and trigonometry were offered in the fourth year (elective for students enrolled in One Foreign Language course). As for students enrolled in the Commercial course, they were only required to take algebra and geometry in the first and second year, respectively.

Over time these requirements shifted and the requirement for students enrolled in New York City public high schools to take algebra and geometry did not continue. A report published by the High School Teachers' Association of New York City (HSTA) suggested the adoption of the "general course" of study which will allow for a "greater flexibility" with the goal of meeting the "tendency in educational theory favoring greater concentration upon fewer high schools' subjects" (High School Teachers' Association of New York City 1914, p. 10). The report's findings resulted in the HSTA's recommendation to move away from the requirement of taking six or more subjects simultaneously, as it "leads to distraction, superficiality, and the violation of many principles of efficiency....[The general course] proposes the five-subject plan per week, each subject allotted four hours per week of required work" (p. 10). Algebra was listed as an elective course for the first-year high school students enrolled in the general course of study (High School Teachers' Association of New York City 1914, p. 10). The curriculum for the second, third, and fourth years required four periods of English per week. A course in mathematics that offered surveying or the elements of engineering "with ample opportunity for field practice and office computations" was listed as one of the elective subjects (High School Teachers' Association of New York City 1914, p. 12). Also, in the event of the adoption of the HSTA's plan, the work required in English, algebra, and foreign language was suggested to "be materially reduced" (High School Teachers' Association of New York City 1914, p. 12).

The adoption of the elective system proposed by the HSTA is confirmed by the *The High Schools of New York City: A Hand-book of Procedure and Personnel* (Walker and Jeffords 1921),[4] which states that in order to graduate in the city of New York, students had to satisfy 17 units of work, and that mathematics classes were not included in the required courses. Further passing of a mathematics examination was not a requirement for successful graduation as prescribed by the New York Board of Education (Walker and Jeffords 1921).

In his 1950 address (subsequently published in *The Mathematics Teacher*) at the 10th summer meeting of the National Council of Teachers of Mathematics, William Betz claimed that the following developments summarized the fate of mathematics during the Progressive Era on the national level (Betz 1950, pp. 382–83):

1. The "cultural" claims of mathematics were increasingly ignored. The subject had to be justified, if at all, in terms of social, utilitarian, and vocational objectives.
2. Academic mathematics became an elective subject in the majority of our states. As a result, the number of students enrolled in algebra or geometry classes shrank at a rapid rate. In typical high schools the number of teachers required to teach secondary mathematics was reduced to fewer than half the former figure.
3. It became current doctrine that only a minority, "the few," had any use for the customary offerings of algebra and geometry. All the rest, "the other 85%," needed a totally different diet, the "mathematics of daily life," or the "mathematics of life situations".

Thus, it can be justly argued that during the Progressive Era, social efficiency started to gain "unstoppable momentum" (Ravitch 2000, p. 86) and shaped the broad purpose of schooling. This movement particularly targeted the subject of mathematics, along with some other subjects:

> Social efficiency called upon the school subject to prove for their right to exist. The so-called practical subjects, it was often assumed, did so by definition. Inevitably, then, the demand fell heavily on those fields by this name identified as traditional, or academic, namely, the foreign languages, mathematics, history, English, and science (Krug 1969, p. 336).

Mathematics educators of the twentieth century were obligated to prove that their educational goals fell within the goals of the ideology of efficient education and that teachers would be "held accountable for meeting [the goals] thoroughly and efficiently" (Kilpatrick 2009, p. 120). Hence, mathematics education that was available to girls should be reviewed in tandem with the goals of education as prescribed by the social efficiency ideology and by vocations that were available and attractive to the female population of the United States.

[4] Hereinafter, this publication will be referred to as the Handbook (Walker and Jeffords 1921).

4 Implications on Mathematics Education for Girls

As differentiated curricula became a dominant feature of high schools, educators often used gender to form curricular structure in secondary schools in the United States during the twentieth century (Graves 1998). In his book *Industrial Training of the Girl*, Professor William A. McKeever describes the increasing trend of differentiated curriculum in high schools as follows:

> The last few years have witnessed marked changes and differentiations in the high-school course of study. At the beginning of the twentieth century there was as a rule only one secondary course of study available for young people, and that consisted of a traditional arrangement of Latin, mathematics, literature, and a smattering of abstract science. But the new high school is succeeding more and more each year in making itself what it pretends to be, namely, an institution for the whole people. Instead of one course as formerly there are now many courses, each one arranged to suit the needs of some class of society. In consideration of what has just been stated, the parent cannot reasonably be satisfied with having merely send his daughter to be enrolled in the high school. He must help her decide what course to pursue, and in doing this he must consult first her individual taste and disposition, and second, her probable destiny as a full-grown woman. No matter how attractive the place, how able the instructor, and how well-equipped the school, the young woman will not make satisfactory advancement in her classes unless she is allowed to pursue some course that appeals enticingly to her inherent interests and desires (McKeever 1914, p. 55).

Thus, in order to review girls' access to mathematics education as well as their experience in mathematics classes during the first decades of the twentieth century, this work will concentrate on two significant issues, the objectives of mathematics education as defined by contemporary educators as well as data on girls' enrollment in mathematics classes.

4.1 Objectives

In order to analyze objectives of female education during the period 1890–1920, as defined by the contemporary educators and writers, this study includes review of publications that contained the advice and opinions of teachers and of public figures who played an influential part in the development of mathematics education of the time.

An analysis of the positions on the mathematical studies for girls expressed by the educators and writers of the Progressive Era (Smith, Snedden, Kilpatrick, and Hall, among others) shows that views on the objective of female mathematics education often diverged. Snedden, believed that the path of a student's education should match the student's strengths and interests as well as future prospects (Drost 1967). Ravitch (2000) recounts the following story about Snedden and his stance on mathematics education for females: "In the closing days of World War I, Snedden argued that mathematics was not necessary for all students, and he referred casually to 'the dead hand of mathematics' and the 'mummified study' of algebra" (p. 85).

According to Ravitch, a professor of mathematics at the University of Cincinnati replied to Snedden that due to their insufficient knowledge of mathematics, many young men failed to become artillery officers, eliciting a reply from Snedden stating that "girls did not, as a rule, become artillery officers or navigators, nor did they aspire to follow the peaceful pursuit of the civil or electrical engineering, machinist, or architect" (Ravitch 2000, p. 85).

Evidence of similar discussions on the connection between girls' interests and a prescribed course of study is present in the contemporary publications of the period under review. For example, an issue of *School Science and Mathematics* contains the article "What Knowledge (of Physics) Is of Most Worth?" by Frank B. Spaulding of Boys' High School (1908). Although the article mostly deals with the teaching of physics, Spaulding also discusses issues of interest and need in the general teaching of science, mathematics included, to boys and girls: "The normal girl's interests and tastes, her fund of observation and her method of approach to many subjects are different from those of the normal boy" (p. 676). Spaulding claims that the school requirements that were prescribed 15 years earlier were based on the assumption that all high school students will go to college. Furthermore, he points out that the committee that worked on the requirements was primarily composed of members who were college professors. Therefore, according to the author, "the course they laid out is obviously a course intended for boys preparing for college" (p. 676). However, the demographics of schools had changed—the majority of students were female and only a small proportion of students intended to go to college. According to Spaulding, it was an injustice to force girls to study the same topics as boys, since girls had no interest in the scientific fields:

> I have had the injustice of the situation brought painfully to my attention recently through association with a pupil of a neighborhood school. From her report and the remarks of the teachers of girls, I judge her case is not an unusual one. With no interest in things scientific, no fund of observational knowledge and no ability to think in mathematical terms, she has tried with most commendable perseverance to remember how to do various "stunts" which are likely to be called for in the coming Regents' examination. To her they will never be anything but "stunts" (p. 677).

Spaulding (1908) goes on to say that he disagrees with the claim that girls receive mental discipline through learning how to deal with such problems: "The present uniform course not only results in dragging some pupils through work for which they are wholly unqualified; it is entirely inadequate to the needs of others" (p. 677). One of Spaulding's suggestions is to have a course that will incorporate preparation for a "future in the office, the factory, the field or the household" (p. 677).

The goal of female mathematics education is addressed in the report *Mathematics in the Public and Private Secondary Schools of the United States* (International Commission on the Teaching of Mathematics, Committees III and IV 1911).[5] The ICTM, which was created under the leadership of Felix Klein, and the American arm of the ICTM, which was chaired by David Eugene Smith, collected a vast

[5] Hereinafter, Committee III (1911).

amount of material which detailed the existing conditions in the field of mathematics education. The report compiled (Committee III 1911), which focused on public secondary schools, stated the following concerning the aim of girls-only public-school mathematics education:

> Nine of the schools[6] were either wholly or in part preparatory for college, and three fitted for technical or normal schools. In the six occupations listed under "vocational aims" one school gave preparation for the occupation of draftsman, two for that of computer, and five for that of accountant or merchant. There were no girls' schools that gave training for the professions of surveyor or civil engineering. In addition to the aims classified three schools directed their mathematical training toward the equipment of the teacher of mathematics (p. 47).

Meanwhile, the same publication stated that all of the boys' schools included in the survey[7] "report that the aim of mathematical teaching is general culture and preparation for college" (p. 39).

Similar observations are made in the Handbook (Walker and Jeffords 1921), finding that Wadleigh[8] High School and Bay Ridge High School, both girls' public high schools located in New York City[9], "while stressing college preparation, [offer] a broadly cultural course which includes home economics and commercial subjects as well as those required for entrance to advanced institutions" (p. 27).

Also, mathematics requirements in girls' schools were not always the same as mathematics in boys' schools. Review of the mathematics offerings (commercial and household arithmetic, algebra, geometry, intermediate algebra, and advanced algebra) in Wadleigh High School and Bay Ridge High School (both girls-only) listed all mathematics courses as electives (Walker and Jeffords 1921).

In comparison, a review of the Stuyvesant High School (boys-only) Student's Guide (1926) specified that the aspect of mathematics was involved in all "characteristic courses" (p. 20). Another boys-only high school that can be used for comparison is Boys' High School, the oldest public high school for boys in New York City. Boys' High School reported the objective of education was primarily concerned with the students' entrance into an institution of higher education; thus, its curriculum was closely matched with college entrance requirements (Walker and Jeffords 1921). Boys' High School required algebra (first year) as well as plane geometry (second year). After the second year, the following mathematics courses are listed as requirements for certain professions (e.g., science, engineering, architecture, medicine): intermediate algebra (half-year) and trigonometry (half-year) to be taken in the third year, followed by solid geometry (half-year) and advanced

---

[6] The report on girls' high schools is based on the results on the replies received from 12 high schools.

[7] Based on the replies received from 10 boys' high schools. "Though the number of schools which sent replies to the questionnaires was not large, they include some of the largest and most important schools in the United States" (Committee III 1911, p. 36).

[8] The first public girls-only high school in New York City.

[9] For a full list of the high schools in New York City, organized by borough and date of first session, see Appendix A.

algebra (half-year) to be taken in the fourth year (Walker and Jeffords 1921). Stuyvesant High School Student's Guide listed similar requirements.

Thus, based on sources cited above, belief that it was unjust to force girls into the same educational path as that of boys, is mirrored in the girls' school mathematics curricula. Mathematics was mostly offered as an elective course in girls-only schools. In comparison, boys' schools' mathematics courses were rigidly fixed to prepare students for higher institutions.

To compare mathematics education offered to girls in coed public high schools, a description of the segregation between girls and boys in mathematics classes should be established in order to analyze how objectives of the girls' mathematics education were different from the one of boys. This analysis[10] was established by Committee III (1911), and with regard to this issue, the report focused on two topics: (1) the advisability of separating girls and boys in mathematics classes and (2) the ability of girls compared with that of boys.

According to Committee III (1911):

> There seems to be no general movement to make such a segregation of sexes in mathematical classes, although it is almost unanimously believed that such a step would be advisable in certain grades of the work. In the schools where a large number of boys are preparing to enter the higher technical schools, the classes in higher mathematics are composed entirely of boys, and much better work is being done than in corresponding mixed classes. Even in the elementary subjects, it has been found in such schools that better results are obtained if the segregation takes place early in the course (p. 50).

Furthermore, as revealed by the teachers' responses, in the majority of cases, as mathematical classes progress to advanced algebra and trigonometry, boys show "ability that is far ahead of a girl" (p. 51).

It must be noted that this opinion was not supported by all members of Committee III (1911):

> In a mixed class this greater ability of the boy, and the greater faithfulness of the girl, react most advantageously on each other. This question of the difference in ability of the girl and the boy is a many-sided one. Every teacher in a mixed school has found in the mathematics class many girls who could do the work as well as the boy, and who by steady application to work has obtained a higher mark than he. But the average girl meets this situation at the start with a prejudice "ex matre," and is therefore handicapped (p. 50).

Furthermore, David Eugene Smith was an outspoken opponent of treating secondary school mathematics as a subject that girls did not need. He addresses this issue in *Problems in the Teaching of Secondary Mathematics: An Address Delivered Before the New England Association of Teachers of Mathematics* (Smith 1913):

> When I hear some man who is skilled in dialectics attack the teaching of algebra to the girl because it does not enter into her immediate life, I am led to wonder if he would have women learn anything whatever except the work of the household and the club. The classics of the world appeal to her immediate needs and interests quite as little as mathematics, and hence these would be dropped. Indeed, we hear not infrequently of some school principal

[10] Analysis performed by the Committee based on several hundred replies from teachers in response to a questionnaire sent by the U.S. Bureau of Education to the teachers of the North Atlantic states.

> who proposes the epoch-making plan of studying literature only from current magazines. If we agree to drop all noble literature, all knowledge of history except what comes from the newspapers, all study of good music and art except by those who expect to be musicians or painters or designers, all science except such as relates to the simple chemistry of foods, all astronomy, all physics, everything except what concerns the science of good physical living, then we may let mathematics go with the rest. Perhaps this is not what these agitators mean—one never knows, for they themselves never seem to have any exact ideas upon the subject. ... Hence the problem seems to me not so much to decide whether or not the girl should study algebra, as to decide how we shall so teach the subject to her that she will know of its beauties, of its purposes, and of the feeling of mastery that comes from its pursuit. Such a problem may well occupy the attention of associations like the one I have the honor to address (pp. 13–14).

This support was based on the principle that career opportunities should be expanded for women and that a lack of demand for mathematics among female students, often cited by those who resisted the expansion, should not be a factor in whether schools provided such education to girls.

While sources listed above show varied opinions on the purpose of the girls' mathematics education, there no doubt that during the first decades of the twentieth century the notion that girls did not need the same education as boys was present and expressed among educators. This stance was influenced by different factors, ranging from gender-related societal beliefs and expectations for students' future careers: "Where does trigonometry apply in a good woman's life? Will it contribute anything toward peace, happiness, and contentment in the home? Will it bake any bread, sew on any buttons, or rock any cradles?" (Krug 1969, p. 118). These words belonged to Professor William A. McKeever, Kansas State Agricultural College, and were published in the college bulletin for parents and students (Ravitch 2000).

Though generalization of teachers' beliefs is not the goal of this work, the opinions held by teachers and administrators played an important role in the demand for mathematics education when high school curricula allowed for election of classes. The Handbook (Walker and Jeffords 1921), published by the High School Teachers' Association of New York City, states:

> [The] fundamental principle of diversity of talent, aptitude, and aim in pupils can be met only through a flexible group of electives extending through the entire course, so arranged as to avoid the forced repetition by pupils of subjects for which they have proved themselves unfitted. Selections from each group are made by pupils individually under the expert guidance of teachers, with whom the pupils individually are brought into intimate and sympathetic contact. Pupil effort is thus concentrated where it will do the most good, and desultory or haphazard choosing of subjects discouraged (p. 23).

This quote demonstrates that teachers' guidance played a large role in the structure of students' curricula. And, with curricula becoming tightly connected to careers, which were presumed to be different for women and men, meeting the individual differences of student needs entailed enabling students to follow the different class offerings provided by the differentiated curricula. Informal gender differentiation ranged from teacher expectations to future career possibilities and imposed norms and rules on girls that might have been as powerful as the policies implemented by schools in governing their actions toward education (Hansot and Tyack 1988). Thus, in reflecting on the goals of education during the Progressive Era, presence of differentiated curricula and the ability to avoid taking mathematics during high school brings up the next query of this research—enrollment of girls in mathematics classes.

4.2 Enrollment

Statistics collected by the United States Department of Health, Education, and Welfare (Brown and Osbourn 1957) show a decline in the percentage of high school students who enrolled in algebra, geometry, and trigonometry classes during the period 1890–1934, as presented in Table 1.

The statistics displayed in Table 1 are not separated by gender. However, the findings of the National Survey of Secondary Education (U.S. Bureau of Education 1933) on the distribution of high school courses taken by students in 1890, 1900, 1910, 1920, and 1930 show that the number of mathematics courses taken by high school students positively correlated with the number of science and foreign language courses taken and negatively correlated with the number of courses taken in nonacademic fields. These correlations can be explained by the fact that students who were preparing for college entrance found it necessary to "take work in all three of these so-called elective academic fields in order to meet the academic requirements set up by colleges. The nonacademic field was largely neglected by such pupils because colleges do not require work in it, and in some cases, colleges refuse to allow entrance credit for nonacademic work" (p. 263). The percentage of girls seeking admission to college during the reviewed period is an important factor in understanding girls' mathematics education, especially with respect to their enrollment numbers in mathematics courses. And although the number of girls in coeducational high schools generally exceeded the number of boys, girls made up less than 25% of the students who applied to college (College Entrance Examination Board 1921).

Table 1 Offerings and percentage of students enrolled in science and mathematics in public high schools 1890–1934

Year	Algebra (%)	Geometry (%)	Trigonometry (%)
1890	45.4	21.3	1.9
1900	56.3	27.4	1.9
1910	56.9	30.9	1.5
1915	48.8	26.5	1.5
1922	40.2	22.7	1.5
1928	35.2	19.8	1.3
1934	30.4	17.1	1.3

Note. Adapted from *Offerings and Enrollments in Science and Mathematics in Public High Schools*, (Brown and Osbourn 1957)

This finding is pertinent to what educators of the Progressive Era were seeing in schools and the following excerpt from the bulletin *High Points in the Work of High Schools of New York City* (New York Board of Education 1921) reveals that the presence of elective curricula had altered girls' enrollment in mathematics classes:

> Now that mathematics has become an elective subject we find that many girls have a tendency to leave it out of their programs. The word that mathematics is "hard" is passed on from one generation of students to the next, and it is only human nature that the girls should look for an "easy" subject to elect (New York Board of Education 1921, p. 12).

The same publication confirms that except for those who planned to go to college, girls were not required to take algebra classes. Given the presence of a changing environment in relation to the demands of employment, many vocational opportunities became available to women in the first decades of the twentieth century and required schooling to efficiently "alphabetize records or to speak properly" (Rury 1991, p. 91). Therefore, study of the presence of vocational education in public high schools and how it affected female mathematics education proves to be appropriate for this research.

5 Commercial Education and Its Influence on Female Mathematics Education

In addition to the general course of study which still dominated in the public high schools in the first decades of the twentieth century (Ravitch 2000), a growing number of schools started to offer vocational education as part of the curricula. And even though vocational education included commercial, industrial, homemaking, and agricultural studies (Committee on Vocational Education and Vocational Guidance 1916), vocational education other than commercial[11] failed to keep up with the demands of the labor market in New York City (Rury 1991). In general, in the United States, the absolute number of students in commercial courses of study increased by nearly 2000% between 1890 and 1920, from fewer than 15,000 to 300,000 (Rury 1984, p. 149). Furthermore, girls were attracted to commercial education for its demand in the labor market and salaries.[12] Historians acknowledge that although commercial courses were available to both boys and girls, commercial education was largely selected by females, and educators were aware of this trend (Krug 1969; Rury 1984; Fennema 1974). Therefore, given the focus of this research, study of the rise of the vocational education in public high schools is limited in this

[11] Commercial education provided teaching of bookkeeping, stenography, or salesmanship (U.S. Department of the Interior, Bureau of Education 1921, p. 109).

[12] At the time, the average annual income of a stenographer in New York City was $954. By comparison, the average annual income of a nurse was $760 (U.S. Department of Commerce and Labor 1911, p. 414).

work to the commercial branch, particularly mathematics curricula offered to girls as part of commercial education.

As per the Handbook (Walker and Jeffords 1921), by 1916, although all New York City traditional high schools offered nonvocational courses, all public schools except for Boys' High School and Stuyvesant High School (both boys-only) offered commercial courses. The graduation requirements for the commercial program followed a pattern similar to those of general courses of study, with the required completion of 19.7 units of work, of which 12.7 were required courses and 7 were electives. Distribution of the required 12.7 units lists algebra and geometry as elective courses and business calculations as a required course (Walker and Jeffords 1921).

However, mathematics classes included in the commercial education available to girls were not necessarily similar to those offered as part of boys' commercial education. Review of the curricula for Julia Richman High School, one of the predominantly commercial high schools for girls, shows that algebra and geometry were not listed among the offered subjects, even as elective courses. Commercial arithmetic is listed as an elective (Walker and Jeffords 1921).

By contrast, according to the same source, the course of study for the Commercial High School (boys-only school) offered arithmetic as a required course, and algebra and geometry as electives.

Presence of the widespread belief that even if both genders elected commercial education, the curricula offered to girls and boys did not need to be the same can be seen in the *Biennial Survey of Education 1916–1918* (U.S. Department of the Interior Bureau of Education 1921). The chapter "School Surveys and Commercial Education" highlights surveys that "bear out the position of the Commissioner of Education Report of 1915–16" (p. 111) on the differentiation of curricula based on gender. Although these survey results show support for commercial education in schools, the authors of the work criticize the structure of commercial education that is simultaneously geared toward boys and girls:

> [The] Cleveland survey shows what any survey invariably does - that commercial education in our public high schools pursues a policy quite independent of the business needs and conditions of the community under consideration. Commercial education has a been a thing of school credits and academic standards conducted in accordance with college entrance requirements or with abstract scholastic procedure. Commercial educators have neither seen nor apparently cared for the actual conditions of employment into which their graduates may go. The school prepares a certain product which business must take or leave just as it chooses. The Cleveland survey makes a distinct point of the fact that boys and girls are trained alike for the same kinds of commercial occupations, whereas an actual survey of business conditions shows that the sex conditions of employment are quite different, males being in excess in certain commercial employments and females in other kinds of commercial employment (p. 112).

Other evidence of this opinion is found in "Vocationalism for Home and Work: Women's Education in the United States, 1880–1930," where John Rury (1984) quotes a 1917 survey of 66 high school principals which found that two thirds of the respondents believed that commercial training for boys and for girls should be different. Rury (1984) also refers to another survey of the time on the occupational status of men and women, which states: "Regardless of the position in which boys and girls started in life, boys worked into administrative positions, while women remained secretaries and clerks until they left office to get married" (p. 13). Thus, as the future careers of the students were often dependent on their gender, commercial education and its mathematics course offerings were often different.

6 Conclusion

This study illustrates that during the Progressive Era the purpose of a high school education shifted. Before the turn of the twentieth century, secondary education was available only to a very few, and the aim was to prepare students for college; the curriculum included a predetermined set of requirements. Furthermore, high schools were plagued by high drop-out rates. But during the first decades of the twentieth century, public high schools aimed to offer education to the majority of American teens. And, as high schools experienced rapid expansion in enrollment, the objective and direction of education changed. The theme of the period was recognition of the fact that not all boys and girls are alike and that they differ in their skills and aspirations; therefore, it was believed that the curriculum must be adjusted to their needs. Thus, the move toward differentiated curricula, with the categorizing of elective subjects, became a sensible solution for serving different groups of students. With this move, mathematics fell into the category of subjects not required for certain trajectories.

In his book, *The American High School* (Brown 1909), in which he examined the state and problems of American high schools of the time, John Franklin Brown, PhD, a professor of education and inspector of high schools for the State University of Iowa, provides an important summary of the purpose of education of the reviewed period:

> The educational value of a subject depends not so much upon its content as upon the reaction which the study of it causes in the life of the student; hence it is extremely important to consider the question from the standpoint of the individual. ... The same subject has very different educational values when judged from the viewpoint of different pupils (p. 102).

Indeed, the differentiation of high school curricula in the early twentieth century for the most part was not based on a perceived difference between girls and boys in aptitude and abilities, though this opinion is also seen in historical records. The majority of records show that this differentiation existed mostly based on the institutions' aims and educators' beliefs that girls and boys needed different skills upon graduation. All of these factors played a role in the students' decisions and guidance they received from educators when it came to elective classes in mathematics. Additionally, many vocations that became available and popular among females during the Progressive Era did not require college education, which in turn meant that girls did not need to take mathematics to graduate from high school, and more importantly, were often not offered mathematics courses beyond arithmetic. This work demonstrates that allowing female students to exclude mathematics subjects from their curriculum proved to lower their enrollment in such classes.

Appendix A: High Schools of New York City

School	Borough	Date of first session	Type (Boys/Girls/Coed)
DeWitt Clinton	Manhattan	Sept. 13, 1897	Boys
Wadleigh	Manhattan	Sept. 13, 1897	Girls
High School of Commerce	Manhattan	Sept. 8, 1902	Boys
Washington Irving	Manhattan	Sept. 8, 1902	Girls
Stuyvesant	Manhattan	Sept. 12, 1904	Boys
Julia Richman	Manhattan	Sept. 8, 1913	Girls
George Washington	Manhattan	Feb. 2, 1920	Coed
Haaren	Manhattan	Sept. 13, 1920	Coed
Morris	Bronx	Sept. 13, 1897	Coed
Evander Childs	Bronx	Sept. 8, 1913	Coed
Theodore Roosevelt	Bronx	Feb. 4, 1918	Coed
Boys' High School	Brooklyn	Jan. 5, 1891	Boys
Girls' High School	Brooklyn	Jan. 5, 1891	Girls
Manual Training	Brooklyn	Nov. 14, 1893	Coed
Erasmus Hall	Brooklyn	Sept. 14, 1896	Coed
Commercial	Brooklyn	Oct. 9, 1899	Boys
Eastern District	Brooklyn	Feb. 1, 1912	Coed
Bushwick	Brooklyn	Sept. 12, 1910	Coed
Bay Ridge	Brooklyn	Feb. 1, 1912	Girls
New Utrecht	Brooklyn	Sept. 13, 1915	Boys
Girls Commercial	Brooklyn	Feb. 2, 1920	Girls
Bryant	Queens	Feb. 1, 1905	Coed
Jamaica	Queens	Nov. 27, 1907	Coed
Flushing	Queens	Dec. 1, 1907	Coed
Richmond Hill	Queens	Feb. 23, 1909	Coed
Newtown	Queens	Feb. 1, 1910	Coed
Far Rockaway	Queens	Sept. 17, 1920	Coed
Curtis	Richmond	Feb. 1, 1904	Coed

Note. Adapted from *High Schools of New York City: A Hand-Book of Procedure and Personnel*, edited by C. F. Walker and Jeffords, 1921

References

Betz, William. 1908. The Teaching of Geometry in Its Relation to the Present Educational Trend. *School Science and Mathematics* 8 (8 November): 625–633. https://onlinelibrary.wiley.com/doi/abs/10.1111/j.1949-8594.1908.tb01283.x.

———. 1950. Five Decades of Mathematical Reform—Evaluation and Challenge. *The Mathematics Teacher* 43 (8): 377–387. www.jstor.org/stable/27953630.

Boorstin, Daniel J. 1973. *The Americans: The Democratic Experience*. New York: Random House.

Brown, John Franklin. 1909. *The American High School*. New York: Macmillan. https://catalog.hathitrust.org/Record/102299632.

Brown, Kenneth E., and Ellsworth S. Osbourn. 1957. *Offerings and Enrollments in Science and Mathematics in Public High Schools*. Washington, DC: U.S. Office of Education. https://babel.hathitrust.org/cgi/pt?id=osu.32435029541513&view=1up&seq=1.

College Entrance Examination Board. 1921. Twenty-first Annual Report of the Secretary. *New York College Entrance Examination Board.*. https://babel.hathitrust.org/cgi/pt?id=osu.32435052586476&view=1up&seq=35.

Committee on Vocational Education and Vocational Guidance. 1916. *Vocational Secondary Education*. Washington, DC: Department of the Interior, Bureau of Education. https://babel.hathitrust.org/cgi/pt?id=mdp.39015063898046&view=1up&seq=51.

Cremin, Lawrence A. 1961. *The Transformation of the School: Progressivism in American Education, 1876–1957*. New York: Alfred A. Knopf.

Department of Education, City of New York. 1901. *Third Annual Report of the city Superintendent of Schools to the board of education for the Year Ending July 31, 1901*. New York: Department of Education, City of New York. https://babel.hathitrust.org/cgi/pt?id=njp.32101059128163&view=1up&seq=5.

Drost, Walter H. 1967. *David Snedden and Education for Social Efficiency*. Madison, WI: University of Wisconsin Press.

Eccles, Jacquelynne. 1984. Sex Differences in Mathematics Participation. In *Women in Science: Advances in Motivation and Achievement 2*, edited by M. L. Maehr and M. W. Steinkamp, 93-137. Greenwich, CT: JAI Press Inc.

Eisenmann, Linda. 1998. *Historical Dictionary of Women's Education in the United States*. Westport, CT: Greenwood Press.

Fennema, Elizabeth. 1974. Mathematics Learning and the Sexes: A Review. *Journal for Research in Mathematics Education* 5 (3): 126–139. https://doi.org/10.2307/748949.

———. 1979. Women and Girls in Mathematics: Equity in Mathematics Education. *Educational Studies in Mathematics* 10 (4): 389–401. http://www.jstor.org/stable/3481825.

Fennema, Elizabeth, and Penelope L. Peterson. 1985. Autonomous Learning Behavior: A Possible Explanation of Sex-Related Differences in Mathematics. *Educational Studies in Mathematics* 16 (3): 309–311. http://www.jstor.org/stable/3482624.

Furinghetti, Fulvia. 2003. Mathematical Instruction in an International Perspective: The Contribution of the Journal L'Enseignement Mathématique. In *One hundred years of L'Enseignement Mathématique, Monographie n.39 de L'Enseignement Mathématique*, ed. D. Coray, F. Furinghetti, H. Gispert, B.R. Hodgson, and G.G. Schubring, 19–46.

Graves, Karen. 1998. *Girls' Schooling During the Progressive Era: From Female Scholar to Domesticated Citizen*. New York/London: Garland Publishing.

Hansot, Elisabeth, and David Tyack. 1988. Gender in American Public Schools: Thinking Institutionally. *Signs* 13 (4): 741–760. http://www.jstor.org/stable/3174110.

High School Teachers' Association of New York City. 1914. *Efficiency in High Schools: Studies, 1911–14, in the Application of the Principles of Scientific Management to High School Problems Made by the High School Teachers' Association of New York City*. New York: Brown Bros. https://archive.org/details/efficiencyinhigh00high.

International Commission on the Teaching of Mathematics, Committees III and IV. 1911. *Mathematics in the Public and Private Secondary Schools of the United States (Bulletin*

16). Washington, DC: U.S. Bureau of Education. https://babel.hathitrust.org/cgi/pt?id=hvd.32044089263487&view=1up&seq=8.

Karp, Alexander. 2014. The History of Mathematics Education: Developing a Research Methodology. In *Handbook on the History of Mathematics Education*, ed. Alexander Karp and Gert Schubring, 9–24. New York: Springer.

Kilpatrick, Jeremy. 2009. The Social Efficiency Movement in the United States and Its Effects on School Mathematics. In *Dig Where You Stand: Proceedings of the Conference "Ongoing Research in the History of Mathematics Education"*, ed. K. Bjarnadottir, F. Furinghetti, and G.G. Schubring, 113–122.

King, Patricia. 1981. The Campaign for Higher Education for Women in 19th-Century Boston. *Proceedings of the Massachusetts Historical Society* 93: 59–79. www.jstor.org/stable/25080888.

Kliebard, Herbert M. 2004. *The Struggle for the American Curriculum, 1893–1958*. 3rd ed. New York: Routledge Falmer.

Knoll, Michael. 2009. From Kidd to Dewey: The Origin and Meaning of Social Efficiency. *Journal of Curriculum Studies* 41 (3): 361–391. https://doi.org/10.1080/00220270801927362.

Krug, Edward A. 1969. *The Shaping of the American High School, 1880–1920*. Madison: University of Wisconsin Press.

Leder, Gilah C. 2001. Pathways in Mathematics Towards Equity: A 25-Year Journey. In *Proceedings of the 25th Conference of the International Group for the Psychology of Mathematics Education 1 (July 12–17, 2001)*, ed. M. van den Heuvel-Panhuizen, 1–54. Utrecht, NL: Freudenthal Institute. https://archive.org/details/ERIC_ED466950/page/n99.

McKeever, William A. 1914. *Industrial Training of the Girl*. New York: Macmillan.

National Committee on Mathematical Requirements. 1920. *The Reorganization of the First Courses in Secondary School Mathematics: A Preliminary Report by the National Committee on Mathematical Requirements*. Washington, DC: Government Printing Office.

New York Board of Education. 1921. *Bulletin of High Points in the Work of the High Schools of New York City* III, no. 7 (September). New York: Board of Education. https://babel.hathitrust.org/cgi/pt?id=mdp.39015035445686&view=1up&seq=1093

Powers, Jane Bernard. 1992. *The "Girl Question" in Education: Vocational Education for Young Women in the Progressive Era*. New York, NY: Routledge.

Ravitch, Diane. 1974. *The Great School Wars: A History of the New York City Public Schools*. New York: Basic Books.

———. 2000. *Left Back: A Century of Failed School Reforms*. New York: Simon & Schuster.

Rury, John L. 1984. Vocationalism for Home and Work: Women's Education in the United States, 1880–1930. *History of Education Quarterly* 24 (1): 21–44. https://doi.org/10.2307/367991.

———. 1991. *Education and Women's Work: Female Schooling and the Division of Labor in Urban America, 1870–1930*. Albany, NY: State University of New York Press.

Schubring, Gert. 2014. On Historiography of Teaching and Learning Mathematics. In *Handbook on the History of Mathematics Education*, ed. Alexander Karp and Gert Schubring, 3–8. New York: Springer.

Smith, David Eugene. 1913. *Problems in the Teaching of Secondary Mathematics: An Address Delivered Before the New England Association of Teachers of Mathematics*. Boston Ginn and Company. https://babel.hathitrust.org/cgi/pt?id=uc1.$b66978&view=1up&seq=5.

Spaulding, Frank. B. 1908. What knowledge (of physics) is of most worth? School of Science and Mathematics, no. 8 (November), 674–679. https://babel.hathitrust.org/cgi/pt?id=mdp.39015057094040&view=1up&seq=719

Stanic, George M.A. 1986. Mental Discipline Theory and Mathematics Education. *For the Learning of Mathematics* 6 (1): 39–47. http://www.jstor.org/stable/40247802.

U.S. Bureau of Education. 1933. *National Survey of Secondary Education*. Washington, DC: Government Printing Office. https://babel.hathitrust.org/cgi/pt?id=mdp.39015024249586&view=1up&seq=715.

U.S. Department of Commerce and Labor. 1911. *Twenty-fifth Annual report of the Commissioner of Labor 1910*. Washington, DC: Government Printing Office. https://babel.hathitrust.org/cgi/pt?id=msu.31293028404006&view=1up&seq=7.

U.S. Department of the Interior, Bureau of Education. 1920. *Bulletin 1: The Problem of Mathematics in Secondary Education: A Report of the Commission on the Reorganization of Secondary Education, Appointed by the National Education Association*. Washington, DC: Government Printing Office. https://files.eric.ed.gov/fulltext/ED543366.pdf.

———. 1921. *Biennial Survey of Education, 1916–1918, Bulletin 1, no.88*. Washington, DC: Government Printing Office. https://ia600201.us.archive.org/11/items/ERIC_ED541459/ERIC_ED541459.pdf.

Walker, Claude F., and Clyde R. Jeffords, eds. 1921. *High Schools of New York City: A Hand-book of Procedure and Personnel*. New York: High School Teachers' Association of New York City. https://archive.org/details/highschoolsofnew00high/page/n3.

Woody, Thomas. 1929. *A History of Women's Education in the United States*. New York: Science Press.

Building an American Mathematical Community from the Ground Up: Artemas Martin and *the Mathematical Visitor*

Sian E. Zelbo

Abstract The Mathematical Visitor, a nineteenth-century American journal, was published regularly from 1877 to 1887, at an inflection point for the American mathematical community. The journal was started a year before James Joseph Sylvester started the American Journal of Mathematics at Johns Hopkins University, an event recognized as marking the beginning of research mathematics in the United States, which until this point had lagged behind Europe. The Mathematical Visitor is often mentioned merely as a step in the evolution of American journals. In fact, the Mathematical Visitor was never meant to produce research mathematics. Its purpose, instead, was education. The Mathematical Visitor was an early effort at outreach by the country's top mathematical figures, including Benjamin Peirce and James Joseph Sylvester, to cultivate mathematical interest and talent in an era when the country's schools were not yet equipped for that task. Viewing the journal in this light allows for a deeper understanding of the development of the American mathematics community and also situates this research within a body of international research about outreach efforts by mathematicians to influence and support pre-college mathematics education in their countries.

Keywords History of mathematics education · Nineteenth century periodicals · History of higher education · Mathematics journals

S. E. Zelbo (✉)
The Brearley School, New York, NY, USA

Stern College for Women, Yeshiva University, New York, NY, USA
e-mail: sez2105@tc.columbia.edu

A. Karp (ed.), *Advances In The History Of Mathematics Education*,
International Studies in the History of Mathematics and its Teaching,
https://doi.org/10.1007/978-3-030-95235-8_9

1 Introduction

The American mathematical community was at a pivotal moment in 1877 when Artemas Martin published the first issue of the *Mathematical Visitor*. Before the late nineteenth century, the systems of education in the United States were unable to educate American students to participate in the research mathematics that was active in Europe. American secondary schools in the 1870s were, according to historian John Thelin, "few in number and uneven in their offerings," which limited the ability of American institutions of higher learning (colleges) at this time to be academically selective or demanding (Thelin 2011, p. 88). America produced mathematical figures of note before the 1870s, such as Benjamin Banneker, Nathaniel Bowditch and Benjamin Peirce, but they (of necessity) were self-taught, and while there were arguably individual instances of original mathematical discovery before this point, there was no mathematical research community (Parshall and Rowe 1994). Colleges taught algebra and geometry, with occasional classes in calculus, but there were few programs focusing on pure mathematics, and no program dedicated to mathematical research until 1876, when Johns Hopkins University opened in Baltimore, Maryland.

The opening of Johns Hopkins University set a new standard for American colleges and for American mathematics. The university hired English mathematician James Joseph Sylvester to found "America's first school of mathematical research" (Parshall and Rowe 1994, p. 53), and by 1878 Sylvester had started the country's first mathematical research journal, the *American Journal of Mathematics*. In his 1890 work, the *Teaching and History of Mathematics in the United States*, Florian Cajori noted the importance of the founding of the *American Journal of Mathematics* and dismissed the journals that came before it, as "ephemeral." Other than a few examples of original work, he wrote, earlier journals contained mostly problem posing and problem solving, one of the "lowest forms of mathematical work" (Cajori 1890, p. 97). Cajori was correct that the *Mathematical Visitor* did not compare to the *American Journal of Mathematics* in this regard, and it was never meant to. The *Mathematical Visitor* was not a research journal but is significant instead because it was an effort by the country's top mathematical figures to educate and support mathematics students outside of the university systems. They set out to create a pipeline to support the country's nascent departments of mathematical research just at the moment mathematical research was gaining a foothold in the United States, and before American schools were able to step in and fill that need.

2 How the Topic Fits Into Existing Research

Much research has been done about aspects of American mathematics education during this same period including school mathematics (Stanic and Kilpatrick 2003), mathematics textbooks (e.g., Van Sickle 2011), and the mathematics at particular institutions (e.g., Nadis and Yau 2013). When conducting research in this field, as with any historical field, it is important also to broaden the perspective and see how the pieces fit into the larger, dynamic social processes of which they necessarily form a part (Karp 2014). While the major people, institutions, and textbooks of this period are well documented, the identities and experiences of the ordinary teachers and students who engaged in mathematics education at the ground level are underexplored.

Some of the research noted above—particularly Stanic and Kilpatrick's (2003)—contextualizes artifacts of mathematics education by uncovering details about the mathematical experiences of ordinary teachers and students. This author's prior research on the recreational mathematics activities of ordinary nineteenth-century Americans was another effort to contribute to this broader understanding (Zelbo 2019b). The present study also seeks to connect mathematics education to the broader social system but in a novel way. While some studies examine how a particular curriculum, an institution, or a professional role has changed over time, the current study seeks instead to focus on moment in time and examine a particular tie that connected people involved in all aspects of mathematics education at that moment, including the profession's leading figures, college professors, school teachers, school administrators, students, and the laypeople who participated in the publication and consumption of the *Mathematical Visitor*. The study further seeks to understand the role that the journal may have played in the personal and professional lives of its contributors.

Nineteenth-century educational periodicals are the subject of a separate and well-explored area of research. Paul Ringel, for example, in his book *Commercializing Childhood*, traces the evolution of American educational periodicals from their emergence in the 1820s and 1830s through their peak of influence in the late nineteenth century and their ultimate decline (2015). These children's periodicals contained stories that tended to be gently didactic, with the journal taking on the role of a "Companion," "Friend," or "Visitor" (words that appeared in the titles) and telling stories that were meant to guide children's moral and social development (Ringel 2015, p. 7). Similarly, in the United Kingdom, there was a proliferation of children's literary magazines in the late nineteenth-century ranging in quality from sensationalist pulp to serious literature (Lang 1980). A small subset of the periodicals was more explicitly academic. These academic children's magazines contained stories, grammar exercises, sheet music, puzzles, and sometimes elementary mathematics. While the *Mathematical Visitor* was not a children's magazine, it was apparently influenced by this popular genre, and the word "Visitor" in the title is one reflection of that influence.

Periodicals with a mathematical focus were less common, of course. The few that existed before the *Mathematical Visitor* were published only sporadically. The elementary *Mathematical Correspondent* (1804–1806) has been cited as the first American mathematics periodical (Hogan 1985, p. 246). Later followed eight issues of *Mathematical Miscellany*, published by Charles Gill, a mathematics teacher on Long Island in New York from 1836 to 1839. Like most mathematical publications of the time, it focused on problems and solutions but was more advanced than the publications that preceded it and included contributions from prominent mathematics figures of the day such as Benjamin Peirce (Hogan 1985). No regular mathematical periodical was published in the United States until the *Analyst*, which was founded by Joel Hendricks in Des Moines, Iowa in 1874. The *Analyst* was still focused on problems and solutions rather than research but for the first time at least reflected an awareness of the research that had been happening in Europe (Cajori 1890, p. 280). The *Mathematical Visitor* was founded around this same time (between the first issues of the *Analyst* in 1874 and the *American Journal of Mathematics* in 1878) and was meant to fill its own educational niche.

In contrast to the state of mathematical journals in United States, the United Kingdom had had the *Ladies' Diary* since 1704, an almanac that published recreational mathematics problems and puzzles that attracted serious amateurs (Costa 2002). The English pedagogical journal *Educational Times* also ran a mathematics column from 1847 to 1915 that was supported by contributions from important figures such as, among others, J. J. Sylvester, Charles Dodgson, and W.W. Rouse Ball (Tattersall 2004). By the mid-nineteenth century, Great Britain was also making progress toward developing a culture of research mathematics, well ahead of the United States. The *Cambridge Mathematical Journal* was founded as an educational journal in 1837, with problems, mathematical memoirs, and essays meant to educate Cambridge's students and prepare them for exams but then evolved into a journal of research and laid the foundation for the research journals that followed (Crilly 2004).

This study fits also into a body of comparative research about the relationship between professional mathematicians and mathematics education in various countries throughout history. There have been numerous historical examples of individual mathematicians who were deeply involved in pre-college mathematics education, such as Leonhard Euler (see Polyakova 2010, pp. 12–15), as well as Felix Klein, Hans Freudenthal, and George Polya. There are also various times throughout history when entire communities of mathematicians became involved in pre-college education. Sometimes this has happened through formal education systems, such as the mathematicians of the *École Polytechnique* in the eighteenth and nineteenth centuries who exerted a profound influence on the way mathematics was taught in France (Belhoste 2001). Mathematicians have also worked from outside of formal educational institutions and reached students directly. Alexey Sossinsky described, for example, a deep culture of "service to mathematics education" in Russian history, in which mathematicians organized "math circles" and mathematics competitions (Sossinsky 2010). Similarly, in Hungary, an organization of scientists and mathematicians began the Eötvös mathematics competition for high school students, whose top prize was intended to help identify and nurture "the future creative mathematician" (Wieschenberg 1990, p. 288). The participation of the American mathematics community in the *Mathematical Visitor* can be understood in this way, as another example of an early effort by a country's community of mathematicians to fill an unmet need that they perceived, to identify, educate, and motivate promising students in mathematics by reaching them directly.

This study collects and analyzes available evidence about the intended and actual audience for the *Mathematical Visitor*. It seeks to answer the following questions:

- What was the purpose of Artemas Martin and other leading American mathematical figures of the time in collaborating on the *Mathematical Visitor*?
- Who was the audience for the *Mathematical Visitor*, and what does their participation reflect about the state of American mathematics education at this time?
- Did the *Mathematical Visitor* accomplish its purposes?

3 Methodology

This study uses primary sources and contemporaneous secondary sources to uncover and collect information about the 171 individuals named as contributors to the *Mathematical Visitor* over the years 1877 to 1887.[1] The individuals contributed in various ways, mostly by writing problems for the journal or offering solutions to published problems. Many individuals did both. In some cases, college professors solved the published problems, and there were also college students who wrote problems for the journal. Because the roles of contributors were not fixed, for purposes of this study, the contributors were treated as a single, heterogeneous group.

For each individual contributor, it was typical for the journal to publish his or her full name, the title or degree held by the individual (if any), the academic institution they were affiliated with (if any), and the town where they lived. With the information published about these individuals, it was possible in 147 of the 171 cases to identify the individual with reasonable certainty and find additional information from primary and secondary sources. (In some cases the individuals had common names or lived in large cities, which made it impossible to identify them with certainty.) Once the individuals were identified, this study used census records, birth records, marriage records, city directories, college catalogs, and other available public records to put together basic biographical data about the contributors, and where possible, a fuller picture of their lives. The Library of Congress' newspaper archives were used as an additional resource, and in some cases, the individuals were prominent enough that it was possible to learn details about their lives through these and other contemporaneous secondary sources.

The United States Census is conducted every ten years, and this study made heavy use the 1880 census, which was the one closest in time to most of the contributions to the *Mathematical Visitor*. These census documents tell, among other things, each individual's age, exact location, occupation, and country of birth, as well as whether they attended school that year. The census was conducted house by house, and so it is also evident from the census records, for example, whether an individual was the head of a household, lived with their parents, or lived in a

[1] A final, short edition of the *Mathematical Visitor* came in 1894, seven years later, and is not included here.

boarding house. It was also possible to see the ages and professions of the other members of the family or household to understand each contributor's role in his or her household.

Once this information was collected, it was organized. The study then analyzed the data, looking both for commonalities of the contributors (such as their professions, level of education, and geographical locations) and well as specific details of any notable outliers. A table of the contributors with the information compiled is available on the author's website (Zelbo n.d.).

While the circumstances of 2020–2021 made it difficult to do in-person research in library archives, the author was able to access certain additional documents from the Artemas Martin archives, held at American University in Washington, D.C., with the help of the archivists there. American University holds, among other things, volumes of Martin's handwritten subscription lists and accounting notes from the *Mathematical Visitor*. Additional contextual information was taken from these documents about the subscriber base and reach of the journal.

4 Findings

4.1 The Founding of the Mathematical Visitor

Artemas Martin was a prominent, self-taught, amateur mathematical figure who supported himself for most of his life through farming. He was born on a farm in Steuben County, New York in 1835. Census records show that when Martin was fifteen, the family had moved to a farm in Venago County, Pennsylvania, where Martin also attended school (U.S. Census Bureau 1850). This is apparently where he received his only formal education (Artemas Martin 1918). Martin continued to live on his family's farm as an adult, and around the age of thirty-five,[2] Martin moved again with his parents and sister to a farm in Erie County, Pennsylvania. Martin would live there until the age of fifty, and this is where he published both the *Mathematical Visitor* and, later, the *Mathematical Magazine* (U.S. Census Bureau 1870; Artemas Martin 1918). In 1885 Martin finally left farming to take a position as Librarian at the United States Coast and Geodetic Survey where he stayed until his death in 1918 (Artemas Martin 1918).

As a young man and self-directed student of mathematics, Martin was a prolific contributor to elementary magazines and puzzle columns, and before he became a publisher himself, he took over as the Mathematical Editor of a magazine for young people called *Our Schoolday Visitor* (later *Schoolday Magazine*) in 1870 (Cajori 1890, pp. 281–282). Before Martin took over at *Our Schoolday Visitor*, the entire math column consisted of a few simple problems such as rate problems or

[2] Martin's registration for the draft (for the U.S. Civil War) lists him still as a resident of Venago County in 1863, and so the family moved sometime between 1863 and 1870 (*Artemus Martin* 1863).

elementary algebra problems with an occasional, more difficult problem contributed by Martin himself (e.g., *Our Schoolday* 1868, p. 25). Martin's own column in *Our Schoolday Visitor*, by contrast, included problems involving logatirthms, trigonometry, and even calculus, with contributions from some of the same leading mathematical figures who would later contribute to the *Mathematical Visitor* (e.g., *Schoolday Magazine* 1872, pp. 332–333). It is not clear who the audience was for this column; the level of the problems stands in contrast to the elementary material in the rest of the magazine and also to the illustrations of tiny children on the same page who are climbing a stairway and are urged to "go up higher" by working the problems in the column. In 1871, Martin also edited a sixteen-page mathematical section of an almanac for young people that he compared explicitly to the *Ladies' Diary*. In his introduction to the Mathematical Department of the almanac (called *Our Schoolday Visitor Mathematical Almanac and Annual*), Martin cited the *Ladies' Diary*. He wrote that "a mathematical annual similar to those published in Europe would cover ground unoccupied in this country" and that the value of these journals was in "kindling and cherishing a pure and lasting love of mathematical science" (Martin 1871, p. 31).

While Martin never held a position in higher education, he went on to receive recognition and respect in the international mathematics community (Allaire and Cupillari 2000), including various honorary degrees from American universities and memberships in learned societies abroad. When Martin received an honorary degree from Yale University in 1877, the *Philadelphia Inquirer* described him as a "self-educated bachelor of forty-two" who "raises vegetables which he sells through the streets of [Erie, Pennsylvania]" while also making voluminous contributions to various American and English journals (Personal and Political 1877).

Even though Martin was occupied with his own mathematical explorations as an adult, he never lost interest in elementary mathematical publications as a vehicle for reaching young people. Martin launched the *Mathematical Visitor* in the spring of 1877 and later the *Mathematical Magazine* in 1882[3] (Cajori 1890, p. 282; Artemas Martin 1918) from his family's farm in Pennsylvania and did the much of the editing, typesetting, and publishing himself. In the first issue of the *Mathematical Visitor*, Martin declared his purpose in starting the journal:

> In England and Europe periodical publications have contributed much to the diffusion of mathematical learning, and some of the greatest scientific characters of those countries commenced their mathematical career by solving the problems proposed in such works. . . . Similar publications have produced like results in this country. Not a few of our ablest teachers and mathematicians were first inspired with a love of mathematical science by the problems and solutions published in the mathematical department of some unpretending periodical (Martin 1881b, p. 1).

[3] The *Mathematical Magazine* was meant to be more elementary than the *Mathematical Visitor* and to reach a wider audience. In the first issue of the *Mathematical Magazine*, Martin explains that while the *Analyst*, the *Mathematical Visitor*, and the *American Journal of Mathematics* were available in the United States to serve the needs of advanced learners of mathematics, a more elementary journal was needed to serve the needs of average teachers and students (Martin 1882, p. 1).

It is noteworthy that Martin mentions both teachers and mathematicians in this opening. Martin saw his journal as a method both to "diffus[e] mathematical learning" and to "inspire" those who would become the country's "ablest teachers and mathematicians." To understand Martin's goals, it is also significant to note that Martin had in mind the "greatest" and the "ablest." Martin did not see his journal as serving the needs of the masses but instead was aiming to create new pathways for those who would rise to the top of their professions in mathematics-related fields. Martin went on in the same introduction to tell the story of a "young mathematician of great promise" who learned what he knew about probabilities "by studying the problems and solutions in the *Schoolday Magazine*" (Martin 1881b, p. 1). Martin may well have been talking about himself since he began his mathematical studies with the same elementary magazine and went on to serve as its Mathematical Editor. Martin, like nearly all of the leading figures of his profession in the United States at the time, learned much of what he knew about mathematics outside of school.

The *Mathematical Visitor*, like the journals that preceded it, was mostly a journal of problems and solutions, arranged in no particular order.[4] Mathematical articles were not the focus of the journal, but they did appear in some issues. The few articles were educational in focus and did not present original research. E.P. Thompson, a teacher, engineer and inventor, wrote a piece about the binomial theorem (Thompson 1881). Frank Freeland, a professor of mechanics at the University of Pennsylvania, explained how to determine the day of the week on any particular date (Freeland 1881). Martin himself proved a simple theorem in number theory—that the sum of the squares of three consecutive integers cannot be a square number (Martin 1881a).

The journal organized the problems loosely by level, into the Junior and Senior Departments. The Junior Department contained mathematics content that students at the time might have studied in school or in college, from arithmetic to basic algebra, trigonometric ratios, and geometry. Some junior problems were similar to textbook problems that would have been common at the time, including practical problems involving rates or computing interest. Some were less routine, such as a problem by Daniel Kirkwood, a mathematics professor at Indiana State University:

> The number of diagonals that can be drawn in a certain polygon is equal to seven times the number of sides. How many sides has the polygon? (Martin 1881b, p. 43)

The journal also contained a "Senior Department" with problems that were more difficult and that required more advanced content to solve, including integral calculus. Some issues contained a separate section of "Unsolved Problems," and some issues also offered "prize problems" in the Junior and Senior Departments with first and second place prizes of free issues offered to contributors with the best solutions.

Artemas Martin and E. B. Seitz of Greenville, Ohio wrote many problems for the journal, and both seemed to prefer problems that were simple to state but difficult to

[4] It is curious that this journal and others like it had a primarily pedagogical purpose and yet made no attempt to organize the material in a way that would make it more useful to self-directed learners.

solve. Geometry, probability, and number theory (particularly Diophantine equations) were common topics. In a "Senior Problem," Seitz, for example, asks for the average area of a triangle formed by taking three points at random from within a sphere (Martin 1881b, p. 5). In one Senior Problem Martin asks for the least integral values[5] of x and y that will solve the equation: $x^2 - 9781y^2 = 1$ (Martin 1881b, p. 4). In an "Unsolved Problem" Martins asks, "If four dice be piled up at random on a horizontal plane, what is the probability that the pile will not fall down?" (Martin 1881b, p. 7). Many of the problems Martin himself proposed were computation-heavy, such as one where the challenge was to find the side lengths of the 100th triangle in a series of right triangles of integral side lengths, where the lengths of the two legs were consecutive integers (such as the 3, 4, 5 triangle or 20, 21, 29 triangle). The hypotenuse given in the solution had 77 digits (Martin 1881b, p. 156). James Joseph Sylvester of Johns Hopkins and Benjamin Peirce of Harvard each contributed problems to the Senior Department. Sylvester wrote various problems on Diophantine equations for the journal, while Peirce contributed problems in probability and geometry, as well as the following problem that reflects his background in astronomy.

> Prove that if two bodies revolve about a center, acted upon by a force proportional to the distance from the center, and independent of the mass of the attracted body, each will appear to the other to move in a plane, whatever may be their mutual attraction (Martin 1881b, p. 146).

This problem of Peirce's was solved by two readers, including DeVolson Wood, an engineering professor from New Jersey, and a mathematical hobbyist, Walter Siverly of Ohio.

The journal published readers' solutions to problems, and it published sometimes multiple solutions even to the most elementary problems, which must have encouraged young readers to see mathematics as creative and flexible—contrary to the impression of mathematics they might derive from the rigid, rule-based textbooks that were common in nineteenth-century America. The *Mathematical Visitor* ran regularly for only a few years. From 1877 to 1881, the six issues of Volume I comprising a total of 198 pages were published. Three issues of Volume II were published from January 1882 to January 1887, with a final, short issue in January 1894. Altogether this volume contained 122 pages. While the journal was successful, Martin slowed down because of an illness that interfered with his work[6] and later because of his move to Washington, D.C. to take the position as the librarian for the United States Coast and Geodetic Survey office (Artemas Martin 1918).

From 1877 to 1887 the *Mathematical Visitor* reached hundreds of people around the country and even some in Mexico, Canada, Europe, and India. In these years,

[5] While the problem does not clarify, it is apparent from the solution that Martin meant the smallest *positive* integer values.

[6] Martin explains cryptically at the end of issue dated January 1883, "This No. has been delayed a long time in consequence of the ill health and sickness of the Editor. The date on the title-page is January, 1883, but these items [meaning the editorial notes at the end of the volume] were written July 16, 1885" (Martin 1883, p. 64).

171 contributors participated actively in this mathematical community by writing problems and offering solutions. Others subscribed and participated without recognition by reading the journal and by working through its problems on their own. Martin nurtured this developing community by including his own notes in each issue with personal and professional news about *Mathematical Visitor* contributors. He celebrated their professional accomplishments, he thanked them for their help with the journal, and he eulogized them when they passed away. Martin also advertised the journals and math columns of his peers, and some of these men in turn, contributed to the *Mathematical Visitor*.[7]

Artemas Martin kept volumes of handwritten subscription lists, one for each year he published the journal. An early subscription list, perhaps his first, begins with a list of approximately 80 "free" subscribers, which included various prominent individuals (such as Benjamin Peirce and James Joseph Sylvester), educational journals (such as the *New England Journal of Education* and *Scientific American*), and the Library of Congress. What follows are several pages containing a total of 146 paying subscribers, with many subscribers ordering multiple copies of the journal (Martin 1881c). A later list of subscribers contains institutions including the University of Pennsylvania, the Library at the U.S. Naval Academy, a public library in San Francisco, the *Erie Gazette* (Martin's local newspaper), and the Smithsonian Institution (Martin 1881c). The subscription lists suggest that the reach of the *Mathematical Visitor* likely went well beyond the individuals listed on the pages to their families, communities, and institutions.

4.2 The Mathematical Visitor Community, An Overview

The 171 contributors to the Mathematical Visitor came from all over North America, and several from overseas. Twenty-six of the (then) thirty-eight United States were represented, and while the contributors were concentrated in a handful of states (many from Pennsylvania, Ohio, Indiana, and New York), there were contributors scattered throughout the United States, as well as from Mexico, Canada, Scotland, England, and India. Five of the 171 contributors were women. The contributors ranged in age from a nineteen-year-old student to an eighty-three-year-old surveyor, and the great majority were in mid-career (Fig. 1).

At least sixty-five of the contributors to the *Mathematical Visitor* had some identifiable connection to the field of education at or before the time of their contributions. This included thirty individuals who taught mathematics at a college, university, or normal school,[8] and thirty-five who were either career teachers or

[7] This included Joel Hendricks of the *Analyst*, Samuel H. Wright of the *Yates County Chronicle*, Hudson A. Wood of the *National Educator*, and William Hoover of the *Wittenberger*.

[8] Normal schools at this time were sometimes classified with secondary schools and sometimes as a special category of higher education. Many eventually became comprehensive colleges (Thelin 2011, pp. 84–85).

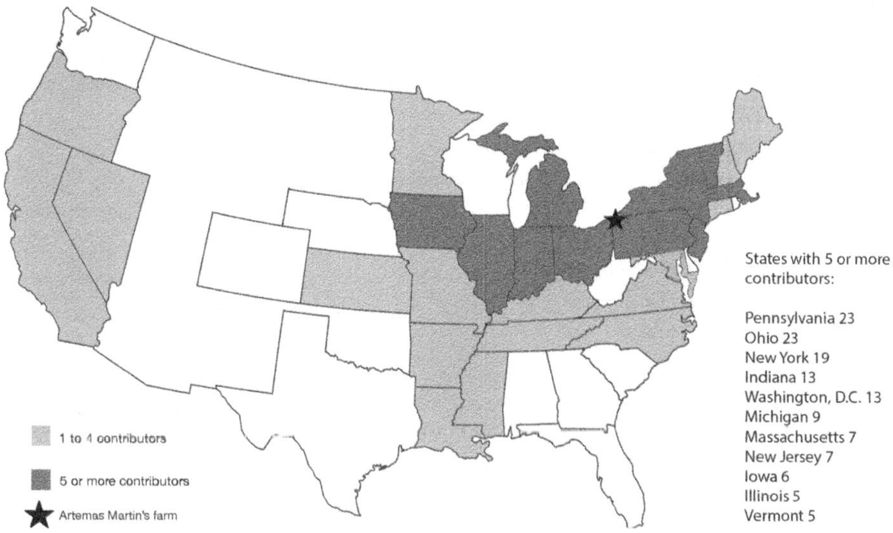

Fig. 1 Map of the United States in the years 1878 to 1887 showing locations of contributors to the Mathematical Visitor

school administrators. These categories were not as clear cut in the nineteenth century before the paths to the teaching profession became more formalized, however, and the movement of many contributors in and out of teaching jobs, made individuals hard to categorize by profession. Some individuals were classified, for example, as a farmer or minister if that was their primary profession, even though some also taught school at some point before contributing to the journal. It is also likely that some additional unclassified individuals in this study also worked as teachers, but that information was not always captured in a decennial census because those documents are a mere snapshot in time.

The participation of serious amateur mathematicians who supported themselves through farming or other work, also undermined an effort to categorize contributors by blurring the line between mathematics as a profession and as a hobby. Some contributors who spent much of their free time doing mathematics, but were not as accomplished as Martin, were not classified as mathematicians for purposes of this study. A contributor named Samuel Cagwin of New York, was one such individual. He worked as a farmer his whole life while he did mathematics in his leisure time. A biographical sketch of Cagwin in *American Mathematical Monthly* says that he had little more than a common school education and yet had an "insatiable love for mathematics" and a "predilection for astronomy" that would drive him to stay up past midnight many nights working problems, even though he woke before dawn every day to milk his cows. Walter Siverly, a petroleum broker from Pennsylvania, was another example. Siverly was mentioned above as one of the men who solved Peirce's astronomy/geometry problem. Siverly was clearly talented. He was in fact often the only contributor to solve some of the hardest problems, sometimes even

Table 1 Classification of contributors by profession

Professions	Number of individuals
Professors at colleges and universities	24
Professors or administrators at normal schools	6
Accomplished mathematical figures without formal degrees or positions in higher education	6
Mathematical professionals outside of education	19
Pre-college teachers and administrators	35
College students	10
Other professions and jobs	47
Unidentified contributors	24
Total	171

This category includes people who were not professors, mathematicians, or math professionals but who worked at some point at or before the time of contribution in the field of pre-college education

solving every problem in an issue, and Martin often complimented and published his solutions. Yet, Cagwin and Siverly were not mathematicians, even by the lower standards of nineteenth-century America. Neither did their own mathematical investigations and, in fact, did not even write problems. Nor did either have honorary degrees or memberships in mathematical societies.

Approximately nineteen contributors worked in mathematics-related professions outside of education, including surveyors, engineers, accountants, and an astronomer. Six of those contributors worked for the United States Coast and Geodetic Survey Office in Washington, D.C. (where Artemas Martin and Benjamin Peirce had worked at different times as well). Other contributors had professions that were squarely outside the field of mathematics, such as doctors, judges, journalists, and did mathematics as a hobby. Several contributors were farmers. An additional ten contributed as college students (Table 1).

4.3 Participation of the Country's Leading Mathematical Figures

Two of United States' leading nineteenth-century mathematical figures, James Joseph Sylvester of Johns Hopkins University and Benjamin Peirce of Harvard, published problems in the *Mathematical Visitor* from its beginning. An additional twenty-two mathematics and engineering professors, listed below, contributed from colleges and universities around the country. An additional six professors, also listed below, either taught at or ran normal schools.

College Mathematics Professors Who Contributed to the Mathematical Visitor

Edward Bowser (Rutgers College, New Jersey)
Wooster Beman (U. of Michigan)
LG Barbour (Central U. in Richmond, Kentucky)
L.G. Carpenter (State Agricultural College, Michigan)
Edgar Edmunds (Southern U., Louisiana)
Joseph Ficklin (U. of Missouri)
Frank T Freeland (U. of Pennsylvania)
TN Haun (Tusculum College, Tennessee)
Arthur Edwin Haynes (Hillsdale College, Michigan)
Christian Hornung (Heidleburg College, Ohio)
William Hoover (Ohio University, Ohio)
William Woolsey Johnson (St. John's College, Maryland)
Joseph H Kershner (Mercersburg College, Ohio)
Daniel Kirkwood (Indiana State U.)
WA Kite (Tusculum College, Tennessee)
HTJ Ludwig (North Carolina College)
Alexander MacFarlane (Edinburgh College, Scotland)
J. Dunlap McAdam (Washington and Jefferson College, Pennsylvania)
JW Nicholson (Louisiana State University)
Benjamin Peirce (Harvard College, Massachusetts)
PH Philbrick (State U. of Iowa)
James Joseph Sylvester (Johns Hopkins College, Maryland)
JFW Scheffer (Mercersburg College, Pennsylvania)
DeVolson Wood (Stevens Institute of Technology, New Jersey)

Teachers and Administrators of Normal Schools

Edward Brooks (Pennsylvania State Normal School, Pennsylvania)
Joseph C. Corbin (Branch Normal College, Arkansas)
Albert Frank (Pennsylvania State Normal School, Pennsylvania)
Hanumanta Rau (Government Normal School, Madras, India)
Enoch Beery Seitz (North Missouri State Normal School, Kirksville, Missouri)
Hudson A. Wood (Keystone State Normal School, Pennsylvania)[9]

Other contributors included prominent mathematical figures without academic affiliations, who are listed below. With the exception of British mathematician Samuel Roberts, these men were all self-taught in mathematics. They belonged to mathematical societies and held honorary degrees and operated within the same professional circles as Sylvester and Peirce, like Artemas Martin himself. This includes Joel Hendricks (founder of the *Analyst*), Samuel Hart Wright (mathematical editor of the *Yates County Chronicle*), William James Wright, (member of the London Mathematical Society) and Sylvester Robins whose extensive

[9] A short autobiographical sketch of Hudson A. Wood is included in the *Yates County Chronicle.*

mathematical correspondence with the period's major figures is archived at Princeton University.

Important Mathematical Figures Without Academic Positions
Joel Hendricks[10]
Aretmas Martin
Samuel Roberts (British)
Sylvester Robins
David Trowbridge
Samuel Hart Wright

The participation of these accomplished figures in an elementary journal would have been, as it was for Martin, a form of outreach to build and nurture the mathematical community and to create pathways for young people of talent. Being mostly self-taught, these mathematicians would have had a special appreciation of the role of mathematical publications in reaching young people who lacked access to a high level of mathematics education. David Trowbridge was one such contributor. Trowbridge was a wagon maker from Waterburg, New York, who scraped together money to buy mathematics books and set aside several hours every day to do his reading, correspondence, and investigations in mathematics and astronomy. Trowbridge received an honorary degree from Indiana State University but lamented at the age of forty-six that he had "accomplished little" because of his lack of access to an education earlier in his life:

> I see now that I have labored under great disadvantages in acquiring what I have, and if I could live my life again, starting with my present knowledge of what might be accomplished if one is directed right, I should expect to be able to master three or four times as much as I have up to this period in my life (Trowbridge 1875, p. 4).

He further remarked that when he was young, he did not know of a single person within his reach that "could enlighten me in regard to things that I wanted to know, and I had to grope my way along as well as I could" (Trowbridge 1875, p. 4). It must have occurred to Trowbridge that a journal like the *Mathematical Visitor* would have made a difference in his own life.

The mathematicians who contributed their time and talent to the *Mathematical Visitor* did so in various ways. Many of course contributed problems, but they were also active in solving each other's problems. They contributed in other ways too, showing that this was indeed a collaborative effort. In one issue, Martin thanked Enoch Beery Seitz "for valuable assistance in correcting the proof sheets," (Martin 1881b, p. 195). He thanked others in various issues—Peirce, Siverly, and Wood for their "invaluable assistance"—and he thanked at least ten other individuals for their help in procuring subscribers to the journal. An inscription on the cover of the bound issues of the *Mathematical Visitor* also shows that the issues were received and held in the Harvard College Library contemporaneously with their publication. Benjamin Peirce was sixty-nine when the first issue of the *Mathematical Visitor* came out and had been at Harvard already for nearly fifty years. Given that Peirce used his

[10]A short biography of Hendricks is given in the American Mathematical Monthly (Colaw 1894).

valuable time to contribute problems to the journal, he may also have had a hand in ensuring that Harvard students would have access to the journal by arranging for it to be in the college's library.

4.4 The Journal's Reach to Young People

One of Martin's stated purposes in founding the *Mathematical Visitor* was to reach talented young people, expose them to mathematics, and spur their interest in mathematics. The journal may indeed have accomplished this purpose, but of course this is difficult to document because some young people who read the journal and worked problems likely did not publish their work. However, among the 171 contributors, there are some notable examples of young contributors. Several were young teachers or college students who went on to have noteworthy careers in mathematics related fields. While it is not clear what the full scope of the journal's impact was among young people, these individual examples serve as case studies to illustrate particular careers the journal may have impacted.

Edwin Place was one young contributor who went on to have an impactful career. Place was a twenty-one-year-old student in electrical engineering at Cornell University when he wrote to the *Mathematical Visitor* for the first time, and he went on after graduation to become an "Edison Pioneer," one of the associates who worked with Thomas Alva Edison, the inventor of the lightbulb, in the early days of his company (Miller 1931, p. 307). William T. Magruder was a nineteen-year-old engineering student at the Stevens Institute of Technology in New Jersey when he first contributed to the *Mathematical Visitor*. Upon graduation Magruder entered Johns Hopkins for graduate study (Board of Trustees 1886) and went on to have a long career as a professor of mechanical engineering at Ohio State University. Lon C. Walker was a twenty-one-year-old farmer in Ohio when he first contributed to the *Mathematical Visitor* (U.S. Census Bureau 1880d). Walker later worked as a local teacher (Teacher's Institute 1884). Twenty years after his first contribution, Walker was still contributing to journals and in approximately 1901 entered Stanford University for graduate study in mathematics (Notes 1902, p. 272). Abram Bullis was a Cornell student when he first contributed to the journal in 1880 (U.S. Census Bureau 1880b) and in 1891 was admitted to the American Mathematical Society (American Mathematical Society 1896). In each of these cases, the individuals had access to college-level mathematics even without the *Mathematical Visitor*, but the journal was, at a minimum, a way for them to connect to the wider mathematical community and perhaps also to push themselves to learn mathematics beyond what was available to them in college. And for some individuals, like Walker, the *Mathematical Visitor* and other journals would have been a lifeline that helped them prepare for and enter graduate school.

Christine Ladd is another example of a contributor whose career was aided by her work in mathematics journals, including the *Mathematical Visitor*. Christine

Ladd graduated from Vassar College in 1869. She worked as a teacher and pursued the study of mathematics independently for ten years, in part through work in elementary mathematics and education journals (Parshall and Rowe 1994, p. 84; Wiggins 1926, p. 13). Ladd was a twenty-eight-year-old science teacher in Union Springs, New York, when she contributed problems to the first issue of the *Mathematical Visitor* in 1877 (signed "Miss Christine Ladd, Cayuga Co., N.Y."). Ladd applied for graduate study at Johns Hopkins, and when James Joseph Sylvester saw her application, he reportedly remembered a submission she had made to a British educational journal (Wiggins 1926, p. 13). By the time the third issue of the *Mathematical Visitor* was published in January 1879, Ladd was signing her name, "Miss Christine Ladd, B.A., Baltimore, Maryland." Ladd had moved to Baltimore to begin her studies, working directly with James Joseph Sylvester. Ladd continued to contribute to the journal while in school, and Martin celebrated her achievements, writing in January 1880:

> Our talented lady contributor, Miss Christine Ladd, has been deservedly honored by the Johns Hopkins University. The trustees have voted her an honorary stipend and an invitation to continue her mathematical studies at that institution. Miss Ladd is a graduate of Vassar College, and is probably the best mathematician of the gentler sex in this country (Martin 1881b, p. 118).

When Ladd became the first American woman to be elected to the London Mathematical Society, Martin announced that, too (Martin 1882, p. 28). Ladd went on to become an accomplished mathematician, logician, and scientist (Woodworth 1930, p. 307). While Ladd was certainly not typical of the journal's contributors, Ladd's achievement must have served as confirmation to Martin that his work with the journal served its intended purpose of encouraging and educating young mathematicians.

4.5 *Teachers and the Mathematical Visitor*

Martin's other stated purpose for starting the journal was to provide mathematical knowledge and inspiration to teachers. As noted above, at least thirty-six of the contributors were teachers or administrators at the pre-college level and another twenty-nine contributors taught mathematics at colleges and normal schools. To understand why the *Mathematical Visitor* may have been so valuable to those in the teaching profession, it must be remembered that at the time the *Mathematical Visitor* was published, many of those working in the teaching profession did not enter the profession though the formalized pathways that later became common. Normal schools for training teachers were opening all over the United States in the last quarter of the nineteenth century (Thelin 2011, pp. 84–85; Ogren 2005, pp. 1–2), and college education was becoming more common, and yet many *Mathematical Visitor* contributors who worked as teachers had at most a high school education. One school principal from Oneida Castle, New York, Benjamin Burleson, had only a common school education, taught himself mathematics in part by working through problems in the *Mathematical Visitor* and other journals, and worked as a teacher while he also held a position as "postmaster" of his town (Notes 1897, p. 324). Other teacher-contributors similarly spent much of their lives doing work unrelated to teaching, typically farming. A contributor named John Rea, for example, was a thirty-eight-year-old teacher living on his family's farm in Ohio in 1880 (U. S. Census Bureau 1880c), but an earlier census document listed his occupation as "farmer" and a later one as a "bank clerk." For some of the contributors, changes in the signature lines of their letters to the *Mathematical Visitor* over the years reflected changes in jobs as they moved in and out of teaching positions.

Even for those who had attended college, opportunities up to this point for the advanced study of mathematics were rare in American colleges. Johns Hopkins' new program was an outlier, and while standards in American colleges were raised over the course of the nineteenth century, it was "well into the twentieth century" before other programs caught up to Johns Hopkins (Thelin 2011, p. 90). Even the country's top mathematical schools, like the John C. Green School of Sciences at Princeton, required only arithmetic, basic algebra, and plane geometry to enter in 1890. Students studied algebra and calculus as part of this special scientific course of study, and by senior year could take elective graduate courses in differential equations and the "theory of functions." (May 1972, p. 9). The mathematics professors teaching in the 1870s when the *Mathematical Visitor* was published had been educated in earlier decades when academic standards were lower. Many of the contributors who held the title of "Mathematics Professor" at the college level had, at most, a bachelor of arts degree with little mathematics beyond basic algebra and may well have taught English, Latin, or some other subject at another point in their careers. Mathematics was not an area of expertise for these college mathematics professors until they began to teach the subject. Even those who included the titles "M.A.," "L.L.D.," and "Ph.D." in their signatures tended to hold honorary degrees received later in their careers. This included prominent mathematical figures like Joseph Ficklin, Edward Brooks, and Enoch Beery Seitz.

The professional path of Louis Green Barbour to mathematics professor was typical for this time period. Barbour was a professor of mathematics at Central University in Kentucky and contributor to the journal. Barbour's degrees were in divinity studies, and he was, alternately, a "professor of pure and applied mathematics and astronomy" as well as "professor of history and the Bible" at the same university (Lewis 1899, pp. 197–202). Similarly, Enoch Beery Seitz grew up on a farm, had only two years of college education, and became a professor of mathematics at the state normal school of Missouri (Finkel 1894). According to Cajori, he received his training through elementary journals and had sufficient talent that, if he had not died so young, "might have done good original work" (Cajori 1890, p. 284). Even Wooster Beman, a prominent mathematics professor at the University of Michigan, had no specialized university training in mathematics. Beman began his career with a Bachelor of Arts degree from the University of Michigan, and upon graduating, he taught Greek and mathematics at Kalamazoo College. Beman only later specialized and became a Professor of Mathematics at the University of Michigan and eventually became a prominent member of his field, authoring and translating several mathematics textbooks with David Eugene Smith of Columbia University (*e.g.*, Beman and Smith 1895). His expertise was acquired through his professional activities, not through his schooling. For a mathematics educator in the late nineteenth century, even at the college level, the *Mathematical Visitor* would have been an important means of professional development, allowing teachers to develop their skills at mathematics and connect with other members of their profession.

4.6 *The Reach of the Mathematical Visitor Beyond Its Intended Audience*

While it was not Martin's stated intention, his journal was also read recreationally by people in remote parts of the country who had little access to education beyond the local town schools. Only about twenty percent of the contributors came from in or near large cities (with 100,000 or more residents) such as Philadelphia, New York, Boston, Chicago, or Washington, D.C. The great majority came from small towns, and many contributors lived on their family farms, with the closest cities hundreds of miles away. Many of these individuals were already middle aged and likely did not aspire to be mathematicians or educators, but the *Mathematical Visitor* allowed them to connect with an intellectual community in a way that was not possible in their daily lives. One remote contributor was Frederick S. Samuels, who sent in a single problem from a tiny mining town, Cerro Gordo, California: "What is the volume of a chip cut at an angle of 45 degrees to the center of a round log, radius r?" (Martin 1881b, p. 82). (The problem was published in the "Senior Department" because it required integral calculus to solve.) Cerro Gordo was in the Inyo mountains on the edge of the Death Valley desert, 275 miles by horse from Los Angeles. Various other contributors held jobs in skilled or unskilled labor, such as Ruben Knecht, a fifty-seven year old photographer from Pennsylvania at the time of his contribution (U. S. Census Bureau 1880e) or "S. C. Gould" from Manchester, New Hampshire whose occupation was listed as a "baggage master" in the 1880 census (U.S. Census Bureau 1880f). The participation of people from such disparate walks of life and from remote areas suggests a desire for education that was unavailable to them when they were younger.

While it was not Martin's explicit intention, it is notable that his journal also reached individuals who were marginalized because of their race, sex, or disability and therefore, despite whatever ability they may have had, did not have the same opportunities for mathematical schooling and careers. One such example was George M. Day, a physically disabled thirty-three year old man living on his brother's farm in Lockport, New York at the time of his contributions. (According to the 1880 census, Day was unemployed and classified as "Maimed, Crippled [or] Bedridden" (U. S. Census Bureau 1880a). Whatever his abilities might have been, it would have been impossible for Day to have a college education and a career, and the *Mathematical Visitor* provided him some access to the education that was otherwise unavailable to him.

At this time, of course, college education—particularly higher-level mathematics—was not widely available to women.[11] It was mentioned above that Christine Ladd, one of the female contributors, secured a fellowship at Johns Hopkins. As accomplished as Ladd was, in spite of completing the requirements for a Ph.D. including a dissertation, Ladd was not awarded a Ph.D. from the university until 1926, after the university changed its policy to allow the conferral of degrees on women (Wiggins 1926, p. 13). At least four other women contributed to the journal—mostly young teachers who contributed elementary problems and solutions.

At least two African-American men contributed to the journal. Both were highly accomplished men who attended college in spite of severe limitations at the time to African-American education. Joseph Carter Corbin graduated from Ohio University in Athens, Ohio in 1853, and E. J. Edmunds, a French-speaking Afro-Creole from New Orleans, graduated from the *École Polytechnique* in Paris in 1873. Both men spent most of their careers in the newly formed Black institutions that sprung up in the South after the Civil War. Corbin founded the Branch Normal College of the Arkansas Industrial University in 1875, a school meant to train teachers for the Black schools, and he served as its first principal (Finney 2017). Edmunds was the first mathematics professor at three important Black institutions, the first high school, the first normal school, and one of the first universities (Southern University) for Black students in Louisiana (Zelbo 2019a). While both men were accomplished in their fields, their race prevented them from being accepted into white academic circles in the South, and in the case of Edmunds in particular, he suffered vicious racist threats and attacks. For all of these marginalized individuals, the *Mathematical Visitor* must have provided a source of intellectual stimulation and also a sense of belonging, giving them a community of peers that was otherwise unavailable to them.

5 Discussion

This study shows that at the same time the United States' leading mathematical figures were working to elevate American mathematics departments to the standards of Europe, some were also taking active steps to help create a population of students who would be prepared to do mathematical work at this new, higher standard. Martin, Peirce, and others collaborated through their work on the *Mathematical Visitor* to write and publish problems and thereby educate the most promising students by bypassing schools and reaching those students directly. Martin evidently had a professional relationship with many of the country's prominent figures, and in

[11] Thelin points out that by 1860, there were at least forty-five colleges that granted degrees to women, many were not equivalent to degrees from men's colleges. Some were nothing more than finishing schools or schools for vocational training (Thelin 2011, p. 83).

turn they must have believed in his work, because Martin was able to recruit these figures to assist him in the work of putting together the journal, but also to use their own influence to spread the reach of the journal.

In a narrow sense, Martin's effort showed some success. While it is not known how many young people were inspired by or learned from the journal, this study was able to identify at least a handful of college-student contributors who went on to have impactful careers in mathematics and mathematics education. Many teachers also contributed to the journal, and it seems likely that their work on the problems in the *Mathematical Visitor,* and the community they found through their participation, had a positive impact on their careers—especially given how few opportunities many of them had to study mathematics beyond basic arithmetic and algebra in school.

Martin apparently did not anticipate how many people from outside of the community of mathematics educators and students would value his journal, and this too reflects something about the time. The fact that so many men who held jobs in skilled and unskilled labor spent their free time submitting to a mathematics journal reflects what is known from other sources—that American systems of pre-college education in the last quarter of the nineteenth century lagged far behind the need for quality education. The women, African-Americans, and one disabled man in the rolls of the journal provide examples of people from groups marginalized by American schools and American society and in similarly in need of a an education or intellectual community that was not available to them in their normal lives.

America's education landscape rapidly changed in the last quarter of the 1800s, with the rise of widely available public education and the rise of American universities. The *Mathematical Visitor* did not last long enough to see these changes, and if it had, its subscriber and contributor base would surely have been different. The *Mathematical Visitor* was unique to its time period—a journal by and for people who largely educated themselves in mathematics at a time when other options were scarce.

References

Allaire, Patricia R., and Antonella Cupillari. 2000. Artemas Martin: An amateur mathematician of the nineteenth century and his contribution to mathematics. *The College Mathematics Journal* 31 (1): 22–34.

American Mathematical Society. 1896. *Annual report.* New York: American Mathematical Society.

Belhoste, Bruno. 2001. The École Polytechnique and mathematics in nineteenth-century France. In *Changing images in mathematics: From the French Revolution to the new millennium*, ed. Umberto Bottazzini and Amy Dahan Dalmedico, 15–30. London and New York: Routledge.

Beman, Wooster W., and David E. Smith. 1895. *New plane and solid geometry*. Boston: Ginn & Co.

Board of Trustees. 1886. Preliminary register of officers and students. *Johns Hopkins University Circulars* VI (52) Retrieved from https://jscholarship.library.jhu.edu/bitstream/handle/1774.2/32885/I052.PDF.

Cajori, Florian. 1890. *The teaching and history of mathematics in the United States*. Washington, D.C.: Washington Government Printing Office.

Colaw, John M. 1894. Biography: Dr. Joel E. Hendricks, A. M. *American Mathematical Monthly* 1 (3): 64–67. https://doi.org/10.2307/2971637.

Costa, Shelley. 2002. The "Ladies' Diary": Gender, mathematics, and civil society in early-eighteenth-century England. *Osiris* 17 (2nd series): 49–73.

Crilly, Tony. 2004. The Cambridge Mathematical Journal and its descendants: the linchpin of a research community in the early and mid-Victorian Age. *Historia Mathematica* 31(4): 455–497.

Finkel, Benjamin F. 1894. E. B. Seitz. *The American Mathematical Monthly* 1 (1): 3–6. https://doi.org/10.2307/2968643.

Finney, Gladys T. 2017. *Joseph Carter Corbin: Educator extraordinaire and founder of the University of Arkansas at Pine Bluff.* Little Rock, AR: Butler Center for Arkansas Studies.

Freeland, Frank. 1881. The day of the week for a given date. In *The Mathematical Visitor, Volume 1, 1877–1881*, ed. Artemas Martin, 155–156. Erie, PA: Artemas Martin. Retrieved from https://babel.hathitrust.org/.

Hogan, Edward R. 1985. *The Mathematical Miscellany* (1836–1839). *Historia Mathematica* 12: 245–257. Retrieved from. https://doi.org/10.1016/0315-0860(85)90022-9.

Karp, Alexander. 2014. The history of mathematics education: Developing a research methodology. In *Handbook on the history of mathematics education*, ed. Alexander Karp and Gert Schubring, 9–24. New York: Springer-Verlag.

Lang, Marjory. 1980. Childhood's champions: Mid-Victorian children's periodicals and the critics. *Victorian Periodicals Review* 13 (1/2): 17–31. Retrieved from https://www.jstor.org/stable/20081920.

Lewis, Alvin F. 1899. *History of higher education in Kentucky*. Washington, D.C.: Government Printing Office.

Martin, Artemas 1863. *U.S. Civil War Draft Registration Records 1863–1865 [database online]*. Ancestry.com.

———. 1871. Mathematical department. In *Our Schoolday visitor mathematical almanac and annual*. Philadelphia: Daughaday & Becker. Retreived from: https://archive.org/details/ourschooldayvisi1871unse.

———. 1881a. Demonstration of two Diophantine theorems. In *The Mathematical Visitor, Volume 1, 1877–1881*, ed. Artemas Martin, 156. Erie, PA: Artemas Martin. Retrieved from https://babel.hathitrust.org/.

———. 1881b. *The Mathematical Visitor, Volume 1, 1877–1881*. Erie, PA: Artemas Martin. Retrieved from https://babel.hathitrust.org/.

———. 1881c. *Subscription lists to Mathematical Visitor, 1878–1881*. Artemas Martin Papers: University Archives and Special Collections, American University Library.

———. 1882. *The Mathematical Visitor II(1)*. Erie, PA: Artemas Martin. Retrieved from https://babel.hathitrust.org/.

———. 1883. *The Mathematical Visitor II(2)*. Erie, PA: Artemas Martin. Retrieved from https://babel.hathitrust.org/.

———. 1918. *Science* 48 (1247): 508.

May, Kenneth O., ed. 1972. *The Mathematical association of America: Its first fifty years*. Washington, D.C.: Mathematical Association of America.

Miller, Francis T. 1931. *Thomas A. Edison: Benefactor of mankind*. Philadelphia: John C. Winston Co.

Nadis, Steve, and Shing-Tung Yau. 2013. *A history in sum: 150 years of mathematics at Harvard (1825–1975)*. Cambridge, MA: Harvard University Press.

Notes. 1897. *The American Mathematical Monthly* 4(12), 323-324, DOI: https://doi.org/10.1080/00029890.1897.11999044

———. 1902. *The American Mathematical Monthly* 9(11), 272-274, DOI: https://doi.org/10.1080/00029890.1902.12000881

Ogren, Christine A. 2005. *The American state normal school: An instrument of good*. New York: Palgrave Macmillan.

Our Schoolday Visitor XII(1). 1868. Available on the author's website: https://sites.google.com/view/sian-zelbo-research/home/artemas-martin-and-the-mathematical-visitor/early-projects-of-artemas-martin

Parshall, Karen H., and David E. Rowe. 1994. In *The emergence of the American mathematical research community 1876-1900*, ed. J.J. Sylvester, Felix Klein, and E.H. Moore. Providence, RI: American Mathematical Society.

Personal and political. 1877. *The Philadelphia Inquirer*, 4.

Polyakova, Tatiana. 2010. Mathematics education in Russia before the 1917 revolution. In *Russian mathematics education: History and world significance*, ed. Alexander Karp and Bruce Vogeli, 1–42. Hackensack, NJ: World Scientific Publishing.

Ringel, Paul B. 2015. *Commercializing childhood: Children's magazines, urban gentility, and the ideal of the child consumer in the United States, 1823–1918*. Amherst, MA: University of Massachusetts Press.

Schoolday Magazine XVI(12). 1872. Available on the author's website: https://sites.google.com/view/sian-zelbo-research/home/artemas-martin-and-the-mathematical-visitor/early-projects-of-artemas-martin

Sossinky, Alexey. 2010. Mathematicians and mathematics education: A history of involvement. In *Russian mathematics education: History and world significance*, ed. Alexander Karp and Bruce Vogeli, 187–222. Hackensack, NJ: World Scientific Publishing.

Stanic, George, and Jeremy Kilpatrick. 2003. *A history of school mathematics*. Reston, VA: National Council of Teachers of Mathematics.

Tattersall, James J. 2004. Using the *Educational Times* in the classroom. *Mediterranean Journal or Research in Mathematics Education* 3 (1–2): 103–114.

Teachers' Institute. 1884. *The Eaton Democrat*, 3. Ohio: Eaton.

Thelin, John. 2011. *A history of American higher education*. 2nd ed. Baltimore, MD: Johns Hopkins University Press.

Thompson, Edward P. 1881. Expansion of polynomials. In *The Mathematical Visitor, Volume 1, 1877–1881*, ed. Artemas Martin, 151–155. Erie, PA: Artemas Martin. Retrieved from https://babel.hathitrust.org/.

Trowbridge, David. 1875. *Biographical sketches: Yates County Chronicle*, 4.

United States Census Bureau. 1850. Pennsylvania, Sandy Creek: Venago County.

———. 1870. Pennsylvania, Millcreek: Erie County.

———. 1880a. New York, Lockport: Niagara County.

———. 1880b. New York, Ithaca: Thompkins County.

———. 1880c. Ohio, Liberty Township: Adams County.

———. 1880d. Ohio, Harrison: Darke County.

———. 1880e. Pennsylvania, Easton: Northampton County.

———. 1880f. 129. Vermont, Hillsborough County: Manchester, E.D..

Van Sickle, Jenna. 2011. *A history of trigonometry education in the United States: 1776–1900 (doctoral dissertation)*. New York, NY: Columbia Teachers College.

Wieschenberg, Agnes A. 1990. The birth of the Eötvös competition. *The College Mathematics Journal* 21 (4): 286–293.

Wiggins, DuBois K. 1926. Husband and wife famous as scholars, *Brooklyn Daily Eagle*, (*Sunday Eagle Magazine* supplement). Retrieved from https://www.newspapers.com.

Woodworth, Robert S. 1930. Christine Ladd-Franklin. *Science* 71 (1838): 301.

Zelbo, Sian. 2019a. E. J. Edmunds, school integration, and white supremacist backlash in reconstruction New Orleans. *History of Education Quarterly* 59 (3): 379–406. https://doi.org/10.1017/heq.2019.26.

———. 2019b. The recreational mathematics activities of ordinary nineteenth century Americans: A case study of two mathematics puzzle columns and their contributors. *British Journal for the History of Mathematics* 34 (3): 155–178. https://doi.org/10.1080/26375451.2019.1646522.

———. n.d. *Research in 19th Century Mathematics Education* [website]. https://sites.google.com/view/sian-zelbo-research/home

Author Index

A. Karp (ed.), *Advances In The History Of Mathematics Education*, International Studies in the History of Mathematics and its Teaching, https://doi.org/10.1007/978-3-030-95235-8

Subject Index

A. Karp (ed.), *Advances In The History Of Mathematics Education*, International Studies in the History of Mathematics and its Teaching, https://doi.org/10.1007/978-3-030-95235-8

The manufacturer's authorised representative in the EU is Springer Nature Customer Service Centre GmbH, Europaplatz 3, 69115 Heidelberg, Germany. If you have any concerns regarding our products, please contact ProductSafety@springernature.com

Printed and bound by CPI Group (UK) Ltd, Croydon, CR0 4YY
30/09/2026
02225245-0008